A Guide to Safe Material and Chemical Handling

Scrivener Publishing
3 Winter Street, Suite 3
Salem, MA 01970

Scrivener Publishing Collections Editors

James E. R. Couper	Richard Erdlac
Rafiq Islam	Pradip Khaladkar
Norman Lieberman	Peter Martin
W. Kent Muhlbauer	Andrew Y. C. Nee
S. A. Sherif	James G. Speight

Publishers at Scrivener
Martin Scrivener (martin@scrivenerpublishing.com)
Phillip Carmical (pcarmical@scrivenerpublishing.com)

A Guide to Safe Material and Chemical Handling

Nicholas P. Cheremisinoff
Anton Davletshin

Scrivener

Copyright © 2010 by Scrivener Publishing LLC. All rights reserved.

Co-published by John Wiley & Sons, Inc. Hoboken, New Jersey, and Scrivener Publishing LLC, Salem, Massachusetts
Published simultaneously in Canada

No part of this publication may be reproduced, stored in a retrieval system, or transmitted in any form or by any means, electronic, mechanical, photocopying, recording, scanning, or otherwise, except as permitted under Section 107 or 108 of the 1976 United States Copyright Act, without either the prior written permission of the Publisher, or authorization through payment of the appropriate per-copy fee to the Copyright Clearance Center, Inc., 222 Rosewood Drive, Danvers, MA 01923, (978) 750-8400, fax (978) 750-4470, or on the web at www.copyright.com. Requests to the Publisher for permission should be addressed to the Permissions Department, John Wiley & Sons, Inc., 111 River Street, Hoboken, NJ 07030, (201) 748-6011, fax (201) 748-6008, or online at http://www.wiley.com/go/permission.

Limit of Liability/Disclaimer of Warranty: While the publisher and author have used their best efforts in preparing this book, they make no representations or warranties with respect to the accuracy or completeness of the contents of this book and specifically disclaim any implied warranties of merchantability or fitness for a particular purpose. No warranty may be created or extended by sales representatives or written sales materials, The advice and strategies contained herein may not be suitable for your situation. You should consult with a professional where appropriate. Neither the publisher nor author shall be liable for any loss of profit or any other commercial damages, including but not limited to special, incidental, consequential, or other damages.

For general information on our other products and services or for technical support, please contact our Customer Care Department within the United States at (800) 762-2974, outside the United States at (317) 572-3993 or fax (317) 572-4002.

Wiley also publishes its books in a variety of electronic formats. Some content that appears in print may not be available in electronic formats. For more information about Wiley products, visit our web site at www.wiley.com.

For more information about Scrivener products please visit www.scrivenerpublishing.com.

Cover design by Russell Richardson.

Library of Congress Cataloging-in-Publication Data:

ISBN 978-0-470-625828

Printed in the United States of America

10 9 8 7 6 5 4 3 2 1

Contents

Preface	ix
Author Biographies	xi
List of Tables	xiii

1. **Corrosion** — 1
   1.1 General Information — 1
   1.2 Types of Corrosion — 1
   1.3 Materials Evaluation and Selection — 6
   1.4 Corrosion Data — 11

2. **Material Properties and Selection** — 43
   2.1 General Properties and Selection Criteria — 43
   2.2 Cast Irons — 45
      2.2.1 Gray Cast Iron — 46
      2.2.2 White Cast Iron — 47
      2.2.3 Malleable Cast Iron — 47
      2.2.4 Nodular Cast Iron — 47
      2.2.5 Austenitic Cast Iron — 48
      2.2.6 Abrasion Resistance — 48
      2.2.7 Corrosion Resistance — 48
      2.2.8 Temperature Resistance — 50
      2.2.9 Welding Cast Iron — 52
   2.3 Steels — 52
      2.3.1 Low Carbon Steels (Mild Steel) — 53
      2.3.2 Corrosion Resistance — 53
      2.3.3 Heat Resistance — 54
      2.3.4 Low Temperatures — 54
      2.3.5 High-Carbon Steels — 54
      2.3.6 Low-Carbon, Low-Alloy Steels — 54

	2.3.7	Mechanical Properties	54
	2.3.8	Corrosion Resistance	55
	2.3.9	Oxidation Resistance and Creep Strength	55
	2.3.10	Low-Temperature Ductility	55
	2.3.11	High-Carbon, Low-Alloy Steels	56
	2.3.12	High-Alloy Steels	56
		2.3.12.1 Chromium Steels (400 Series), Low-Carbon Ferritic (Type 405)	56
		2.3.12.2 Medium Carbon Martensitic	56
		2.3.12.3 Medium Carbon Ferrule	57
		2.3.12.4 Chromium/Nickel Austenitic Steels (300 Series)	57
	2.3.13	Precipitation Hardening Stainless Steels	58
2.4	Materials Properties Data Tables		58

3. Property Tables of Various Liquids, Gases, and Fuels — 107

3.1	General Properties of Hydrocarbons		116
	3.1.1	General Information	116
	3.1.2	Isomers	119
	3.1.3	Alkenes	121
	3.1.4	Alkynes	125
	3.1.5	Straight-Chain Hydrocarbon Nomenclature	125
	3.1.6	Aromatic Hydrocarbons	127
	3.1.7	Hydrocarbon Derivatives	129
	3.1.8	Halogenated Hydrocarbons	130
	3.1.9	Alcohols	133
	3.1.10	Ethers	135
	3.1.11	Ketones	136
	3.1.12	Aldehydes	137
	3.1.13	Peroxides	137
	3.1.14	Esters	138
	3.1.15	Amines	138
3.2	Fuel Properties		138
	3.2.1	Crude Oil	139
	3.2.2	Gasoline	154
	3.2.3	Bioethanol and ETBE	155
	3.2.4	Diesel Oil, Kerosene, Jet A1, and Biodiesel	156
	3.2.5	Fuel Oil	158

Contents

        3.2.6 Natural Gas, Biogas, LPG and Methane Hydrates   159
        3.2.7 Hydrogen   161

4. General Guidelines on Fire Protection, Evacuation, First Responder, and Emergency Planning   165

    4.1 Flammability Properties   165
            4.1.1.1 General Information   165
            4.1.1.2 Flammability Designation   170
        4.1.2 Ignition Temperature   173
        4.1.3 Flammability Limits   175
        4.1.4 Vapor Density   175
        4.1.5 Specific Gravity   178
        4.1.6 Water Solubility   179
        4.1.7 Responding to Fires   179
        4.1.8 Firefighting Agents   184
            4.1.8.1 Water   184
            4.1.8.2 Foam   186
            4.1.8.3 Alcohol-Resistant Foams   188
            4.1.8.4 High Expansion Foams   189
            4.1.8.5 Other Extinguishing Agents   189
            4.1.8.6 Carbon Dioxide   190
        4.1.9 Electrical Fire Prevention   190
        4.1.10 Firefighting Guidance   192
            4.1.10.1 Types   192
            4.1.10.2 Firefighting Agents and Extinguishers   193
            4.1.10.3 Vehicles   196
            4.1.10.4 Firefighting Gear   196
        4.1.11 Specialized Rescue Procedures   198
        4.1.12 First Responder to Electrical Fire Incidents   199
        4.1.13 Evacuation Planning   201
            4.1.13.1 Designated Roles and Responsibilities   202
            4.1.13.2 Preparation & Planning for Emergencies   202
        4.1.14 Evacuation Procedure   203
        4.1.15 General   204
        4.1.16 Template for Emergency Evacuation Plan   204

5. Chemical Data	207
6. Chemical Safety Data	263
7. Recommended Safe Levels of Exposure	371
8. Fire and Chemical Reactivity Data	399

Preface

Though the world has seen a great deal of change in industry and science over the last two decades, the growth of new technologies, and the rise of new industries, the most important information for engineers has not changed. There is still an intense need for an easy reference for anyone working with materials and chemicals. That is one of the aims of this volume.

There is a second, equally important, reason for this work, namely, to offer a useful contribution to the industrial health and safety literature, for which there is also still an intense, or growing, need. Even with all we have learned, there are still explosions, chemical spills, and other incidents that could have been avoided by following basic standards. Our hope, with this volume, is to prevent such accidents by providing the engineer with the information necessary for a safer, incident-free environment.

This volume is intended as a general reference of useful information for engineers, technologists and students. It is a compilation of general data that have been collected over the years and serves as a quick reference for information that can be consulted for general properties values of materials, chemicals and safety measures when handling industrial chemicals.

The volume provides short discussions that are introductory notes and then provides data in the form of tables. There are five chapters that cover the subjects of corrosion, material properties and selection information, properties of various liquids, gases and fuels, properties of hydrocarbons and fuel properties, guidelines on fire protection, and chemical safety data.

To get maximum use of the volume the user should first refer to the List of Tables at the beginning of the volume to find the topical areas of interest.

The authors wish to thank Scrivener Publishing for the fine production of this volume.

Nicholas P. Cheremisinoff
Anton Davletshin

Author Biographies

Nicholas P. Cheremisinoff is a consultant to industry, international lending institutions and donor agencies on pollution prevention and responsible environmental care practices. With a career spanning more than 30 years, he is also the author, co-author or editor of more than 150 technical books and hundreds of state of the art review and scientific articles. He received his B.Sc., M.Sc., and Ph.D. degrees in chemical engineering from Clarkson College of Technology.

Anton Davletshin is a Construction Management student at the Virginia Polytechnic Institute where he has gained education and experience in design, construction and management of commercial and industrial infrastructure investments. He has taken several internships during his undergraduate studies and most recently joined N&P Ltd, an environmental services group that focuses on pollution prevention and environmental management. Mr. Davletshin has focused on green building practices and technologies and has assisted N&P Ltd on a number of site investigations including the Tennessee Valley Authority flyash spill.

List of Tables

Chapter 1	Corrosion	
Table 1.1	Corrosion rates of zinc and steel panels	11
Table 1.2	Chemical corrosion data	14

Chapter 2	Material Properties and Selection	
Table 2.1	Mechanical properties of cast iron	45
Table 2.2	Properties of white iron	47
Table 2.3	Properties of spheroidal cast irons	49
Table 2.4	Properties of flake cast irons	50
Table 2.5	Properties of graphite casts irons	51
Table 2.6	Aluminum alloys generally used in cryogenic applications	58
Table 2.7	General properties of titanium, tantalum and zirconium as reported in british standard code	59
Table 2.8	Mechanical properties of titanium and alloys	60
Table 2.9	General properties of carbon and graphite	60
Table 2.10	General properties of precious metals	61
Table 2.11	Effect of temperature on tensile strength (n/mm^2) of titanium alloys	62
Table 2.12	Chemical corrosion resistance of tantalum and platinum	62
Table 2.13	Properties and uses of common thermoplastics	63
Table 2.14	Mechanical properties of common thermoplastics	75
Table 2.15	Average physical properties of plastics	76
Table 2.16	Hydrostatic design pressures for thermoplastic pipe at different temperatures	81
Table 2.17	Properties of nylons	81
Table 2.18	Properties of common engineering plastics	82
Table 2.19	Properties of fiberglass reinforced plastics	82
Table 2.20	Filler materials and properties imparted to plastics	83
Table 2.21	Chemical resistance of epoxy resin coatings	84

Table 2.22	Properties of cements	86
Table 2.23	Recommended materials for pumping different liquids	88

Chapter 3 — Property Tables of Various Liquids, Gases, and Fuels

Table 3.1	Properties of gases at standard temperature and pressure (273°k, 1 atm)	107
Table 3.2	Thermodynamic properties	109
Table 3.3	Properties of common organic solvents	111
Table 3.4	Properties of chlorinated solvents	113
Table 3.5	Volatility propertied of glycerine water solutions	115
Table 3.6	Typical properties of alkanes	120
Table 3.7	Typical properties of alkenes	124
Table 3.8	A listing of common radicals	127
Table 3.9	Comparison of benzene and some of its derivatives	129
Table 3.10	Approximate heating values of various common fuels	140
Table 3.11	Heating values of various fuel gases	141
Table 3.12	Wood heating value by species	142
Table 3.13	Fuel economy cost data	143
Table 3.14	Ultimate analysis of coal	146
Table 3.15	Typical syngas compositions	146
Table 3.16	Calorific value and compositions of syngas	147
Table 3.17	Glass transition and decomposition temperature of some materials	152
Table 3.18	Typical biomass compositions	153

Chapter 4 — General Guidelines on Fire Protection, Evacuation, First Responder, and Emergency Planning

Table 4.1	Examples of petroleum liquids that are combustible	166
Table 4.2	Classes of flammable and combustible liquids	167
Table 4.3	Explosive limits of hazardous materials	173
Table 4.4	Limits of flammability of gases and vapors, % in air	176

List of Tables

Chapter 5 **Chemical Data**
Table 5.1   Physical properties data   207
Table 5.2   Chemical synonyms   224

Chapter 6 **Chemical Safety Data**
Table 6.1   Recommended emergency response to   263
            acute exposures and direct contact

Chapter 7 **Recommended Safe Levels of Exposure**
Table 7.1   Recommended safe levels of exposure   371

Chapter 8 **Fire and Chemical Reactivity Data**
Table 8.1   Fire and chemical reactivity data   399

1
Corrosion

1.1 General Information

Corrosion occurs in various forms and is promoted by a variety of causes, all related to process operating conditions to which equipment and support structures are subjected. It is a continuous problem that can lead to contaminated process streams. This subsequently leads to poor product quality and unscheduled equipment shutdowns, the consequence of which is reduced production, high maintenance costs, and equipment replacement costs. Minimizing corrosion is a key consideration for the designer and can be accomplished in two ways: (1) proper material selection for apparatus, and (2) preventive maintenance practices. Both of these approaches must be examined.

1.2 Types of Corrosion

Corrosion is characterized by the controlling chemical-physical reaction that promotes each type. Each of the major types is described below.

Uniform corrosion is the deterioration of a metal surface that occurs uniformly across the material. It occurs primarily when the surface is in contact with an aqueous environment, which results in a chemical reaction between the metal and the service environment. Since this form of corrosion results in a relatively uniform degradation of apparatus material, it can be accounted for most readily at the time the equipment is designed, either by proper material selection, special coatings or linings, or increased wall thicknesses.

Galvanic corrosion results when two dissimilar metals are in contact, thus forming a path for the transfer of electrons. The contact may be in the form of a direct connection (e.g., a steel union joining two lengths of copper piping), or the dissimilar metals may be immersed in an electrically conducting medium (e.g., an electrolytic solution). One metal acts as an anode and consequently suffers more corrosion than the other metal, which acts as the cathode.

The driving force for this type of corrosion is the electrochemical potential existing between two metals. This potential difference represents an approximate indication of the rate at which corrosion will take place; that is, corrosion rates will be faster in service environments where electrochemical potential differences between dissimilar metals are high.

Therogalvanic corrosion is promoted by an electrical potential caused by temperature gradients and can occur on the same material. The region of the metal higher in temperature acts as an anode and thus undergoes a high rate of corrosion. The cooler region of the metal serves as the cathode; hence, large temperature gradients on process equipment surfaces exposed to service environments will undergo rapid deterioration.

Erosion corrosion occurs in an environment where there is flow of the corrosive medium over the apparatus surface. This type of corrosion greatly accelerates when the flowing medium contains solid particles. The corrosion rate increases with velocity. Erosion corrosion generally manifests as a localized problem due to maldistributions of flow in the apparatus. Corroded regions are often clean, due to the abrasive action of moving particulates, and occur in patterns or waves in the direction of flow.

Concentration cell corrosion occurs in an environment in which an electrochemical cell is affected by a difference in concentrations in the aqueous medium. The most common form is crevice corrosion. If an oxygen concentration gradient exists, usually at gaskets and lap joints, crevice corrosion often occurs. Larger concentration gradients cause increased corrosion due to the larger electrical potentials present.

Cavitation corrosion occurs when a surface is exposed to pressure changes and high-velocity flows. Under pressure conditions, bubbles form on the surface. Implosion of the bubbles causes large enough pressure changes to flake off microscopic portion of the metal from the surface. The resulting surface roughness promotes further bubble formation, thus increasing the rate of corrosion.

Fretting corrosion occurs where there is friction, generally caused by vibrations, between two metal surfaces. The debris formed by fretting corrosion accelerates the initial damage done by contact welding. Vibrations cause contact welds to break with subsequent surface deterioration. Debris formed acts to accelerate this form of corrosion by serving as an abrasive. Fretting corrosion is especially prevalent in areas where motion between surfaces is not foreseen.

If allowances for vibration are not made during design, fretting corrosion may be a strong candidate.

Pitting corrosion is a form of localized corrosion in which large pits are formed in the surface of a metal usually in contact with an aqueous solution. The pits can penetrate the metal completely. The overall appearance of the surface involved does not change considerably; hence, the actual damage is not readily apparent. Once a pit forms, it acts as a local anode. Conditions such as debris and concentration gradients in the pit further accelerate degradation. There are several possible mechanisms for the onset of pitting corrosion. Slight damage or imperfections in the metal surface, such as a scratch or local molecular dislocation, may provide the environment necessary for the beginning of a pit.

Exfoliation corrosion is especially prevalent in aluminum alloys. The grain structure of the metal determines whether exfoliation corrosion will occur. In this form of corrosion, degradation propagates below the surface of the metal. Corrosion products in layers below the metal surface cause flaking of the metal.

Selective leaching occurs when a particular constituent of an alloy is removed. Selective leaching occurs in aqueous environments, particularly acidic solutions. Graphitization and dezincification are two common forms of selective leaching. Dezincification is the selective removal of zinc from alloys containing zinc, particularly brass. The mechanism of dezincification of brass involves dissolving the brass with subsequent plating back of copper while zinc remains in solution. Graphitization is the selective leaching of iron or steel from gray cast irons.

Intergranular corrosion occurs selectively along the grain boundaries of a metal. This is an electrochemical corrosion in which potential differences between grain boundaries and the grain become the driving force. Even with relatively pure metals of only one phase, sufficient impurities can exist along grain boundaries to allow for intergranular corrosion. Intergranular corrosion is generally not visible until the metal is in advanced stages of deterioration. These advanced stages appear as rough surfaces with loose debris, or dislodged grains. Welding can cause local crystal graphic changes which favor intergranular corrosion. It is especially prevalent near welds.

Stress corrosion cracking is an especially dangerous form of corrosion. It occurs when a metal under a constant stress (external, residual or internal) is exposed to a particular corrosive environment. The effects of a particular corrosive environment vary for different metals. For example, Inconel-600 exhibits stress corrosion cracking in high-purity water with only a few parts per million of contaminants at about 300°C. The stress necessary for this type of corrosion to occur is generally of the residual or internal type. Most external stresses are not sufficient to induce stress corrosion cracking. Extensive cold working and the presence of a rivet are common stress providers. Corrosion products also can build up to provide stress sufficient to cause stress corrosion cracking. The damage done by stress corrosion cracking is not obvious until the metal fails. This aspect of stress corrosion cracking makes it especially dangerous.

Corrosion fatigue is caused by the joint action of cyclically applied stresses (fatigue) and a corrosive medium, generally aqueous. Metals will fail due to cyclic application of stress. The presence of an aqueous corrosive environment causes such failure more rapidly. The frequency of the applied stress affects the rate of degradation in corrosion fatigue. Ordinary fatigue is generally not frequency dependent. Low-frequency applied stresses cause more rapid corrosion rates. Intuitively, low frequencies cause extended contact time between cracks and the corrosive medium. Generally, the cracks formed are transgranular.

Hydrogen blistering is caused by bubbling of a metal surface due to absorbed hydrogen. Monatomic hydrogen can diffuse through metals, whereas diatomic hydrogen cannot. Ionic hydrogen generated by chemical processes, such as electrolysis or corrosion, can form monatomic hydrogen at a metal surface. This hydrogen can diffuse through the metal and combine on the far side of the metal forming diatomic hydrogen. The diffusion hydrogen also can combine in voids in the metal. Pressure within the void increases until the void actually grows visibly apparent as a blister and ultimately ruptures, leading to mechanical failure.

Hydrogen embrittlement is due to the reaction of diffused hydrogen with a metal. Different metals undergo specific reactions, but the result is the same. Reaction with hydrogen produces a metal that is lower in strength and more brittle.

Decarburization results from hydrogen absorption from gas streams at elevated temperatures. In addition to hydrogen blistering, hydrogen can remove carbon from alloys. The particular mechanism depends to a large extent on the properties of other gases present. Removal of carbon causes the metal to lose strength and fail.

Grooving is a type of corrosion particular to environmental conditions where metals are exposed to acid-condensed phases. For example, high concentrations of carbonates in the feed to a boiler can produce steam in the condenser to form acidic condensates. This type of corrosion manifests as grooves along the surface following the general flow of the condensate.

Biological corrosion involves all corrosion mechanisms in which some living organism is involved. Any organism, from bacteria and fungi to mussels, which can attach themselves to a metal surface, can cause corrosion. Biological processes may cause corrosion by producing corrosive agents, such as acids. Concentration gradients also can be caused by localized colonies of organisms. Some organisms remove protective films from metals, either directly or indirectly, leaving the actual metal surface vulnerable to corrosion. By selective removal of products of corrosion, biological organisms also can cause accelerated corrosion reactions. Some bacteria also directly digest certain metals (e.g., iron, copper or aluminum). Microorganisms also may promote galvanic corrosion by removing hydrogen from the surface of a metal, causing potential irregularities between different parts of metal.

Stray current corrosion is an electrolytic degradation of a metal caused by unintentional electrical currents. Bad grounds are the most prevalent causes. The corrosion is actually a typical electrolysis reaction.

Gaseous corrosion is a general form of corrosion whereby a metal is exposed to a gas, usually at elevated temperatures. Direct oxidation of a metal in air is the most common cause. Cast iron growth is a specific form of gaseous corrosion in which corrosion products accumulate onto the metal surface, particularly at grain boundaries, to the extent that they cause visible thickening of the metal. The entire metal thickness may succumb to this before loss of strength causes failure.

Tuberculation occurs in aqueous solutions. Mounds form over metal surfaces providing for concentration differences, favorable environments for biological growth, and an increase in acidity leading to hydrogen formation.

Deposit attack occurs when there is non-uniform deposition of a film on a metal surface. The most common form appears as unequal scale deposits in an aqueous environment. Unequal film provides for concentration cells, which degrade the metal by galvanic means.

Impingement is corrosion caused by aerated water streams, constricting metal surfaces. It is similar to erosion corrosion in which air bubbles take the place of particles. The pits formed by impingement attack have a characteristic tear drop shape.

Liquid metal corrosion occurs when a metal is in contact with a liquid metal. The main type of corrosion with highly pure liquid metals is simple solution. The solubility of the solid metal in the liquid metal controls the rate of damage. If a temperature gradient exists, a much more damaging form of corrosion takes place. Metal dissolves from the higher temperature zone and crystallizes in the colder zone. Transfer of solids to liquid metal is greatly accelerated by thermal gradients. If two dissimilar metals are in contact with the same liquid metal, the more soluble metal exhibits serious corrosion. The more soluble metal dissolves along with alloys from the less soluble metal. Metal in solutions may move by gross movement of the liquid metal or by diffusion. Depending on the system, small amounts of impurities may cause corrosive chemical reactions.

High-temperature corrosion is induced by accelerated reaction rates inherent in any temperature reaction. Layers of different types of corrosion on one metal surface is one phenomenon that occurs frequently in heavy oil-firing boilers.

Causes of corrosion are the subject of extensive investigation by industry. Almost any type of corrosion can manifest itself under widely differing operating conditions. Also, different types of corrosion can occur simultaneously. It is not uncommon to see crack growth from stress corrosion to be accelerated by crevice corrosion, for example.

1.3 Materials Evaluation and Selection

Materials evaluation and selection are fundamental considerations. When done properly, and in a systematic manner, considerable time and cost can be saved in design work, and design errors can be avoided. The design or specification of any equipment must be unified and result in a safe functional system. Materials used for each apparatus should form a well coordinated and integrated entity, which should not only meet the requirements of the apparatus' functional utility, but also those of safety and product purity.

Materials evaluation should be based only on actual data obtained at conditions as close as possible to intended operating environments.

Prediction of a material's performance is most accurate when standard corrosion testing is done in the actual service environment. Often it is extremely difficult in laboratory testing to expose a material to all of the impurities that the apparatus actually will contact. In addition, not all operating characteristics are readily simulated in laboratory testing. Nevertheless, there are standard laboratory practices that enable engineering estimates of the corrosion resistance of materials to be evaluated.

Environmental composition is one of the most critical factors to consider. It is necessary to simulate as closely as possible all constituents of the service environment in their proper concentrations. Sufficient amounts of corrosive media, as well as contact time, must be provided for test samples to obtain information representative of material properties degradation. If an insufficient volume of corrosive media is exposed to the construction material, corrosion will subside prematurely.

The American Society for Testing Materials (ASTM) recommends 250 ml of solution for every square inch of area of test metal. Exposure time is also critical. Often it is desirable to extrapolate results from short time tests to long service periods. Typically, corrosion is more intense in its early stages before protective coatings of corrosion products build up. Results obtained from short-term tests tend to overestimate corrosion rates, which often results in an overly conservative design.

Immersion into the corrosive medium is important. Corrosion can proceed at different rates, depending on whether the metal is completely immersed in the corrosive medium, partially immersed or alternately immersed and withdrawn. Immersion should be reproduced as closely as possible since there are no general guidelines on how this affects corrosion rates.

Oxygen concentration is an especially important parameter to metals exposed to aqueous environments. Temperature and temperature gradients should also be reproduced as closely as possible. Concentration gradients and mixing conditions should be reproduced as closely as possible, and careful attention should be given to any movement of the corrosive medium.

The condition of the test metal is important. Clean metal samples with uniform finishes are preferred. The accelerating effects of surface defects lead to deceptive results in samples. The ratio of the area of a defect to the total surface area of the metal is much higher in a sample than in any metal in service. This is an indication of the inaccuracy of tests made on metals with improper finishes. The sample metal should have the same type of heat treatment as the metal to be used in service. Different heat treatments have different effects on corrosion. Heat treatment may improve or reduce the corrosion resistance of a metal in an unpredictable manner. For the purpose of selectivity, a metal stress corrosion test may be performed. General trends of the performance of a material can be obtained from such tests; however, it is difficult to reproduce the stress that actually will occur during service.

For galvanic corrosion tests, it is important to maintain the same ratio of anode to cathode in the test sample as in the service environment. Evaluation of the extent of corrosion is no trivial matter. The first step in evaluating degradation is the cleaning of the metal. Any cleaning process involves removal of some of the substrate. In cases in which corrosion products are strongly bound to the metal surface, removal causes inaccurate assessment of degradation due to surface loss from the cleaning process. Unfortunately, corrosion assessments involving weight gain measurements are of little value. It is rare for all of the corrosion products to adhere to a metal. Corrosion products that flake off cause large errors in weight gain assessment schemes.

The most common method of assessing corrosion extent involves determining the weight loss after careful cleaning. Weight loss is generally considered a linear loss by conversion. Sometimes direct measurement of the sample thickness is made. Typical destructive testing methods are used to evaluate loss of mechanical strength. Aside from inherent loss of strength due to loss of cross section, changes brought about by corrosion may cause loss of mechanical strength. Standard tests for tensile strength, fatigue, and impact resistance should be run on test materials.

There are several schemes for nondestructive evaluation. Changes in electrical resistance can be used to follow corrosion. Radiographic techniques involving X-rays and gamma rays have been applied. Transmitted radiation and back-scattered radiation have been used.

Radiation transmission methods, in which thickness is determined by (measured as) the shadow cast from a radioactive source, are limited to pieces of equipment small enough to be illuminated by small radioactive sources. There are several schemes for highlighting cracks. If the metal is appropriate, magnetic particles can be used to accentuate cracks. Magnetic particles will congregate along cracks too small to be seen normally. An alternate method involves a dye. A dye can be used that will soak into cracks preferentially.

Because of the multitude of engineering materials and the profusion of material-oriented literature, it is not possible to describe specific engineering practices in detail in a single chapter. However, we can outline general criteria for parallel evaluation of various materials that can assist in proper selection. The following is a list of general guidelines that can assist in material selection:

1. Select materials based on their functional suitability to the service environment. Materials selected must be capable of maintaining their function safely and for the expected life of the equipment and at reasonable cost.
2. When designing apparatus with several materials, consider all materials as an integrated entity. More highly resistant materials should be selected for the critical components and for cases in which relatively high fabrication costs are anticipated. Often, a compromise must be made between mechanically advantageous properties and corrosion resistance.
3. Thorough assessment of the service environment and a review of options for corrosion control must be made. In severe, humid environments it is sometimes more economical to use a relatively cheap structural material and apply additional protection, rather than use costly corrosion-resistant ones. In relatively dry environments, many materials can be used without special protection, even when pollutants are present.
4. The use of fully corrosion-resistant materials is not always the best choice. One must optimize the relation between capital investment and cost of subsequent maintenance over the entire estimated life of the equipment.
5. Consideration should be given to special treatments that can improve corrosion resistance (e.g., special welding methods, blast peening, stress relieving, metallizing, sealing of welds). Also, consideration should be given to fabrication methods that minimize corrosion.
6. Alloys or tempers chosen should be free of susceptibility to corrosion and should meet strength and fabrication requirements. Often a weaker alloy must be

selected over one that cannot be reliably heat-treated and whose resistance to a particular corrosion is low.
7. If, after fabrication, heat treatment is not possible, materials and fabrication methods must have optimum corrosion resistance in their as-fabricated form. Materials that are susceptible to stress corrosion cracking should not be employed in environments conducive to failure. Stress relieving alone does not always provide a reliable solution.
8. Materials with short life expectancies should not be combined with those of long life in irreparable assemblies.
9. For apparatuses for which heat transfer is important, materials prone to scaling or fouling should not be used.
10. For service environments in which erosion is anticipated, the wall thickness of the apparatus should be increased. This thickness allowance should secure that various types of corrosion or erosion do not reduce the apparatus wall thickness below that required for mechanical stability of the operation. Where thickness allowance cannot be provided, a proportionally more resistant material should be selected.
11. Nonmetallic materials should have the following desirable characteristics: low moisture absorption, resistance to microorganisms, stability through temperature range, resistance to flame and arc, freedom from out-gassing, resistance to weathering, and compatibility with other materials.
12. Fragile or brittle materials whose design does not provide any special protection should not be employed under corrosion-prone conditions.

Thorough knowledge of both engineering requirements and corrosion control technology is required in the proper design of equipment. Only after a systematic comparison of the various properties, characteristics, and fabrication methods of different materials can a logical selection be made for a particular design. Design limitations or restrictions for materials might include:

- size and thickness
- velocity

- temperature
- composition of constituents
- bimetallic attachment
- geometric form
- static and cyclic loading
- surface configuration and texture
- special protection methods and techniques
- maintainability
- compatibility with adjacent materials

1.4 Corrosion Data

Table 1.1 provides data on zinc and steel panels exposed to varying environmental conditions over 2 year exposure periods. Table 1.2. provides corrosion data for different materials exposed to various industrial chemicals.

Table 1.1 Corrosion rates of zinc and steel panels

No.	Location	mils/yr Steel	mils/yr Zinc	Environment[a]
1.	Norman Wells, NWT, Canada	0.06	0.006	R
2.	Phoenix, AZ	0.18	0.011	R
3.	Saskatoon, Sask., Canada	0.23	0.011	R
4.	Vancouver Island, BC, Canada	0.53	0.019	RM
5.	Detroit, Mi	0.57	0.053	I
6.	Fort Amidor, Panama C.Z.	0.58	0.025	M
7.	Morenci, MI	0.77	0.047	R
8.	Ottawa, Out., Canada	0.78	0.044	U
9.	Potter County, PA	0.81	0.049	R
10.	Waterbury, CT	0.89	0.100	I

Table 1.1 (cont.) Corrosion rates of zinc and steel panels

No.	Location	mils/yr Steel	mils/yr Zinc	Environment[a]
11.	State College, PA	0.90	0.045	R
12.	Montreal, Que., Canada	0.94	0.094	U
13.	Melbourne, Australia	1.03	0.030	I
14.	Halifax, NS, Canada	1.06	0.062	U
15.	Durham, NH	1.08	0.061	R
16.	Middletown, OH	1.14	0.048	SI
17.	Pittsburgh, PA	1.21	0.102	I
18.	Columbus, OH	1.30	0.085	U
19.	South Bend, PA	1.32	0.069	SR
20.	Trail, BC, Canada	1.38	0.062	I
21.	Bethlehem, PA	1.48	0.051	I
22.	Cleveland, OH	1.54	0.106	I
23.	Miraflores, Panama C.Z.	1.70	0.045	M
24.	London (Battersea), England	1.87	0.095	I
25.	Monroeville, PA	1.93	0.075	SI
26.	Newark, NJ	2.01	0.145	I
27.	Manila, Philippine Islands	2.13	0.059	U
28.	Limon Bay, Panama C.Z.	2.47	0.104	M
29.	Bayonne, NJ	3.07	0.188	I
30.	East Chicago, IN	3.34	0.071	I
31.	Cape Kennedy, FL	3.42	0.045	M
32.	Brazos River, TX	3.67	–0.072	M
33.	Piisey Island, England	4.06	0.022	IM
34.	London (Stratford), England	4.40	–0.270	I

Table 1.1 (cont.) Corrosion rates of zinc and steel panels

No.	Location	mils/yr Steel	mils/yr Zinc	Environment[a]
35.	Halifax, NS, Canada	4.50	0.290	I
36.	Cape Kennedy, FL 180 ft.	5.20	0.170	M
37.	Kure Beach, NC 800 ft.	5.76	0.079	M
38.	Cape Kennedy, FL 180 ft.	6.52	0.160	M
39.	Daytona Beach, FL	11.7	0.078	M
40.	Widness, England	14.2	0.400	I
41.	Cape Kennedy, FL 180 ft.	17.5	0.160	M
42.	Dungeness, England	19.3	0.140	IM
43.	Point Reyes, CA	19.8	0.060	M
44.	Kure Beach, NC 80 ft.	21.2	0.250	M
45.	Galatea Point, Panama C.Z.	27.3	0.600	M

[a] R- Rural, SI- Semi-industrial, M- Marine, IM- Industrial Marine, RM- Rural Marine, SR- Semi-rural, I- Industrial, U- Urban

Table 1.2 Chemical corrosion data

	Aluminum	Brass	Carbon Steel	Ductile Iron / Cast Iron	316 Stainless Steel	17-4PH	Alloy 20	Monel	Hastelloy C	Buna N (Nitrile)	Delrin	EPDM/EPR	Viton	Flexible Graphite	Teflon-Reinforced or
Acetaldehyde	B	C	C	C	A		A	A	A	D	A	B	C		A
Acetamine	B	B	B	B	B				A	A					A
Acetate Solvents	A	B	A	B	A			A	A	D	D		D		A
Acetic Acid, Aerated	B	F	D	D	A			A	A	C	D		C	A	A
Acetic Acid, Air Free	B	B	D	D	A	A	A	A	A	C	D		D	A	A
Acetic Acid Glacial					A			A	D		B	C	A	A	
Acetic Acid, pure	C	C	D	D	A	A	A	D	A	D	D		D	A	A
Acetic Acid 10%	C	C	C	C	A	A	A	B	A	D	B	B	D	A	A
Acetic Acid 80%	C	C	C	C	A	A	A	B	A	D	D	C	D	A	A
Acetic Acid Vapors	B	D			D	D	B	C	A	D				A	A
Acetone	A	A	A	A	A	A	A	A	A	D	A	A	D	A	A
Other Ketones	A	A	A	A	A	A	A	A	A	D	A	D	D		A
Acetyl Chloride	D	A		C	C			B	A	D	D	D	D		A
Acetylene	A	B	A	A	A	A	A	A	A	B	A	A	A		A
Acid Fumes	B	D	D	D	B		B			C	D				A
Air	A	A	A	A	A		A	A	A	A	A	A			A
Alcohol, Amyl	B	B	B	C	A		B	B	B	C	A	A	B	A	A
Alcohol, Butyl	B	B	B	C	A		A	A	B	A	C	A	A	A	A
Alcohol, Diacetone	A	A	A	A	A		A	B	A	D	A	B	D	A	A
Alcohol, Ethyl	B	B	B	B	B		A	B	A	A	A	A	A	A	A
Alcohol, Isopropyl	B	B	B	B	B		A	B	B	C	A	A	A	A	A
Alcohol, Methyl	B	B	B	B	A		A	A	A	B	A	A	C	A	A

Corrosion

Table 1.2 (cont.) Chemical corrosion data

	Aluminum	Brass	Carbon Steel	Ductile Iron / Cast Iron	316 Stainless Steel	17-4PH	Alloy 20	Monel	Hastelloy C	Buna N (Nitrile)	Delrin	EPDM/EPR	Viton	Flexible Graphite	Teflon-Reinforced or
Alcohol, Propyl	A	A	B	B	A		A	A	A	B	A	A	A	A	A
Alumunia	A	A						A	A	A				A
Aluminum Acetate	C	D		D	A	B	B	C	B	D	D	A	D		A
Aluminum Chloride Solution	C				D	C	B	B	B	B	D		A	A	A
Aluminum Fluoride	C		D	D	C			B	A	A	C	A	A		A
Aluminum Hydroxide	A	A	D	D	A	B	B	B	A	A	C	A	A		A
Aluminum Nitrate	D	D		D	C		B	C	B	B	D	B	D		A
Aluminum Oxalate	B						A	B							A
Alum (Aluminum Sulfate)	C	C	D	D	B	A	B	C	A	A	D	A	A	A	A
Amines	B	B	B	C	A	A	A	B	B	D	C	C	D		A
Ammonia Alum	C				A		A		A	B	C			A	A
Ammonia, Anhydrous Liquid	A	D	A	B	A	A	A	B	A	B	D	B	D	A	A
Ammonia, Aqueous	B	C	C	C	A		A	A	A	D	A	B	C		A
Ammonia Liquor	A	B	A	B	A			A	A	D	D		D		A
Ammonia Solutions	B	F	D	D	A			A	A	C	D		C	A	A
Ammonium Acetate	B	B	D	D	A	A	A	A	A	C	D		D	A	A
Ammonium Bicarbonate	B	B	C	B	B		B	B		B	A	A	A		A
Ammonium Bromide 5%	D				B		B	B			A				A

Table 1.2 (cont.) Chemical corrosion data

	Aluminum	Brass	Carbon Steel	Ductile Iron / Cast Iron	316 Stainless Steel	17-4PH	Alloy 20	Monel	Hastelloy C	Buna N (Nitrile)	Delrin	EPDM/EPR	Viton	Flexible Graphite	Teflon-Reinforced or
Ammonium Chloride	D	D	D	D	C	C	B	B	B	B	C	A	A		A
Ammonium Hydroxide 28%	C	D	C	C	B	A	A	F	B	B	D	B	A	A	A
Ammonium Hydroxide Concentrated	C	D	C	C	A	A	A	C	B	C	D	A	A	A	A
Ammonium Monosulfate	D			B		B	B	B		D					A
Ammonium Nitrate	B	D	D	D	A	A	B	D	B	A	D	A	A		A
Ammonium Persulfate	C	C			A		A	D		D	D	B	B		A
Ammonium Phosphate	C	D	D	D	B		B	C	A	C	C	A	A		A
Ammonium Phosphate Di-basic	B	C	D	D	B		B	C	A	A	A		A		A
Ammonium Phosphate Tri-basic	C	C	D	D	B		B	C	A	A	A		A		A
Ammonium Sulfate	C	C	C	D	B	B	B	B	A	B	B	A	B	A	A
Ammonium Sulfide	C	D	D	D	B		B	B	A	A	A	A	D		A
Ammonium Sulfite	C	C	C	C	A		B	D		B	A	B	A		A
Amyl Acetate	B	B	C	C	B	A	A	B	A	D	A	B	D		A
Amyl Chloride	D	B		B	A		A	B	B	D	A	D	D		A
Aniline	C	D	C	C	B		A	B	B	D	D	C	C	A	A
Anline Dyes	C	C	C	C	A		A	A		C	A	C	B		A

Corrosion

Table 1.2 (cont.) Chemical corrosion data

	Aluminum	Brass	Carbon Steel	Ductile Iron / Cast Iron	316 Stainless Steel	17-4PH	Alloy 20	Monel	Hastelloy C	Buna N (Nitrile)	Delrin	EPDM/EPR	Viton	Flexible Graphite	Teflon-Reinforced or
Apple Juice	B	C	D	D	B		A	A		A	A	B	A		A
Aqua Regia (Strong Acid)	D	D	D	D	B		B			D	D	D	D	D	A
Aromatic Solvents	A	A	C	B	A		A	B		D	A	D			A
Arsenic Acid	D	D	D	D	B		B	D	B	A	D	B	A	A	A
Asphalt Emulsion	C	A	B	B	A		A	A	A	D	A	D	A		A
Asphalt Liquid	C	A	B	B	A		A	A	A	C	A	D	A		A
Barium Carbonate	C	B	B	B	B		B	B	A	B	A	A	A		A
Barium Chloride	D	B	C	C	B	B	C	B		A	A	A	A		A
Barium Cyanide	D	C		C	B		B	D		B	A	B	B		A
Barium Hydrate	D	D			A		A	B			A				A
Barium Hydroxide	D	C	C	B	B	A	A	B		A	A	B	A		A
Barium Nitrate	B				A		A				A				A
Barium Sulfate	D	C	C	C	A		A	B		A	A	B	A		A
Barium Sulfide	D	D	C	D	B		B	C		A	A	A	A		A
Beer	A	B	D	D	A	A	A	A		B	A	B	A		A
Beet Sugar Liquors	A	A	B	B	A		A	A		A	A	B	A		A
Benzaldehyde	A	A	A	C	A		A	B	B	D	A	A	D		A
Benzene (Benzol)	B	B	B	B	B	B	A	A	B	D	C	D	B	A	A
Benzoic Acid	B	B	D	D	B	A	B	B	A	C	A	D	B		A
Berryllium Sulfate	B	B		B	B		A	B		B	A	B	B		A

Table 1.2 (cont.) Chemical corrosion data

	Aluminum	Brass	Carbon Steel	Ductile Iron / Cast Iron	316 Stainless Steel	17-4PH	Alloy 20	Monel	Hastelloy C	Buna N (Nitrile)	Delrin	EPDM/EPR	Viton	Flexible Graphite	Teflon-Reinforced or
Bleaching Powder Wet		B			C		B	A	D	D	B	B	B		A
Blood (meat juices)	B	B		D	A	A	A		B	A	B	B	B		A
Borax (Sodium Borate)	C	D	C	C	A			A	B	A	A	A	A		A
Bordeaux Mixture					A		A		A						A
Borax Liquors	C	A	C	C	B		A	B		A	A	A	A		A
Boric Acid	B	C	D	D	B		B	B	A	B	A	B	A	A	A
Brake Fluid	B	B		B	B	A		B		D	B	B	D		A
Brines, Saturated	C	B	D	C	B		B	B	A	A	A	A	A		A
Bromine, dry	C	B	D	D	D		B	A	A	D	D	D	B	B	A
Bunker Oils (Fuel)	A	B	B	B	A		A	A		B	A		A		A
Butadiene	B	C	B	B	A		A	C	B	C	A	C	B		D
Butane	A	A	B	B	A		A	B	A	B	A	D	A		A
Butter					A		A			B	A				A
Buttermilk	A	D	D	D	A		A	D		A	A	B	A		A
Butyl Acetate	B	B		B	B		A	B	B	D	B	D	D		A
Butylene	A	A	A	A	A		A	A		D	A	D	D		A
Butyric Acid	B	C	D	D	B		B	B	A	C	A	C	C		A
Calcium Bisulfite	C	C	D	D	B		B	D	B	A	D	D	A		A
Calcium Carbonate	C	C	D	D	B		B	B	B	A	A	B	A		A
Calcium Chlorate	B	D		C	B		B	B		B	D	B	B	B	A

Corrosion

Table 1.2 (cont.) Chemical corrosion data

	Aluminum	Brass	Carbon Steel	Ductile Iron / Cast Iron	316 Stainless Steel	17-4PH	Alloy 20	Monel	Hastelloy C	Buna N (Nitrile)	Delrin	EPDM/EPR	Viton	Flexible Graphite	Teflon-Reinforced or
Calcium Chloride	C	B	C	C	B	B	B	B	A	A	A	B	A		A
Calcium Hydroxide	D	C	C	C	B		B	A	A	A	A	A	A		A
Calcium Nitrate	B				B		B			B	C	B			A
Calcium Phosphate	D	C		C	B		B			B	B	B	B		A
Calcium Silicate	D	C		C	B		B			B	A	B	B		A
Calcium Sulfate	B	C	C	C	B	B	B	B	B	A	A	B	A		A
Caliche Liquor			B		A		A			B	A				A
Camphor	C	C		C	B		C	C		B	A	B	B		A
Cane Sugar Liquors	A	B		B	A		A	B		B	A	B	B		A
Carbonated Beverages	B	B	D	B	B	B	B	C		B	A	B	B	A	A
Carbonated Water	A	B	B	A	A	B	A	B		A	A	A	A	A	A
Carbon Bisulfide	A	C	B	B	B		B	B		D	A	D	A		A
Carbon Dioxide, dry	A	A	A	B	A	A	A	A		C	A	B	B	A	A
Carbonic Acid	A	D	D	D	B	B	A	B		B	A	B	A	A	A
Carbon Monoxide	A	A		B	A	A	A	A	A	B	A	B	B		A
Carbon Tetrachloride, dry	B	C	B	C	A	A	A	A	A	D	A	D	B	A	A
Carbon Tetrachloride, wet		D	D	D	B		B	B	B	D	B	D	B	A	A
Casein	C	C		C	B		B	C		B	A	B	B		A
Caster Oil	A	A	B	B	A		A	A	A	A	A	B	A		A

Table 1.2 (cont.) Chemical corrosion data

	Aluminum	Brass	Carbon Steel	Ductile Iron / Cast Iron	316 Stainless Steel	17-4PH	Alloy 20	Monel	Hastelloy C	Buna N (Nitrile)	Delrin	EPDM/EPR	Viton	Flexible Graphite	Teflon-Reinforced or
Caustic Potash				A		A	B		B	D					A
Caustic Soda	D		B	B	A		A	A		C	D	B	B		A
Cellulose Acetate	B	B		B	B			B	B	D	C	B	D		A
China Wood Oil (Tung)	A	C	C	C	A		A	A	A	A	A	D	A		A
Chlorinated Solvents	D	C	C	C	A		A	B		D	A	D	C		A
Chlorinated Water	C			C	D	A	D	D	B	D		A	B	A
Chlorine Gas, dry	B	C	B	B	B	C	A	A	A	C	D	D	B	A	A
Chlorobenzene, dry	B	B	B	B	A		A	B	B	D	B	D	A		A
Chloroform, dry	D	B	B	C	A	B	A	A	B	D	A	D	B		A
Chlorophyll, dry	B	B		B	B		A	B		B		B	B		A
Chlorosulfonic Acid, dry	B	C	B	B	B		B	B	A	D	D	D	D		A
Chrome Alum	C	C	B	C	A		A	B		B	B	B	B		A
Chromic Acid < 50%	C	D	D	D	C	C	B	C	B	D	D	C	C		A
Chromic Acid > 50%	D	D	D	C	C	D	B	D	B	D	D	C	C		A
Chromium Sulfate	B	C		D	B		C	B		B	C	B	B		A
Cider	B				A		B	A			A				A
Citric Acid	B	C	D	D	B	C	A	B	A	B	A	B	A	A	A
Citrus Juices	C	B	D	D	B		A	A		A	A		A		A
Coca Cola Syrup					A		A			B	A		B		A
Coconut Oil	B	B	C	C	B		A	B		A	A	A	A		A
Coffee	A	A		D	A		A	B		A	A	A	A		B

CORROSION

Table 1.2 (cont.) Chemical corrosion data

	Aluminum	Brass	Carbon Steel	Ductile Iron / Cast Iron	316 Stainless Steel	17-4PH	Alloy 20	Monel	Hastelloy C	Buna N (Nitrile)	Delrin	EPDM/EPR	Viton	Flexible Graphite	Teflon-Reinforced or
Coffee Extracts, hot	A	B	C	C	A		A	A			A				A
Coke Oven Gas	A	C	B	B	A		A	B		C	D	D	B		A
Cooking Oil	B	B	B	B	A		A	A		A	A	D	A		A
Copper Acetate	D	D	D	D	A		A	C	B	C	D	B	D		A
Copper Carbonate	D				A		A				A				A
Copper Cyanide	D	D		D	A		A	C		A	A	B	B		A
Copper Nitrate	D	D	D	D	B		B	D		A	A	B	A		A
Copper Sulfate	D	D	D	D	B	B	B	C	A	A	A	A	A	A	A
Corn Oil	B	B	C	C	B		B	B		A	A	C	A		A
Cottonseed Oil	B	B	C	C	B		B	B		A	A	C	B		A
Cresol				B			B			D	D	D	D		A
Creosote Oil	B	B	B	B	B	B	A	B	B	C	D	D	A		A
Cresylic Acid	C	C	C	D	B		B	B		D	D	D	B		A
Crude Oil, sour	B	C	B	C	A		A	B		A	A	D	A		A
Crude Oil, sweet	A	B	B	B	A		A	A		A	A		A		A
Cupric Nitrate	D				A		A	D			D				A
Cutting Oils, Water Emulsions	A	A	B	B	A		A			A	A		A		A
Cyanide Plating Solution	D	D		D	B		B	D		B	D	B	B		A
Cyclohexane	A	A	A	A	A		A	B	B	C	A	D	A		A
Cyclohexanone	B	B			A		A	B	B	D	A				A
Detergents, Synthetic	B	B		B	B		A	B		B	A	B	A		A

Table 1.2 (cont.) Chemical corrosion data

	Aluminum	Brass	Carbon Steel	Ductile Iron / Cast Iron	316 Stainless Steel	17-4PH	Alloy 20	Monel	Hastelloy C	Buna N (Nitrile)	Delrin	EPDM/EPR	Viton	Flexible Graphite	Teflon-Reinforced or
Dextrin	B	B		B	B		B	B		B	A	B	B		A
Dichloroethane				C	C		B	B		D	D	D			A
Dichloroethyl Ether	B	B		B	B		B			D	D	D	D		A
Diesel Oil Fuels	A	A		A	A		A	A		A	A	D	A		A
Diethylamine	B	B	A	B	A		A	B		B	A	C	D		A
Diethyl Benzene				B			B			D	C	D			A
Diethylene Glycol	B	B	A	A	A		A	B		A	A	A	B		A
Diethyl Sulfate	B	B		B	B		B	B		C	A	C	B		A
Dimethyl Formamide	B	B		B	A		A	B		B	A	D	D		A
Dimethyl Phthalate										B	C		D		A
Dioxane	B	B		B	B		B	B		D	C	C	D	A	A
Dipentane (Pinene)	A	A		A	A		A			B	A	D	B		A
Disodium phosphate	B			B			B	C		B	A		B		A
Dowtherm	A	A	B	B	A		A	A		D	A	D	A	A	A
Drilling Mud	B	B	B	B	A		A	A		B	A	A	A		A
Dry Cleaning Fluids	A	C	B	B	A		A	A		B	D	A	B		A
Drying Oil	C	C	C	B	B		B	B		B	A	A			A
Enamel		A								B	A				A
Epsom Salts (MgSO$_4$)	A	B	C	C	B		B	B		B	A	A	A		A
Ethane	A	B	C	C	B		B	B		A	A	D	A		A
Ethers	A	B	A	B	A	B	A	B		D	C	C	C		A
Ethyl Acetate	A	C	B	C	B	A	B	B	B	D	C	C	D		A

Corrosion

Table 1.2 (cont.) Chemical corrosion data

	Aluminum	Brass	Carbon Steel	Ductile Iron / Cast Iron	316 Stainless Steel	17-4PH	Alloy 20	Monel	Hastelloy C	Buna N (Nitrile)	Delrin	EPDM/EPR	Viton	Flexible Graphite	Teflon-Reinforced or
Ethyl Acrylate	C	B	C	C	A		A	B	A	D	B	C	D		A
Ethyl Benzene							A		A	C	A	D			A
Ethyl Bromide	B	A		B	B		C	B		B	A	B	B		A
Ethyl Chloride, dry	B	B	B	B	A	A	A	B	B	C	A	C	B		B
Ethyl Chloride, wet	D	C	C	D	B		B	B	B	C	A	B	B		A
Ethylene Chloride	C				A		A	B	B	D	A		D		A
Ethylene Dichloride				B			A	B		D	C	D	D	A	A
Ethylene Glycol	A	B	B	B	B	A	A	B	A	A	A	A	A		A
Ethylene Oxide	C	C	B	B	B		B	B	A	D	A	D	D		A
Ethyl Ether		B		C	A		A	A	B	D	A	D	D		A
Ethyl Silicate	B	B		B	B		B	B		B	A	B	B		A
Ethyl Sulfate	A			B			B			B	A	C	A		A
Fatty Acids	B	C	D	D	A		A	B	A	B	A	D	A	A	A
Ferric Hydroxide				A			A	A		B	A				A
Ferric Nitrate	D	D	D	D	C	B	A	D	B	A	A	A	A		A
Ferric Sulfate	D	D	D	D	B	B	A	D		A	A	A	A		A
Ferrous Ammonium Citrate	B			B			B				A				A
Ferrous Chloride	D	B	D	D	D		D	D	D	A	A	A	A	A	A
Ferrous Sulfate	C	B	D	D	B		B	B	B	A	A	A	A	A	A
Ferrous Sulfate, Saturated	C	C	C	C	A		A	B	B	C	A	B	B		A
Fertilizer Solutions	B	C	B	B	B		B	B		B					A

Table 1.2 (cont.) Chemical corrosion data

	Aluminum	Brass	Carbon Steel	Ductile Iron / Cast Iron	316 Stainless Steel	17-4PH	Alloy 20	Monel	Hastelloy C	Buna N (Nitrile)	Delrin	EPDM/EPR	Viton	Flexible Graphite	Teflon-Reinforced or
Fish Oils	C	B	B	B	A		A	A		A	A	D	A		A
Flue Gases	C	B		B	A		A	B		C	C	D	C		A
Fluoboric Acid	B				B		A			A	D				A
Fluorosilicic Acid	D	B	D	D	B		B	A	B	C	C	C	C		A
Formaldehyde, cold	A	A	A	B	A	A	A	A	B	B	A	B	D		A
Formaldehyde, hot	B	B	D	D	C		B	B	B	B	A				A
Formic Acid, cold	D	B	D	D	B	B	A	B	A	D	D		B	A	A
Formic Acid, hot	D	B	D	D	B	D	B	B	B	D	D		A	A	A
Freon Gas, dry	B	B	B	B	A	A	A	A	B	C	A	C	C	A	A
Freon 11, MF, 112, BF	B	B		C	A		A	B	B	C	A	C	D	A	
Freon 12, 12, 32, 114, 115	A	A		B	A		A	B	B	B	A	A	D	A	
Freon 21, 31	B	B		C	A		A	B	B	D	A	D	D	A	
Freon 22	A	A		B			A		B	D	A	D	D	A	
Freon 113, TF	B	B		C	A		A	B	B	B	A	C	C	A	
Freon, wet	D	D		D	C	B	B	B	B	B	A	B	D	A	A
Fruit Juices	B	B	D	D	A		A	B		A	A	A	A		A
Fuel Oil	A	B	B	B	A		A	B		A	A	D	A		A
Fumaric Acid							A			B	A				A
Furfural	A	A	A	B	A	B	A	B	B	D	A	C	D		A
Gallic Acid 5%	A	C	D	D	B		B	B	B	B	A	C	A		A
Gas, Manufactured	B	B	B	B	B		B	A		A	A		A		A

Corrosion

Table 1.2 (cont.) Chemical corrosion data

	Aluminum	Brass	Carbon Steel	Ductile Iron / Cast Iron	316 Stainless Steel	17-4PH	Alloy 20	Monel	Hastelloy C	Buna N (Nitrile)	Delrin	EPDM/EPR	Viton	Flexible Graphite	Teflon-Reinforced or
Gas, Natural	B	B	B	B	A		B	A		A	A	D	A		A
Gas, Odorizers	A	A	B	B	B		A	B		B	A		A		A
Gasoline, Aviation	A	A	A	B	A		A	A	A	C	A		A	A	A
Gasoline, Leaded	A	A	A	A	A		A	B	A	C	A		A	A	A
Gasoline, Motor	A	A	A	B		A	A	A	A	C	A	D	A	A	A
Gasoline, Refined	A	B	B	B	A		A	B	A	C	A	D	A	A	A
Gasoline, sour	A	B	B	B	A		A	C	A	C	A	D	A	A	A
Gasoline, Unleaded	A	A	A	B	A		A	A	A	C	A		A	A	A
Gelatine	A	A	D	D	A		A	B		A	A	A	A		A
Glucose	A	A	B	B	A		A	A	A	A	A	A	A		A
Glue	A	B	A	B	B		A	B	A	A	A	B	A		A
Glycerine (Glycerol)	A	B	C	B	A	A	A	A	A	C	A	A	B	A	
Glycol Amine	C	D		B	B	A			D	A	C	D	D	A	
Glycol	A	B	C	B	B		A	B		B	C	A	A		A
Graphite	B	B		C	B		A	B		B	A	B	B		A
Grease	B	C	A	A	A		A	B		A	A	D	A		A
Helium Gas	B	B		B	A		A	B	A	B	A	B	B		A
Heptane	A	A	B	B	A		A	B	A	A	A	D	A		A
Hexane	A	B	B	B	A		A	B	A	A	A	D	A		A
Hexanol, Tertiary	A	A	A	A	A		A	A	A	A	A	D	B		A
Hydraulic Oilm Petroleum Base	A	B	A	B	A		A	A		A	A	D	A		A

Table 1.2 (cont.) Chemical corrosion data

	Aluminum	Brass	Carbon Steel	Ductile Iron / Cast Iron	316 Stainless Steel	17-4PH	Alloy 20	Monel	Hastelloy C	Buna N (Nitrile)	Delrin	EPDM/EPR	Viton	Flexible Graphite	Teflon-Reinforced or
Hydrazine	C	D		D	B		B	D		C	D	B	D		A
Hydrocyanic Acid	A	D	D	C	A		A	C	B	B	D	B	A		A
Hydrofluosilicic Acid	D	A	D	D	C		C	B		B	A	B	A	A	A
Hydrogen Gas, cold	A	B	B	B	A		A	A		B	A	B	A		A
Hydrogen Gas, hot	C		B		B		A		A	B	A	B			A
Hydrogen Peroxide, Concentrated	A	D	D	D	B		B	D	D	D	D	B	B		A
Hydrogen Peroxide, Dilute	A	C	D	D	B		B	D	D	A	D	B	A		A
Hydrogen Sulfide, dry	A	C	B	B	A	B	B	B	B	C	C	A	A	A	A
Hydrogen Sulfide, wet	B	D	C	D	B		B	C	D	C	C	B	A	A	A
Hypo (Sodium Thiosulfate)	B	C	D	D	B		B	B		C	A	A	A		A
Illuminating Gas	A	A	A	A	A		A	A		A	A	D	A		A
Ink-Newsprint	C	C	D	D	A		A	B		A	A	B	A		A
Iodoform	C	C	B	C	A		A	C			A		A		A
Iso-Butane					B		B			B	A	D			A
Iso-Octane	A	A	A	B	A		B		A	A	A	D	A		A
Isopropyl Acetate					B				D	D	A	D		A	A
Isopropyl Ether	B	A	A	B	A		B	A	C	C	A	D	D	A	A
JP-4 Fuel	A	A	A	B	A		A	A	A	A	A		A		A
JP-5 Fuel	A	A	A	A	A		B	A	B	B	A		A		A

Table 1.2 (cont.) Chemical corrosion data

	Aluminum	Brass	Carbon Steel	Ductile Iron / Cast Iron	316 Stainless Steel	17-4PH	Alloy 20	Monel	Hastelloy C	Buna N (Nitrile)	Delrin	EPDM/EPR	Viton	Flexible Graphite	Teflon-Reinforced or
JP-6 Fuel	A	A	A	A	A		A	A	A	A		A			A
Kerosene	A	A	B	B	A		A	A	A	A	A	D	A	A	A
Ketchup	D	D	D	D	A		A	B		A	A		A		A
Ketones	A	A	A	A	A		A	A		D	A	D	D		A
Laquer (and Solvent)	A	A	C	C	A		A	A		D	A	D	D		A
Lactic Acid Concentrated cold	C	D	D	D	A	D	A	D	A	B	D	B	A	A	A
Lactic Acid Concentrated hot	C	D	D	D	B	D	A	D	B	C	D	B	B	A	A
Lactic Acid Dilute cold	A	D	D	D	A	B	A	C	A	B	D	B	A	A	A
Lactic Acid Dilute hot	B	D	D	D	A	D	A	D	B	C	D		D	A	A
Lactose	B	B		C	B		B	B		B	A	B	B		A
Lard	A	B		A	A		A			B	A	C			A
Lard Oil	B	B	C	C	B		A	B		A	A	B	A		A
Lead Acetate	D	C	D	D	B		B	B		A	A	B	B		A
Lead Sulfate	D	C		D	B		B	B		B	A	B	B		A
Lecithin	C	C		C	B		B	B		D	A	D	B		A
Linoleic Acid	A	B	B	B	A		A	B		B	A	D	B		A
Linseed Oil	A	B	A	A	A		A	B		A	A	D	A		A
Lithium Chloride	D	B		B	B		A	B		B	A	B	B		A
LPG	A	A	B	B	B		B	B		A	A	D	A		A

Table 1.2 (cont.) Chemical corrosion data

	Aluminum	Brass	Carbon Steel	Ductile Iron / Cast Iron	316 Stainless Steel	17-4PH	Alloy 20	Monel	Hastelloy C	Buna N (Nitrile)	Delrin	EPDM/EPR	Viton	Flexible Graphite	Teflon-Reinforced or
Lubricating Oil Petroleum Base	A	B	A	A	A		A	B		A	A	D	A		A
Ludox	D	D		B	B		B	B		B	B	B	B		A
Magnesium Bisulfate	B	B	B	B	A		A	B		B	A	B	B		A
Magnesium Bisulfide	C	D		D	B		B	B		B	A	B	B		A
Magnesium Carbonate	B	A		B	A		A	B		B	A	B	B		A
Magnesium Chloride	C	B	C	D	B	C	B	B	A	A	A	A	A		A
Magnesium Hydroxide	D	B	B	B	A	A	A	B	B	A	A	A	A		A
Magnesium Hydroxide Hot	D	D	B	B	A	A	A	A	B	B	A		A		A
Magnesium Nitrate	B				A		A	B		B	A		B		A
Magnesium Sulfate	B	B	B	B	A	A	A	B	A	A	A	A	A		A
Maleic Acid	B	B	B	C	B		B	B	A	B	A	D	A		A
Maleic Anhydride	B	B		B	B		B	B	B	D	C	D	B		A
Malic Acid	B	B	D	D	B		B	B		A	A		A		A
Malt Beverages					A		B	A		A	A	B	A		A
Manganese Carbonate	B				B		A			B	A				A
Manganese Sulfate	B	B		D	A		A	B		B	A	B	B	A	A
Mayonnaise	D	D	D	D	A		A	B		A	A		A		A
Meat Juices	B	D			A		A			B	A				A
Melamine Resins				D	C		C			B	A				A

Table 1.2 (cont.) Chemical corrosion data

	Aluminum	Brass	Carbon Steel	Ductile Iron / Cast Iron	316 Stainless Steel	17-4PH	Alloy 20	Monel	Hastelloy C	Buna N (Nitrile)	Delrin	EPDM/EPR	Viton	Flexible Graphite	Teflon-Reinforced or
Methanol	B	B		B	A		A	B		B	C	D	B		A
Mercuric Chloride	D	D	D	D	B		B	D	B	A	A	A	A		A
Mercuric Cyanide	D	D	D	D	A		A	C	B	A	A	A	A		A
Mercurous Nitrate	D	D			A		A	D			A		B		A
Mercury	D	D	A	A	A		A	B	B	A	A	A	A		A
Methane	A	A	B	B	A		A	B	A	A			A		A
Methyl Acetate	A	A	B	B	A		A	B	A	D	B	B	D		A
Methyl Acetone	A	A	A	A	A		A	A		D	B	A	D		A
Methylamine	A	D	B	B	A		A	C	B	D	A	B	D		A
Methyl Bromide 100%	C	C		D	B		A	B		B	A	D	B		
Methyl Cellosolve	A	A	B	B	A		A	B	B	C	A	B	D		A
Methyl Cellulose					A		A		B	D	A				A
Methyl Chloride	D	B	B	B	A		A	B		D	A	D	B		A
Methyl Ethyl Ketone	A	A	A	A	A		A	A	B	D	A	B	D	A	A
Methylene Chloride	C	A	B	B	A		A	B	B	D	A	D	C		A
Methyl Formate	C	A	C	C	B		A	B	B	D	A	B	D		A
Methyl Isobutyle ketone				A			A			D	A			A	A
Milk & Milk Products	A	B	D	D	A		A	B		A	A	A	A		A
Mineral Oils	A	B	B	B	A		A	A		A	A	D	A		A
Mineral Spirits	A	B	B	B	B		B	B		A	A		A		A
Mixed Acids (cold)	D	D	C	C	B		B	C		D	D	D	B		A

Table 1.2 (cont.) Chemical corrosion data

	Aluminum	Brass	Carbon Steel	Ductile Iron / Cast Iron	316 Stainless Steel	17-4PH	Alloy 20	Monel	Hastelloy C	Buna N (Nitrile)	Delrin	EPDM/EPR	Viton	Flexible Graphite	Teflon-Reinforced or
Molasses, crude	B	A	A	A	A		A	A		A	A		A		A
Molasses, Edible	A	A	C	C	A		A	A		A	A		A		A
Molybdic Acid					A		A			A					A
Monochloro Benzene dry				B			B	B		D	C			A	A
Morphine	B	B		B	A		A	B		D	A	B	D		A
Mustard	B	A	B	B	A		A	A		A	A		A		A
Naptha	A	B	B	B	B		B	B	A	B	A	D	A		A
Napthalene	B	B	B	B	B		B	B	B	D	A	D	A		A
Natural Gas, Sour	B	B	B	B	A		A	D	A	A	A	D	A		A
Nickel Ammonium Sulfate	D	D	D	D	A		A	C		A	C	B	D		A
Nickel Chloride	D	D	D	D	B		A	B	A	A	D	B	A	A	A
Nickel Nitrate	C	D	D	D	B		A	B		A	C	A	A		A
Nickel Sulfate	D	D	D	D	B		A	B	B	A	C	B	A	A	A
Nicotinic Acid	A	A	B	C	A		A	A		D	C	D	B		A
Nitric Acid 10%	D	D	D	D	A	A	A	D		C	D		A	A	A
Nitric Acid 30%	D	D	D	D	A	D	A	D		C	D	B	A	B	A
Nitric Acid 80%	B	D	D	D	C	D	B	D		D	D	B	B	B	A
Nitric Acid 100%	B	D	D	D	A	D	A	D		D	D	D	B	B	A
Nitric Acid Anhydrous	B	D	D	C	A	D	A	D		D	D	D	A	B	A
Nitrobenzene	C	D	B	B	A		A	B	B	D	B	C	C		A

Table 1.2 (cont.) Chemical corrosion data

	Aluminum	Brass	Carbon Steel	Ductile Iron / Cast Iron	316 Stainless Steel	17-4PH	Alloy 20	Monel	Hastelloy C	Buna N (Nitrile)	Delrin	EPDM/EPR	Viton	Flexible Graphite	Teflon-Reinforced or
Nitrogen	A	A	A	A	A		A	A		A	A	B	A		A
Nitrous Acid 10%	D	D	D	D	B		B	D		C	B		A		A
Nitrous Gases	B	D	B	C	A		A	D			B				A
Nitrous Oxide	C	B	B	C	B		B	D	B	B	A		A		A
Oils & Fats	B				A		A			B	A	D			A
Oils, Animal	A	A	A	A	A		A	B	A	A	A	B	B		A
Oils, Petroleum	A	B	A	A	A		A	A	A	A	A	A	D		A
Oils, Petroleum Sour	A	C	B	C	A		A	A	A	B	A	A	D		A
Oils, Water Mixture	A	A	B	B	A		A			A	A	A			A
Olaic Acid	B			B			B	A	A		D	C			A
Oleic Acid	B	B	C	C	B		A	B	B	B	B	C	D	A	A
Oleum	B	C	B	D	B		B	C	B	D	D	D	C		A
Oleum Spirits	D	D		D	B		B	D		C	D	D	A		A
Olive Oil	B	C	B	B	A		A	A		A	A	B	A		A
Oxalic Acid	C	B	D	D	B	D	B	B		C	C	B	A	A	A
Oxygen	A	A	B	B	A	A	A	A	A	B	D	A	A		A
Ozone, dry	A	A	A	A	A		A	A	A	D	C	A	B		A
Ozone, wet	B	B	C	C	A		A	A	A	D	C	B	B		A
Paints & Solvents	A	A	A	A	A		A	A		D	A	D	B		A
Palmitic Acid	B	B	C	C	B		B	B		B	A	B	A		A
Palm Oil	A	B	C	C	B		A	A		B	A	D	A		A
Paper Pulp	D	B		B	A		A	B		B	A	B	B		

Table 1.2 (cont.) Chemical corrosion data

	Aluminum	Brass	Carbon Steel	Ductile Iron / Cast Iron	316 Stainless Steel	17-4PH	Alloy 20	Monel	Hastelloy C	Buna N (Nitrile)	Delrin	EPDM/EPR	Viton	Flexible Graphite	Teflon-Reinforced or
Paraffin	A	A	B	B	A		A	A	A	A	A	D	A		A
Paraformaldehyde	B	B	B	B	B		B	B		B	A	D			A
Paraldehyde				B			B			B	A	D		A	A
Pentane	A	A	B	B	A		A	B		A	A	D	A		A
Perchlorethylene, dry	B	C	B	B	A		A	B	B	D	B	D	A		A
Petrolatum (Vaseline Petroleum Jelly)	B	B	C	C	B		A	A		A	A		A		A
Phenol	A	B	D	D	A	B	A	A		D	C	D	B		A
Phosphate Ester 10%	D	D	A	A	A		A	A	A	D	A	A			A
Phosphate Acid 10%	D	D	D	D	D	B	B	D		B	D	B	A	A	A
Phosphoric Acid 50% cold	D	D	D	D	B	B	B	C		B	D	B	A	A	A
Phosphoric Acid 50% hot	D	D	D	D	D	D	B	C		B	D	B	A	A	A
Phosphoric Acid 85% cold	D	D	B	B	A	C	B	A		C	D		B	A	A
Phosphoric Acid 85% hot	D	D	C	C	B	D	B			C	D			A	A
Phosphoric Anhydride	A			A			A			D	B		B	A	A
Phophorous Trichloride	D		B	C	A		A			D	D	B	B	A	A
Phthalic Acid	B	B	C	C	B		B	A	B	C	B		A		A

CORROSION 33

Table 1.2 (cont.) Chemical corrosion data

	Aluminum	Brass	Carbon Steel	Ductile Iron / Cast Iron	316 Stainless Steel	17-4PH	Alloy 20	Monel	Hastelloy C	Buna N (Nitrile)	Delrin	EPDM/EPR	Viton	Flexible Graphite	Teflon-Reinforced or
Phthalic Anhydride	B	B	C	C	B		B	A	A	C	A		A		A
Picric Acid	C	C	D	D	B	C	B	D	B	C	D	B	B		A
Pineapple Juice	A	C	C	C	A		A	A		A	A		A		A
Pine Oil	B	B	B	B	A		A	B		A	A	D	A		A
Pitch (Bitumen)					A		A			C	A	D			A
Polysulfide	D	D		B	B		A	B		B	D	B	B		A
Polyvinyl Acetate	B	B		B	B		B	B			A	B			A
Polyvinyl Chloride	B	B		B	B		B	B			A	B			A
Potassium Bicarbonate	A			A		A	B		B	A					A
Potassium Bichromate	A			A	A	A		B	B		B		B		A
Potassium Bisulfate	B			A		A	B		B	A	B	A			A
Potassium Bisulfate	C	C	D	D	B		B	D		A	A	B	A		A
Potassium Bromide	C	C	D	D	A	C	B	B		A	A		A		A
Potassium Carbonate	D	B	B	B	B	A	B	B		A	A	B	A		A
Potassium Chlorate	C	B	B	B	B	B	B	C		A	A	B	A		A
Potassium Chloride	D	C	C	B	B	B	A	B	B	A	A	A	A		A
Potassium Chromate	B	B		B	B		B	B		B	A	B	B		A
Potassium Cyanide	D	D	B	B	B		B	B	B	A	A	A	A		A
Potassium Dichromate	A	D	C	C	B		A	B		A	A	B	A		A

Table 1.2 (cont.) Chemical corrosion data

	Aluminum	Brass	Carbon Steel	Ductile Iron / Cast Iron	316 Stainless Steel	17-4PH	Alloy 20	Monel	Hastelloy C	Buna N (Nitrile)	Delrin	EPDM/EPR	Viton	Flexible Graphite	Teflon-Reinforced or
Potassium Ferricyanide	B	D	C	C	A	B	B	B		A	A	B	A		A
Potassium Ferrocyanide	B	B	C	C	B		B	A		A	A		A		A
Potassium Hydroxide Dilute cold	D	D	A	A	B	B	B	A		A	D		D		A*
Potassium Hydroxide to 70%, cold	D	D	B	B	B	C	B	A		B	D	B	D		A*
Potassium Hydroxide Dilute hot	D	D	B	B	B	C	B	A		B	D	A			A*
Potassium Hydroxide to 70%, hot	D	D	A	B	B	D	B	A		C	D				A*
Potassium Iodide	D	D	C	C	B	B	B	C		A	A	B	A		A
Potassium Nitrate	A	B	B	B	B	B	B	B	B	A	A	B	A		A
Potassium Oxalate	C			A		A				A					A
Potassium Permanganate	B	B	B	B	B	B	B	B	B	A	A	B	A		A
Potassium Phosphate	D	C		C	B		B	B	B	A	A	A	A		A
Potassium Phosphate Di-basic	B	B	A	A	A		A	B	B	A	A	B	A		A
Potassium Phosphate Tri-basic	D		A	A	B		B	B		B		B			A
Potassium Sulfate	A	B	B	C	A	A	A	B		A	A	A	A		A
Potassium Sulfide	B	B	B	B	A		A	C	A	A	A	B	B		
Potassium Sulfite	B	B	B	B	A		A	C	B	B	A	A	B		A

Corrosion

Table 1.2 (cont.) Chemical corrosion data

	Aluminum	Brass	Carbon Steel	Ductile Iron / Cast Iron	316 Stainless Steel	17-4PH	Alloy 20	Monel	Hastelloy C	Buna N (Nitrile)	Delrin	EPDM/EPR	Viton	Flexible Graphite	Teflon-Reinforced or
Producer Gas	B	B	B	B	B	A	B	A		A	A	D	A		A
Propane Gas	A	A	B	B	B	A	A	B	A	A	A	D	A		A
Propyl Bromide	B	B		B	B		A	B		B	A	B	B		A
Propylene Glycol	A	B	B	B	B		B	B		A	C	B	A		A
Pyridine	B			B	B		A			D	D		D		A
Pyrolgalic Acid	B	B	B	B	B	B	A	B		A	A		A		A
Quench Oil	A	B	B	B	A		A			A	A		A		A
Quinine, sulfate, dry					A	B	A	B			A				A
Resins & Rosins	A	A	C	C	A	B	A	A		C	A		A		A
Resorcinol					B		B								A
Road Tar	A	A	A	A	A		A	A		B	A	D	A		A
Roof Pitch	A	A	A	A	A		A	A		B	A		A		A
Rosin Emulsion	A	B	C	C	A		A	A		D	A		B		A
R P-1 Fuel	A	A	A	A	A		A	A		B	A		A		A
Rubber Latex Emulsions	A	A	B	B	A		A				A		A		A
Rubber solvents	A	A	A	A	A		A	A		D	C		D		A
Salad Oil	B	B	C	C	B		A	B		A	A	B	A		A
Salicylic Acid	C	C	D	D	A		B	B		A	A	B	A		A
Salt (NaCl)	B	B	C	C	B		A	A		A	A		A		A
Salt Brine	B	B		D	B		B	B		A	A	B	B		A
Sauerkraut Brine					B		B				C				A

Table 1.2 (cont.) Chemical corrosion data

	Aluminum	Brass	Carbon Steel	Ductile Iron / Cast Iron	316 Stainless Steel	17-4PH	Alloy 20	Monel	Hastelloy C	Buna N (Nitrile)	Delrin	EPDM/EPR	Viton	Flexible Graphite	Teflon-Reinforced or
Sea Water	C	C	D	D	B		B	A		A	A	A	A		A
Sewage	C	C	C	D	B	A	B	B		A	B	B	B		A
Shellac	A	A	A	B	A		A	A		A	A				A
Silicone Fluids	B	B		B	B		B			B	A		B		A
Silver Bromide	D				A	C	A	B			D				A
Silver Cyanide	D	D		D	A		A	B		B	D		B		A
Silver Nitrate	D	D	D	D	A		A	D		C	A	A	A		A
Silver Plating sol.	B				A		A				D				A
Soap Solutions (Stearates)	C	A	A	B	A		A	A		A	A	A	A		A
Sodium Acetate	B	B	C	C	B		B	B	B	B	A	B	A		A
Sodium Aluminate	D	B	C	C	A		B	B	A	A	A	B	A		A
Sodium Benzoate	B				B		B	B			B				A
Sodium Bicarbonate	B	B	C	C	B		A	B		A	B	A	A		A
Sodium Bichromate	A				B		B			D	A				A
Sodium Bisulfate 10%	D	B	D	D	A		A	B		A	D	B	A		A
Sodium Bisulfite 10%	D	B	D	D	A		B	B	B	A	D	B	A		A
Sodium Borate	B	B	C	C	B		B	B		A	A	B	A		A
Sodium Bromide 10%	B	B	C	D	B		B	B		A	A	B	A		A
Sodium Carbonate (Soda Ash)	D	B	B	B	A		A	B	B	A	A	B	A		A
Sodium Chlorate	C	B	C	C	B		B	C	B	A	A	B	A	B	A

Corrosion

Table 1.2 (cont.) Chemical corrosion data

	Aluminum	Brass	Carbon Steel	Ductile Iron / Cast Iron	316 Stainless Steel	17-4PH	Alloy 20	Monel	Hastelloy C	Buna N (Nitrile)	Delrin	EPDM/EPR	Viton	Flexible Graphite	Teflon-Reinforced or
Sodium Chloride	B	B	C	C	B		A	A	B	A	A	B	A	A	A
Sodium Chromate	D	C	B	B	A		B	B		A	A	B	A		A
Sodium Citrate	D				B		B				A				A
Sodium Cyanide	D	D	B	B	A	B	A	B		A	A	B	A		A
Sodium Ferricyanide	A				A		A	B			A				A
Sodium Fluoride	C	C	D	D	B	B	A	B		A	A	B	A		A
Sodium Hydroxide 20% cold	D	A	A	A	A	A	B	A		A	D	B	B	A	A*
Sodium Hydroxide 20% hot	D	A	B	B	A	C	A	A		B	D	B	C	A	A*
Sodium Hydroxide 50% cold	D	A	A	B	A	B	A	A		A	D	B	C	A	A*
Sodium Hydroxide 50% hot	D	A	B	B	A	C	A	B		B	D		C	A	A*
Sodium Hydroxide 70% cold	D	A	A	A	A	B	B	A		B	D	B	C	A	A*
Sodium Hydroxide 70% hot	D	B	B	B	A	C	B	B		D	D	B	C	A	A*
Sodium Hypochlorite (Bleach)	D	D	D	D	D	D	C	D	A		D		A		A
Sodium Hyposulfite	B			B			B	B			A				A
Sodium Lactate	D				A		A	B			A			.	A
Sodium Metaphosphate	A	C	B	C	B	B	B		A	A	B	B		A	

Table 1.2 (cont.) Chemical corrosion data

	Aluminum	Brass	Carbon Steel	Ductile Iron / Cast Iron	316 Stainless Steel	17-4PH	Alloy 20	Monel	Hastelloy C	Buna N (Nitrile)	Delrin	EPDM/EPR	Viton	Flexible Graphite	Teflon-Reinforced or
Sodium Metasilicate cold	B	B	C	C	A		A	A		B	A		B		A
Sodium Metasilicate hot	B	B	D	D	A		A	A	A		A				A
Sodium Nitrate	A	B	B	B	A	B	A	B	B	C	A	B	A		A
Sodium Nitrite	A			B			B	C	B	C	B	A	B		A
Sodium Perborate	B	B	B	B	B	B	B	B	B	C	A	A	A		A
Sodium Peroxide	C	D	C	C	B	B	B	B	B	C	A	A	A		A
Sodium Phosphate	D	C	C	C	B	B	B	B	B	B	B	A	A		A
Sodium Phosphate Di-basic	D	C	C	C	B		B	B	B	A	A	A	A		A
Sodium Phosphate Tri-basic	D	C	C	C	B		B	B	B	A	A	A	A		A
Sodium Polyphosphate				B			B	B	B	B		A			A
Sodium Salicylate				A			A				A	A			A
Sodium Silicate	B	B	B	B	B		B	B		A	A	B	A		A
Sodium Silicate, hot	C	C	C	C	B		B	B			A	B			A
Sodium Sulfate	B	B	B	B	A	B	A	A		A	A	A	A		A
Sodium Sulfide	C	D	B	B	B	A	B	B		A	A	B	A		A
Sodium Sulfite	B	C		A	A	A	A	B	B	A	A	B	B		A
Sodium Tetraborate				A	A		A			A	A	B			A
Sodium Thosulfate	B	B	B	C	B	A	B	B		A	A	A	A		A

Table 1.2 (cont.) Chemical corrosion data

	Aluminum	Brass	Carbon Steel	Ductile Iron / Cast Iron	316 Stainless Steel	17-4PH	Alloy 20	Monel	Hastelloy C	Buna N (Nitrile)	Delrin	EPDM/EPR	Viton	Flexible Graphite	Teflon-Reinforced or
Soybean Oil	B	B	C	C	A		A	A		A	B	B	A		A
Starch	B	B	C	C	B		A	A		A	A	C	A		A
Steam (212°F)	A	A	A	A	A	A	A	B		D	D	B	C	A	A
Stearic Acid	A	C	C	C	B		B	B	A	A	A	B	A	A	A
Styrene	A	A	A	B	A		A	B	A	D	A	D	B		A
Sugar Liquids	A	A	B	B	A		A	A		A	A	B	A		A
Sugar, Syrups & Jam	B	B		C		A	A			A					A
Sulfate, Black Liquor	C	C	C	C	B	A	B	B		C	C	B	C		A
Sulfate, Green Liquor	B	C	C	C	B	A	B	B		C	A		C		A
Sulfate, White Liquor	B	C	C	C	B	B	D	C		C	D		C		A
Sulfur	A	D	C	C	B		A	B		D	A	B	B		A
Sulfur Chlorides	D	B	D	D	D		A	B		D	A	C	A	A	A
Sulfur Dioxide, dry	A	B	B	B	A	A	B	B	A	D	A	A	A	A	A
Sulfur Dioxide, wet	C	D			A	C	B	A	B	D	D	B		A	A
Sulfur Hexafluoride	A	B			A		A				A				A
Sulfur, Molten	A	D	C	B	B		A	D	B	D	D	B	B		A
Sulfur Trioxide		B	B	B	B	B	B		B	D	D		B	D	A
Sulfur Trioxide, dry	A	B	B	B	B	B	B	B	B	D	A	B	A	D	A
Sulfuric Acid 0 to 77%	C	C	D	D	C		B	B		B	D		A	A	A
Sulfuric Acid 100%	D	C	C	B	A	B	A	D		D	D	C	B	D	A
Sulfurous Acid	C	D	D	D	B		B	D	B	C	C	C	A	A	A

Table 1.2 (cont.) Chemical corrosion data

	Aluminum	Brass	Carbon Steel	Ductile Iron / Cast Iron	316 Stainless Steel	17-4PH	Alloy 20	Monel	Hastelloy C	Buna N (Nitrile)	Delrin	EPDM/EPR	Viton	Flexible Graphite	Teflon-Reinforced or
Tall Oil	C	B	B	B	B		B	B	A	B	A	D	A		A
Tannic Acid (Tannin)	C	B	C	C	B	B	B	B	B	B	A	B	A		A*
Tanning Liquors	A			B		B			B	D					A*
Tar & Tar Oils	A	A	A	A	A	A	A	A		C	A	D	A		A*
Tartaric Acid	B	B	D	D	A	A	A	B	B	C	A	B	A		A*
Tetraethyl Lead	B	B	C	C	B		B	A			A				A*
Toluol (Toluene)	A	A	A	A	A		A	A	A	D	C	D	B		A*
Tomato Juice	A	C	C	C	A		A	B		A	A		A		A
Transformer Oil	A	B	A	B	A		A	A		A	A		A		A
Tributyl Phosphate	A	A	A	A	A		A	A		D	A	B	D		A
Trichlorethylene	A	B	B	C	B	A	B	B	A	D	A	D	B	A	A
Trichloroacetic Acid	D	B		D	D		B	B	A	C	D		D		A
Triethanolamine	B			B			B	B	A	C	A	B			A
Triethylamine		B			B		B		A	B	C				A
Trisodium Phosphate	D			B			B		A	A	A	B	B		A
Tung Oil	B	B	B	B	A		A	C	A	A	A	D	A		A
Turpentine	B	B	B	B	B	A	B	D	A	B	A	D	A		A
Urea	B	B	C	C	B		B	B	A	C	A	B	D		A
Uric Acid	D				A		A		A		B				A
Varnish	A	A	C	C	A		A	A	A	C	A	D	B		A
Vegetable Oils	A	B	B	B	A		A	B	A	A	A	D	A		A
Vinegar	C	B	D	D	A		A	B	A	D	B	A	D		A

Table 1.2 (cont.) Chemical corrosion data

	Aluminum	Brass	Carbon Steel	Ductile Iron / Cast Iron	316 Stainless Steel	17-4PH	Alloy 20	Monel	Hastelloy C	Buna N (Nitrile)	Delrin	EPDM/EPR	Viton	Flexible Graphite	Teflon-Reinforced or
Vinyl Acetate	B	B		B	B		B	B	A		D	A		A	A
Water, Distilled	A	A	D	D	A	A	A	A	A	C	A	B	A		A
Water, Fresh	A	A	C	C	A	A	A	A	A	C	A	B	A		A
Water, Acid Mine	D	D	D	D	B	B		D	C	B	A	A	D	A	A
Waxes	A	A	A	A	A		A	A	A	A	C	A			A
Whiskey & Wines	D	B	D	D	A		A	A	A	B	A	A	A		A
Xylene (Xylol), dry	A	A	B	B	A		A	A	A	D	A	D	B		A
Zinc Bromide	D	B		D	B		B	B	A	B	A	B	B	A	A
Zinc Hydrosulfite	D	C	A	B	A		A	B	A	A	A	A			A
Zinc Sulfate	D	B	D	D	B		A	A	A	A	A	A	A		A

(Source: http://www.accutech.com.sg/Images/Products/Reference/ CATALOGE_PDF/ CORROSION_DATA.pdf)

2
Material Properties and Selection

2.1 General Properties and Selection Criteria

Proper material selection is critical to the performance of any machine, structure, refinery, or chemical plant. Among the many parameters that must be considered are structural strength specifications, heat resistance, corrosion resistance, physical properties, fabrication characteristics, composition, and structure of material and cost.

The properties that materials must have for a particular application depend largely on the environment in which they are to be used. Material selection begins from determination of equipment, operating conditions, temperature, pressure, and various components in the process.

No materials have properties that fulfill all requirements. For example, good heat conductivity is a desirable property for the fabrication of heat exchanger surfaces but not for insulation purposes. Obviously, both positive and negative properties can coexist in a single material. A corrosion resistant material may be insufficient for heat resistance or mechanical strength. Strong materials may be too brittle, e.g., ferrosilicon. Also, materials that have good mechanical and chemical properties may be too expensive.

The initial cost of a material does not provide the entire economic picture. At first, strong materials that are expensive may be more favorable than less expensive ones. The cost of processing cheap materials is sometimes high, thus creating abnormally high fabrication costs. For example, the cost of a ton of granite is a dozen times cheaper than that of nickel/chromium/molybdenum steel. However, granite structures are more expensive than steel towers of the same volume because of the high costs associated with processing granite. Furthermore, granite structures are much heavier than the steel ones; therefore, they require stronger, and thus more expensive, foundations.

Because any material may be characterized by some desirable and undesirable properties with respect to a specific application, the selection of materials is reduced to a reasonable compromise. In doing so, one strives to select materials so that properties correspond to the basic demands determined by the function and operating conditions of the equipment, tolerating some of the undesirable properties. The basic requirement for materials intended for fabricating chemical apparatuses or any equipment is mostly corrosion resistance, because this determines the durability of equipment. Often, corrosion data are reported as a weight loss per unit of surface area per unit of time.

Materials must have high chemical resistance as well as durability. For example, if the material dissolves in the product, the product quality may deteriorate, or materials may act as catalysts promoting side reactions and thus decreasing the yield of the primary product. Usually there are several materials suitable for use under the process conditions. In such cases, the material is selected by additional considerations. For example, if a vessel must be equipped with a sight glass, the material for fabricating this item must be transparent and safe. In this case, plexiglas may be used if the vessel operates at low temperatures. Glass is used for higher temperatures; however, glass is brittle and very sensitive to drastic temperature changes. Therefore, the accessories must be designed so that the glass cannot be broken occasionally and the poisonous or aggressive liquid is allowed to escape. In this application, double glasses or valves must be provided for an emergency to shut off the accessory from the working space of the vessel. Consequently, the poor construction property of glass may cause additional complications in the design. At very high temperatures, sight glasses are made from mica. For high-pressure drops, they are made from rock crystal, an excellent but very expensive material.

This section provides discussions and data that will assist in material selection. There are principal construction materials for welded, forged, and cast structures, such as vessels: cast irons, gray cast iron, white cast iron, malleable cast irons, nodular cast iron, austenitic cast iron, high-silicon cast iron, low-carbon steels (mild steel), high-carbon steels, low-carbon/low-alloy steels, high-carbon/low-alloy steels, high-alloy steels (corrosion-resistant, heat resistant and high-temperature), nickel, and nickel alloys. The latter part of this section contains useful information on general properties of materials of construction that have been complied and used by the authors over the years.

2.2 Cast Irons

Three main factors that determine the properties of cast iron are the chemical composition of the cast iron, the rate of cooling of the casting in the mold, and the type of graphite formed.

Most commercial cast irons contain 2.5–4% carbon, and the occurrence of some of this carbon, as free graphite in the matrix, is the characteristic feature of thin material. About 0.8–0.9% carbon is in a bound form as cementite (iron carbide). The cast irons usually have a ferrite-pearlite structure, which determines its mechanical properties. The ferrite content determines the cast iron's viscosity, while the pearlite content determines its rigidity and strength.

Table 2.1 Mechanical properties of cast iron

Material	Specification	Tensile strength (ton$_f$/in.^{2})	Tensile strength (N/mm^2)	Elongation (%)
Gray Cast Iron	BS1452 Grade 10	10	155	
	14	14	215	
	26	26	400	
Nodular Cast Iron	BS2789 SNG 24/17	24	370	17
	32/7	32	500	7
	47/2	47	730	2
Malleable Cast Iron	BS310 B290/6		290	6
	B340/12		340	12
Whiteheart	BS309 W340/3		340	3
	W410/4		410	4
Pearlite	BS3333 P44017		440	7
	P540/5		540	5
	P690/2		690	2

Because cast iron has a carbon content approximately equivalent to its eutectic composition, it can be cast at lower temperatures than steel and flows more readily than steel because of its much narrower temperature solidification range. The presence of the graphite flakes in cast iron decreases its shrinkage on solidification much less than that of steel. These factors contribute to the fabrication of cast iron as sound castings in complex shapes with accurate dimensions at low cost.

The physical properties of cast irons are characterized by the following data:

- Density = 7.25 kg/dm^3
- Melting temperature between 1250–3280°C
- Heat capacity Cp = 0.13 kcal/kg°C
- Heat conductivity A = 22–28 kcal/m°C hr
- Coefficient of linear expansion 11 X 10^{-6}

The cast irons do not possess ductility. They cannot be pressed or forged even while heated; however, their machining properties are considered good. Typical mechanical properties of various types of cast iron are given in Table 2.1.

2.2.1 Gray Cast Iron

Gray cast iron is the most commonly used cast iron and is the least expensive. It is the easiest to cast and machine. The tensile strength of gray cast iron ranges from 155 to 400 N/mm^2 (10 to 26 ton/in.^{2}). The tensile modulus ranges from 70 to 140 kN/mm^2 and the hardness from 130 to 300 DPN.

In nearly all standards for gray cast iron, the grades are designated according to the tensile strength, not composition. In the British standard BS1452, for example, there are seven grades from 155 to 400 N/mm^2 (10 to 26 ton$_f$/in.^{2}). This is the tensile strength measured on a test bar having a diameter of approximately 30 mm. The actual strength of a casting will differ from that of the test bar according to the cross-sectional area.

Castings are designed to be loaded in compression because the compressive strength of gray iron is about three times that of its tensile strength. The recommended maximum design stress in tension is one-quarter the ultimate tensile strength for cast irons a value up to 185 N/mm^2 (12 tonf/ in.^{2}). The fatigue strength is one-half the tensile strength. Notched specimens show the same value as unnotched specimens. For 220 N/mm² (14 tonf/in.²) grades and above, the fatigue strength of unnotched specimens is approximately one-third the tensile strength. There is some notch sensitivity, although much less than is found in steel.

2.2.2 White Cast Iron

White cast iron is very hard (from 400 to 600 DPN) and brittle. All white cast irons are very difficult to machine and usually are finished by grinding. Table 2.2 gives properties of the four principal types of white cast irons.

Table 2.2 Properties of white iron

	Unalloyed White iron	Low-Alloy White Iron	Martensitic White Iron (Ni-hard)	High-Carbon, High-chromium, White Iron
Composition (%) Carbon	3.5	2.6	3.0	2.8
Silicon	0.5	1.0	0.5	0.8
Nickel			3.5	
Chromium		1.0	2.0	27
Hardness (DPN)	600	400	600	500
Tensile strength, (N/mm²)	270	300	330	420

2.2.3 Malleable Cast Iron

This type of cast iron is made by high-temperature heat treatment of white iron castings. They are normally applied to the fabrication of conveyor chain links, pipe fittings, and gears.

2.2.4 Nodular Cast Iron

Nodular cast iron, also referred to as ductile cast iron, is manufactured by inoculating the molten metal with magnesium or cesium.

It is characterized by a homogeneous structure, higher than usual abrasion-resistance and strength for dynamic loads, and by easy machining. A wide variety of grades are available, with typical tensile strengths ranging from 380 to 700 N/mm^2 (25 to 40 tonf/in.^{2}), elongations from 1–2%, and hardness from 150 to 300 DPN (see Table 3.1). The tensile modulus is approximately 170 kN/mm^2. The design stress is half the 0.1% proof stress, and the fatigue design stress is one-third the fatigue limit. The nodular cast iron is used for many applications such as valves in pipelines for petroleum products and underground pipelines.

2.2.5 Austenitic Cast Iron

Austenitic cast irons, either flake graphite irons or nodular graphite irons, are produced by mining in nickel from 13–30%, chromium from 1–5%, and copper from 0.5–7.5 to lower nickel-containing grades to augment the corrosion resistance at lower cost.

The main advantages of austenitic cast irons are corrosion and heat resistance. For corrosion resistance, the flake and nodular are similar, but the mechanical properties of nodular cast irons are superior. Some of the commercially available austenitic cast irons are given in the Tables 2.3 through 2.5.

2.2.6 Abrasion Resistance

The white cast irons and their low alloys have good abrasion resistance properties. White cast irons are used for grinding balls, segments for mill liners, and slurry pumps. They are used for muller tyres and augers in the ceramic industry, for attrition mill plates and chip feeders in the pulp and paper industry, and for balls for grinding pigments in the paint industry.

2.2.7 Corrosion Resistance

The corrosion resistance of unalloyed and low-alloy flake, nodular, malleable and white cast iron is comparable to mild-alloy and low-alloy steel. However, these cast irons have a major advantage over steel, namely, greater cross section or wall thickness than steel. Consequently, they have a longer life although they corrode at the same rates. Although the matrix of a cast iron pipe

Material Properties and Selection

Table 2.3 Properties of spheroidal cast irons

BS3468 Designation	ASTM A439 Designation	Trade Names	C (% max.)	Si (%)	Mn (%)
AUS202 Grade A	D-2	SG Ni-resist type D2	3.0	1.0 to 2.8	0.7 to 1.5
AUS202 Grade B	D-2b	SG Ni-resist type D2b	3.0	1.0 to 2.8	0.7 to 1.5
AUS203	D-2c	SG Ni-resist type D2c	3.0	1.0 to 2.8	1.8 to 2.4
	ASTM		2.6	2.5	3.75 to 4.5
	A571				
AUS204		SG Nicro-silal	3.0	4.5 to 5.5	1.5
AUS205	D-3	SG Ni-resist type D3	2.6	1.5 to 2.8	0.5 max.
	D-4	SG Ni-resist type D4	2.6	5.0 to 6.0	1.0 max.
	D-5	SG Ni-resist type D5, Minovar	2.4	1.0 to 2.8	1.0 max.

may rust, for example, the graphite network prevents disintegration of the pipe and permits its duty for a longer time than a steel pipe. The austenitic cast irons are widely used in many industries: food, pharmaceutical, petroleum, chemical, petrochemical, pulp and paper, etc. in mildly corrosive and erosive situations where the life of unalloyed or low-alloy cast iron or steel is short, but the high cost of stainless steel and nonferrous alloys cannot be justified. Other austenitic cast iron applications can be found in food and dairy production, where the metallic contamination of the product must be eliminated.

Table 2.4 Properties of flake cast irons

BS3468 Designation	ASTM A436 Designation	Trade Names	C (% max.)	Si (%)	Mn (%)
AUS101 Grade 1	Type 1	Ni-resist type 1	3.0	1.0 to 2.8	1.0 to 1.15
AUS101 Grade B	Type lb	Ni-resist type lb	3.0	1.0 to 2.8	1.0 to 1.5
AUS102 Grade A	Type 2	Ni-resist type 2	3.0	1.0 to 2.8	1.0 to 1.5
AUS104	–	Nicrosilal	1.6 to 2.2	4.5 to 5.5	1.0 to 1.5
AUS105	Type 3	Ni-resist type 3	2.6	1.0 to 2.0	0.4 to 0.8
	Type 4	Ni-resist type 4	2.6	5.0 to 6.0	0.4 to 0.8
–	Type 6		3.0	1.5 to 2.5	0.8 to 1.5

2.2.8 Temperature Resistance

The persistent increase in volume of cast iron items in high-temperature situations becomes the limiting factor in the use of unalloyed cast irons, especially in flake graphite castings. The addition in a casting of about 1% of chromium can control the growth in the temperature range from 400 to 600°C. Above 600°C, scaling due to surface oxidation becomes an undesirable phenomenon in the *use of* unalloyed cast irons. For achieving dimensional stability and long life, silicon cast irons (containing 5.5–7% silicon) may be used for temperatures up to 800°C in cases where it is not subjected to thermal shock. For thermal cycling and thermal shock situations of temperatures up to 950°C, the 30% nickel austenitic cast irons are preferred. Above this temperature, where there is no thermal shock, the 28% chromium cast iron is recommended.

Material Properties and Selection

Table 2.5 Properties of graphite casts irons

Ni (%)	Cr (%)	UTS (N/mm² min.)	0.5% Proof Stress (N/mm² min.)	Elongation (% min.)	Hardness (HB)	Elastic Modulus (kN/mm²)
1.80 to 22.0	1.0 to 2.5	370	230	8.0	201 max.	Y (115)
18.0 to 22.0	2.0 to 3.5	370	230	6.0	205 max.	(151)
21.0 to 24.0	0.5 max.	370	230	20.0	170 max.	(110)
21.0 to 24.0	0.20 max.	430		30	170 max.	
18.0 to 22.0	1.0 to 2.5	370	230	10.0	230 max.	
28.0 to 32.0	2.5 to 3.5	370	230	7.0	201 max.	(110)
28.0 to 32.0	4.5 to 5.5	400			202–273	
34.0 to 36.0	0.10 max.	380	210°	20.0	131–185	

Ni (%)	Cr (%)	Cu (%)	UTS (N/mm² min.)	Elongation (% min.)	Hardness (HE)	Elastic Modulus (kN/mm²)
13.5 to 17.5	1.0 to 2.5	5.5 to 7.5	140	2.0	212 max.	(90)
13.5 to 17.5	2.0 to 3.5	5.5 to 7.5	180		248 max.	(105)
18.0 to 22.0	1.0 to 2.5	0.5 max.	140		212 max.	(90)
18.0 to 22.0	1.8 to 4.5	0.5 max.	190	2.0	248 max.	(110)
28.0 to 32.0	2.5 to 3.5	0.5 max.	170		212 max.	(105)
29.0 to 32.0	4.5 to 5.5	0.5 max.	170		149–212	
18.0 to 22.0	1.0 to 2.0	3.5 to 5.5	170		124–174	

Gray cast irons do not have the abrupt ductile to brittle fraction transition down to −40°C that takes place in steels. Special austenitic nodular cast iron, similar to the AUS 203 grade but containing a higher manganese content of about 4%, has been obtained for cryogenic purposes for temperatures down to −253°C.

2.2.9 Welding Cast Iron

Welding is sometimes used to repair broken and defective castings. This process is more difficult than welding steel because the high urban content in cast iron may lead to brittle structures on cooling, thus causing cracking. However, special techniques have been developed for fusion and non-fusion welding. One should consider carefully whether the properties of varieties of cast irons would suit one's demands before specifying more expensive materials.

2.3 Steels

A second group of structural materials in the iron base category is steels. They have obtained an exclusive importance because of their strength, viscosity, and their ability to withstand dynamic loads. Also, they are beneficial for producing castings, forgings, stamping, rolling, welding, machining, and heat treatment works. Steels change their properties over a wide range depending on their composition, heat treatment, and machining. Most steels have a carbon content of 0.1–1%, but in structural steels this does not exceed 0.7%. With higher carbon contents, steel increases in strength but decreases in plasticity and weldability. In the carbon steels designed for welding, the carbon content must not exceed 0.3%; in the alloy steels it must not exceed 0.2%. When the carbon content in the steels exceeds the abovementioned value, they are susceptible to air hardening. Hence, high stresses may be created and hardening fractures in welding zones may be formed. The steels with low carbon content (below 0.2%) are well stamped and stretched, well cemented and nitrated, but badly machined. The physical properties of low-carbon, low-alloy steels are characterized by the following data:

- density = 7.85 kg/dm^3
- melting temperature Tm = 1400–1500°C
- thermal conductivity = 40–50 kcal/m°C hr

2.3.1 Low Carbon Steels (Mild Steel)

Mild steel (< 0.25% carbon) is the most commonly used, readily welded construction material and has the following typical mechanical properties (Grade 43A in BS4360; weldable structural steel):

- Tensile strength, 430 N/mm^2
- Yield strength, 230 N/m^2
- Elongation, 20%
- Tensile modulus, 210 kN/mm^2
- Hardness, 130 DPN

No other steel exceeds the tensile modulus of mild steel. Therefore, in applications in which rigidity is a limiting factor for design (e.g., for storage tanks and distillation columns), high-strength steels have no advantage over mild steel. Stress concentrations in mild steel structures are relieved by plastic flow and are not as critical in other, less-ductile steels. Low-carbon plate and sheet are made in three qualities: fully killed with silicon and aluminum, semi-killed (or balanced), and rimmed steel. Fully killed steels are used for pressure vessels. Most general-purpose structural mild steels are semi-killed steels. Rimming steels have minimum amounts of deoxidation and are used mainly as thin sheet for consumer applications. The strength of mild steel can be improved by adding small amounts (not exceeding 0.1%) of niobium, which permits the manufacture of semi-killed steels with yield points up to 280 N/mm^2. By increasing the manganese content to about 1.5%, the yield point can be increased up to 400 N/mm^2. This provides better retention of strength at elevated temperatures and better toughness at low temperatures.

2.3.2 Corrosion Resistance

Equipment from mild steel usually is suitable for handling organic solvents, with the exception of those that are chlorinated, cold alkaline solutions (even when concentrated), sulfuric acid at concentrations greater than 88%, and nitric acid at concentrations greater than 65% at ambient temperatures. Mild steels are rapidly corroded by mineral acids even when they are very dilute (pH less than 5). However, it is often more economical to use mild steel and include a considerable corrosion allowance on the thickness of the apparatus. Mild steel is not acceptable in situations in which metallic contamination of the product is not permissible.

2.3.3 Heat Resistance

The maximum temperature at which mild steel can be used is 550°C. Above this temperature the formation of iron oxides and rapid scaling makes the use of mild steels uneconomical. For equipment subjected to high loadings at elevated temperatures, it is not economical to use carbon steel in cases above 450°C because of its poor creep strength. Creep strength is time-dependent, with strain occurring under stress.

2.3.4 Low Temperatures

At temperatures below 10°C the mild steels may lose ductility, causing failure by brittle fracture at points of stress concentrations, especially at welds. The temperature at which the transition from ductile to brittle fraction occurs depends not only on the steel composition but also on thickness. Stress relieving at 600–700°C for steels decreases operation at temperatures some 20°C lower. Unfortunately, suitable furnaces generally are not available, and local stress relieving of welds, etc., is often not successful because further stresses develop upon cooling.

2.3.5 High-Carbon Steels

High-carbon steels containing more than 0.3% are difficult to weld, and nearly all production of this steel is as bar and forgings for such items as shafts, bolts, etc. These items can be fabricated without welding. These steels are heat treated by quenching and tempering to obtain optimum properties up to 1000 N/mm^2 tensile strength.

2.3.6 Low-Carbon, Low-Alloy Steels

Low-carbon, low-alloy steels are widely used for fabrication-welded and forged-pressure vessels. The carbon content of these steels is usually below 0.2%, and the alloying elements that do not exceed 12% are nickel, chromium, molybdenum, vanadium, boron, and copper.

2.3.7 Mechanical Properties

The maximum permissible loading of low-alloy steels, according to the ASME code for pressure vessels, is based on proof stress (or yield point), which is applicably superior to those of carbon steels. The cost of a pressure vessel in alloy steel may be more expensive than in carbon steel. However, consideration should be given to other cost savings resulting from thinner-walled vessels, which provide fabrication savings on welding, stress relieving, transportation, erection, and foundation.

2.3.8 Corrosion Resistance

The corrosion resistance of low-alloy steels is not significantly better than that of mild steel for aqueous solutions of acids, salts, etc. The addition of 0.5% copper forms a rust-colored film preventing further steel deterioration. Small amounts of chromium (1%) and nickel (0.5%) increase the rust resistance of copper steels still further. Low-alloy steels have good resistance to corrosion by crude oils containing sulfur.

In operations involving hydrogen at partial pressures greater than 35 kgf/cm^2 and temperatures greater than 250°C, carbon steels are decarbonized and fissured internally by hydrogen. Small additions of molybdenum prevent hydrogen attack at temperatures up to 350°C and pressures up to 56 kgff cm^2. For higher temperatures and pressures chromium/molybdenum steels (2.25 Cr, 0.5 Mo) are used.

2.3.9 Oxidation Resistance and Creep Strength

Chromium is the most effective alloying element for promoting resistance to oxidation. In atmospheres contaminated with sulfur, lower maximum temperatures are necessary. In fractionation columns for petroleum products, where the oxygen content is restricted, higher temperatures can be used without excessive waste of the metal. The creep strength of steels is a factor limiting the maximum temperatures for such high-pressure equipment as shells and stirrers of high temperature reactors. The stress for 1% creep in 100,000 hours, which is a design criterion, is accepted to be not less than two-thirds of the creep stresses.

2.3.10 Low-Temperature Ductility

Nickel is the alloying element used for improving low-temperature ductility. The addition of 1.5% nickel to 0.25% Cr/0.25% Mo steels provides satisfactory application for moderately low temperatures down to about −50°C. Heat treatment by quenching and tempering improves the low temperature ductility of steels, such as 0.5 Cr, 0.5% Mo, 1% Ni Type V. For lower-temperature application (below –196°C), up to 9% nickel is used as the sole alloying element.

2.3.11 High-Carbon, Low-Alloy Steels

High-carbon (about 0.4%), low-alloy steels that are not weldable usually are produced as bars and forging for such items as shafting, high-temperature bolts, and gears and ball bearing components. These steels can be less drastically quenched and tempered to obtain tensile strengths of at least 1500 N/mm^2, thus minimizing the danger of cracking.

2.3.12 High-Alloy Steels

Stainless and heat-resisting steels containing at least 12% by weight chromium and 8% nickel are widely used in industry. The structure of these steels is changed from a magnetic, body-centered cubic, or ferritic crystal, structure to a nonmagnetic, face-centered cubic, or austenitic crystal, structure.

2.3.12.1 Chromium Steels (400 Series), Low-Carbon Ferritic (Type 405)

12–13% Chromium – This type of steel is mainly used for situations in which the process material may not be corrosive to mild steel, yet contamination due to rusting is not tolerable, and temperatures or conditions are unsuitable for aluminum. However, prolonged use of these steels in the temperature range of 450 to 550°C causes low-temperature embrittlement of most ferritic steels with more than 12% chromium.

2.3.12.2 Medium Carbon Martensitic

I3–17% Chromium (Types 403, 410, 414, 416, 420, 431, 440) – These steels resist oxidation scaling up to 825°C but are difficult to weld and, thus, are used mainly for items that do not involve welded joints. They are thermally hardened and useful for items that require cutting edges and abrasion resistance in mildly corrosive situations. However, they should not be tempered in the temperature range of 450 to 650°C. This reduces the wear resistance and hardness and also lowers the corrosion resistance because of the depletion of chromium in solution from the formation of chromium carbides.

2.3.12.3 Medium Carbon Ferrule

17–30% Chromium (Types 430 and 446) – The 17% ferritic steels are easier to fabricate than the martensitic grades. They are used extensively in equipment for nitric acid production. The oxygen-resistant and sulfur-resistant 30% chromium steel can be used at temperatures up to 1150°C but only for lightly loaded and well-supported furnace items because of its poor creep and brittlement properties when equipment is down to ambient temperatures.

2.3.12.4 Chromium/Nickel Austenitic Steels (300 Series)

The excellent corrosion resistance over a wide range of operating conditions and readily available methods of fabrication by welding and other means of shaping metals make these steels the most extensively used throughout the chemical and allied industries. The formation of a layer of metal oxide on the surface of this steel provides better corrosion resistance in oxidizing environments than under reducing conditions. Common steels 304, 304L, 347, 316, and 316L are used for equipment exposed to aqueous solutions of acids and other low-temperature corrosive conditions. For high-temperature regimes involving oxidation, carbonization, etc., the 309 and 310 compositions may be recommended because of their higher chromium content and, thus, better resistance to oxidation.

Type 3W-I9/10 (chromium nickel) provides a stable austenitic structure under all conditions of fabrication. Carbon (0.08% max.) is sufficient to have reasonable corrosion resistance without subsequent corrosion resistance for welded joints.

Type 304 is used for food, dairy, and brewery equipment and for chemical plants of moderate corrosive duties.

Type 304L is used for applications involving the welding of plates thicker than about 6.5 mm. Type 321 is an 18/10 steel that is stabilized with titanium to prevent weld decay or inter-granular corrosion. It has similar corrosion resistance to types 304 and 304L but a slightly higher strength than 304L. Also, it is more advantageous for use at elevated temperatures than 304L.

Type 347 is an 18/11 steel that is stabilized with niobium for welding. In nitric acid it is better than Type 321. Otherwise, it has similar corrosion resistance.

Type 316 has a composition of 17/12/2.5 chromium/nickel/molybdenum. The addition of molybdenum greatly improves the resistance to reducing conditions such as dilute sulfuric acid solutions and solutions containing halides (such as brine and sea water).

Type 316L is the low-carbon (0.03% max.) version of type 316 that should be used where the heat input during fabrication exceeds the incubation period of the 316 (0.08% carbon) grade. It is used for welding plates thicker than 1 cm., for example.

Type 309 is a 23/14 steel with greater oxidation resistance than 18/10 steels because of its higher chromium content.

Type 315 has a composition that provides a similar oxidation resistance to type 309 but has less liability to embrittlement due to sigma formation if used for long periods in the range of 425 to 815°C. (Sigma phase is the hard and brittle intermetallic compound FeCr formed in chromium rich alloys when used for long periods in the temperature range of 650 to 850°.)

Alloy 20 has a composition of 20% chromium, 25% nickel, 4% molybdenum and 2% copper. This steel is superior to type 316 for severely reducing solutions such as hot, dilute sulfuric acid.

2.3.13 Precipitation Hardening Stainless Steels

These steels do not have ANSI numbers and are referred to by trade name. They can be heat-treated to give the following mechanical properties.

2.4 Materials Properties Data Tables

Table 2.6 Aluminum alloys generally used in cryogenic applications

Cryogenic Liquid	Boiling Point (°C)	Material	Minimum Temperature of Use (°C)
Propane	−42	Carbon steels	−50
Carbon Dioxide	−78	2.25% nickel steel	−65
Acetylene	−84	3.5% nickel	−100
Ethylene	−104	Aluminum/ magnesium alloys	−270
Methane	−161	Austenitic stainless steel	−270
Oxygen	−182	Nickel alloys	−270
Argon	−186	Copper alloys	−270

Material Properties and Selection

Table 2.7 General properties of titanium, tantalum and zirconium as reported in british standard code of practice cp3003: lining of vessels and equipment of chemical processes, part 9

	Density (g/cm^2)	Melting Point (°C)	Coefficient of Expansion × 10^{-6}¶ (°C)	Thermal Conductivity (W/m°C)	Yield Strength (N/mm^2)	Tensile Modulus (N/mm^2)	Hardness (DPN)
Titanium	4.5	1668	9	15	345	103,000	150
Tantalum	16.6	2996	6.5	55	240	185,000	170
Zirconium	6.5	1852	7.2	17	290	80,000	180

Table 2.8 Mechanical properties of titanium and alloys

Grade	Alloying Element (% max.)	UTS (N/mm² min.)	Elongation (%)	Hardness (DPN)
1	Oxygen 0.18	240	24	150
2	Oxygen 0.25	345	20	180
4	Oxygen 0.4	550	15	260
5	Aluminum 6 Vanadium 4	900	10	
6	Aluminum 5 Tin 2.5	820	10	
7	Palladium 0.15/0.25 Oxygen 0.25	345	20	180
8	Palladium 0.15/0.25 Oxygen 0.35	450	18	210

(Select data from ASTM/B265/337/338 and Gleekman, L.W., Trends in Materials Applications of Non Ferrous Metals, Chem. Eng. Casebook, 111–118, Oct. 12, 1970)

Table 2.9 General properties of carbon and graphite

	Carbon	Graphite
Density (g/cm³)	1.8	1.8
Tensile Strength (N/mm)	28	10
Compressive Strength (N/mm²)	135	70
Tensile Modulus (kN/mm²)	10	3
Thermal Conductivity (W/m°C)	4	70
Linear Coefficient of Expansion (°C^{-1})	3×10^6	4×10^6

Material Properties and Selection

Table 2.10 General properties of precious metals

Property	Platinum	Gold	Silver	10% Rh/Pt	20% Rh/Pt	10% Ir/Pt	20% Ir/Pt	70% Au 30% Pt 1% Rh
Density (g/cm^3)	21.45	19.3	10.5	20	18.8	21.6	21.7	20
Melting point (°C)	1769	1063	961	1850	1900	1800	1815	1250
Thermal Conductivity (W/m °C)	70	290	418					
Young's Modulus (kN/mm^2)	170	70	70	195	215			
Tensile Strength Annealed (N/mm^2)	140	110	140	325	415	370	695	925
Hardness (DPN) Annealed	40	20	26	75	90	120	200	250

Table 2.11 Effect of temperature on tensile strength (N/mm^2) of titanium alloys

Material ASTM Grade	Room Temperature	100°C	200°C	300°C	400°C	500°C
1	310	295	220	170	130	
2	480	395	295	205	185	
3	545	460	325	250	200	
4	760	585	425	310	275	
5	1030	890	830	760	690	655
6	960	820	690	620	535	460
7	480	395	295	205	185	
8	545	460	325	250	200	

(for more extensive data see Gleekman, L.W. " Trends in Materials Application-Non-Ferrous Metals," Chem. Eng. Casebook, 111–118 (October 12, 1970)).

Table 2.12 Chemical corrosion resistance of tantalum and platinum

Chemical	Tantalum	Platinum
Acetylene	G[a]	NR
Alkalis	NR[b]	G
Bromine (wet or dry)	G	NR
Bromic Acid	G	NR
Cyanides	G	NR
Fluorine	NR	G
Compounds Containing Fluorine	NR	G
Ethylene	G	NR
Lead Salts	G	NR
Lead Oxide	NR	G
Metals (molten)	G	NR
Mercury	G	NR
Mercury Compounds	G	NR
Oleum	NR	G
Phosphoric Acid	NR	G
Sulfur Trioxide	NR	G

[a] G = good
[b] NR = not recommended

Table 2.13 Properties and uses of common thermoplastics

Plastic Material	First Introduced	Strength	Electrical Properties	Acids	Bases	Oxidizing Agents	Common Solvents	Product Manufacturing Methods	Common Applications
Acetal Resins	1960	H[a]		P[b]	P	P	G[c]	Injection, blow or extrusion molded	Plumbing, appliance, automotive industries
Acrylic Plastics	1931		G	P	P	F-P	F-P	Injection, compression extrusion or blow molded	Lenses, aircraft and building glazing, lighting fixtures, coatings, textile fibers
Arc Extinguishing Plastics	1964		E[d]					Injection or compression molded and extruded	Fuse tubing, lightning arrestors, circuit breakers, panel boards

Table 2.13 (cont.) Properties and uses of common thermoplastics

Plastic Material	First Introduced	Strength	Electrical Properties	Acids	Bases	Oxidizing Agents	Common Solvents	Product Manufacturing Methods	Common Applications
Cellulose Plastics Cellulose Acetate	1912	M^e					P	All conventional processes	Exellent vacuum-forming material for blister packages, etc.
Cellulose Acetate Butyrate		H				F	F	Molded with plasticizers	Exellent moisture resistance-metalized sheets and film, automobile industry
Cellulose Nitrate	1889	M						Cannot be molded	Little use today because of fire hazard
Cellulose Propionate		H						All conventional processes	Toys, pens, automotive parts, toothbrushes, handles

Material Properties and Selection

Ethyl Cellulose		H+					All conventional processes	Military applications, refrigerator components, tool handles
Chlorinate Polyether	1959	M+	A^f	VG	VG	VG	Injection, compression, transfer or extrusion molding	Bearing retainers, tanks, tank linings, pipe, valves, process equipment
Fluorocarbon (TFE)	1930	M	A	VG	VG	VG	Molding by a sintering process following preforming	High-temperature wire and cable insulation, motorlead insulation; chemical process equipment

Table 2.13 (cont.) Properties and uses of common thermoplastics

Plastic Material	First Introduced	Strength	Electrical Properties	Acids	Bases	Oxidizing Agents	Common Solvents	Product Manufacturing Methods	Common Applications
Fluorinated Ethylene Propylene (FEP)		M+	A	G	G	G	G	Injection, blow molding, and extrusion and other conventional methods	Autoclavable laboratory ware and bottles
Class-Bonded Mica	1919	M	G	VG	G	G	G	Moldable with inserts like the organic plastics	Arc chutes, radiation generation equipment, vacuum tube components, thermocouples

Material Properties and Selection

Hydro-carbon Resins	1960	M	A	A	A	A	A	Molding with transfer and compression process, costing	Used as lamination resins for various industrial laminates
Methylpentene Polymers (TPX)	1965	M+	E	F	F	F	F	Most conventional processes	Used for electrical and mechanical applications
Parylene (polyparaxylene)	1960							A monomer of the organic compound is vaporized and condensed on a surface to polymerize	Coating material for sensing probes
Phenoxy Plastics	1962	H-M	F	F	F	F	F	Injection, blow and extrusion molding, coatings and adhesives	Adhesives for pipe-bonding compounds, bottles

Table 2.13 (cont.) Properties and uses of common thermoplastics

Plastic Material	First Introduced	Strength	Electrical Properties	Acids	Bases	Oxidizing Agents	Common Solvents	Product Manufacturing Methods	Common Applications
Polyamide Plastics Nylon	1938	H-M	A	P	P	P	VG	Injection, blow and extrusion molding	Mechanical components (gears, cams, bearings), wire insulation pipe fittings
Polycarbonate Plastics	1959	H-M	VG	G-F	G-F	F	F	All molding methods, thermoforming, fluidized bed coating	Street light globes centrifuge bottles, high-temperature lenses, hot dish handles
Polychlorotrifluoroethylene (CTFE)	1938	H	E	VG	VG	VG	VG	Molded by all conventional techniques	Wire insulation, chemical ware, pipe lining, pipe, process equipment lining

Material Properties and Selection

Polyester-Reinforced Urethane	1937	H		G	G	G	G	Compression molded over a wide temperature range	For heavy-duty leather applications-industrial applications
Polyimides	1964	H-M	E	G-VG	P	P	G-VG	Molded in nitrogen atmosphere	Bearings, compressors, valves, piston rings
Polyolefin Plastics Ethlene Vinyl Acetate (EVA)	1940	H		G	G	G-F	G-F	Most conventional processes	Molded appliance and automotive parts, garden hose, vending machine tubing
Polyallomers	1962	H	G-VG	F	F	F	F	Molding processes, all thermoplastic processes	Chemical apparatus, typewriter cases, bags luggage shells, auto trim

Table 2.13 (cont.) Properties and uses of common thermoplastics

Plastic Material	First Introduced	Strength	Electrical Properties	Acids	Bases	Oxidizing Agents	Common Solvents	Product Manufacturing Methods	Common Applications
Polyethylene	1939	H	VG	G-VG	G-VG	P	P	Injection, blow, extrusion and rotational	Pipe, pipe fittings, surgical implants, coatings, wire and cable insulation
Polypropylene Plastics	1954	H-M	VG	VG	VG	F	F	Same as PVC	Housewares, appliance parts, auto ducts and trim, pipe, rope, nets
Polyphenylene Oxide	1964	M	F-G	E	E	VG	VG	Extruded, injection molded, thermoformed and machined	Autoclavable surgical tools, coil forms, pump housing, valves, pipe

Material Properties and Selection

Polysulfone	1965	M	VG	VG	VG	F	P-F	Extrusion and injection molded	Hot water pipes, lenses, iron handles, switches, circuit breakers
Poly-vinylidene Fluoride (VF$_2$)	1961	H	VG+	G-VG	G	G-VG	G-VG	Molded by all process, fluidized bed coatings	High-temperature valve seats, chemical resistant pipe, coated vessels, insulation
Styrene Plastics ABS Plastics	1933	M-H	VG+	G-VG	G-VG	F-G	F	Thermoforming, injection, blow, rotational and extrusion molds	Business machine and camera housings, blowers, bearings, gears, pump impellers

Table 2.13 (cont.) Properties and uses of common thermoplastics

Plastic Material	First Introduced	Strength	Electrical Properties	Acids	Bases	Oxidizing Agents	Common Solvents	Product Manufacturing Methods	Common Applications
Polystyrene	1933	M-H	VG+	G	G	F	P-F	Most molding processes	Jewelry, light fixtures, toys, radio cabinets, housewares, lenses, insulators
Styrene Acrylonitrile (SAN)		H	VG	VG	G	G	G	Most molding processes	Lenses, dishes, food packages, some chemical apparatus, batteries, film
Urethane	1955	M-H+		G-VG	G	G	F-G	Extruded and molded	Foams for cushions, toys, gears, bushings, pulleys, shock mounts

Material Properties and Selection

Vinyl Plastics, Copolymers of Vinyl Acetate and Vinyl Chloride	1835–1912	M+		G	G-F	G	G	All molding processes	Floor products, noise insulators
Polyvinyl Acetate	1928	M		P-G	G	G	P	Coating and adhesives	Adhesives, insulators, paints, sealer for cinder blocks
Polyvinyl Aldehyde	1940	H		VG	VG	VG	VG	Most molding processes	Used for coatings and magnet wire insulation, interlayer of safety glasses
Polyvinyl Chloride (PVC)	1940	M-H		VG	G	G	G	Extrusion, injection, rotational, slush, transfer, compression, blow mold	Pipe conduit and fittings, cable insulation, downspouts, bottles, film

Table 2.13 (cont.) Properties and uses of common thermoplastics

Plastic Material	First Introduced	Strength	Electrical Properties	Acids	Bases	Oxidizing Agents	Common Solvents	Product Manufacturing Methods	Common Applications
PVC Plastisols	1940	M-H		VG	G	G	G	Slush and rotationally molded, foamed, extruded	Used in coating machines to cover paper, cloth and metal
Polyvinylidene Chloride	1940	H		VG	VG	VG	VG	Same as PVC	Auto seat covers, film, bristles, pipe and pipe linings, paperboard coatings

(Source: Cheremisinoff, N.P. and P.N. Cheremisinoff, Fiberglass Reinforced Plastics Deskbook, Ann Arbor Science Publishers Inc., Ann Arbor, Mich, 1978)

[a]H = high [b]P = poor [c]G=good [d]E = excellent [e]M = moderate [f]A = average

Material Properties and Selection

Table 2.14 Mechanical properties of common thermoplastics

	Polyethylene	Polypropylene	PVC[a]	ABS[a]	PTFE[a]	Acrylics	Nylon 66	Acetal
Specific Gravity	0.93	0.9	1.4	1.05	2.2	1.2	1.15	1.4
Tensile Strength (N/mm^2)	10	34	55	34	21	70	83	70
Elongation at Break (%)	500	300	15	50	300	5	100	70
Tensile Modulus (N/mm^2)	170	1360	3100	2100	415	3450	2760	2900
Thermal Expansion (10^{-3}/°C)	15	11	7	10	10	7	10	10

[a]PVC = polyvinychloride; ABS = acrylonitrile butadiene styrene; PTFE = polytetrafluorenethylene

Table 2.15 Average physical properties of plastics

Material	Formula	Density g cm⁻³	Flammability	Limiting oxygen index %	Optical transmission %	Radiation resistance	Refractive index	Resistance to Ultraviolet	Water absorption %	Water absorption equilibrium %	Water absorption – over 24 hours %
Cellulose Acetate	CA	1.3	HB	19	–	Fair	1.49	Fair	–	–	1.9–7.0
Cellulose Acetate Butyrate	CAB	1.20	HB	17	–	Fair	1.478	Good	–	–	0.9–2.2
Ethylene-Chloro-trifluoroethylene copolymer	E-CTFE	1.68	V0	60	–	Fair	–	–	–	–	<0.02
Ethylene-Tetrafluoroethylene Copolymer	ETFE	1.7	V0	30–32	–	Fair	1.403	Excellent	0–0.03	–	–
Fluorinated Ethylene Propylene Copolymer	FEP	2.15	V0	95	–	Poor	1.344	Excellent	0.01	–	–
Polyacrylonitrile butadiene-styrene	ABS	1.05	HB @ 1.5mm	19	–	Fair	–	Poor	–	–	0.3–0.7
Polyamide – Nylon 6	PA 6	1.13	HB	25	–	Fair	1.53	Poor	–	>8	2.7
Polyamide – Nylon 6,6	PA 6,6	1.14	HB	23	–	Fair	1.53	Poor	–	8	2.3

Material Properties and Selection

Polyamide – Nylon 6,6 30% Carbon Fiber Reinforced	PA 6, 6–30% CFR	1.28	HB	22	–	–	–	–	–	<0.1
Polyamide – Nylon 6,6 30% Glass Fiber Reinforced	PA 6, 6 30% GFR	1.4	HB	22	–	–	–	–	–	1–5
Polyamide – Nylon 12	PA 12	1.02	HB-V2	21	–	Fair	–	Fair?	1.6	–
Polyamide/imide	PAI	1.42–1.46	V0	44–45	–	Good	–	Good	3–4	0.3
Polybenzimidazole	PBI	1.3	Does not burn	58	–	Good	–	–	–	0.4
Polybutylene terephthalate	PBT	1.31	HB	25	–	Good	–	Fair?	–	0.1
Polycarbonate	PC	1.2	V0-V2	25–27	–	Fair	1.584–6	Fair	0.35	0.1
Polycarbonate – 30% Glass Fiber Filled	PC–30% GFR	1.43	V-1	–	–	–	–	–	0.11	–
Polycarbonate – Conductive	PC	1.28–1.35	–	–	–	–	–	–	0.28	–
Polychlorotrifluoro-ethylene	PCTFE	2.10–2.14	V-0	–	–	–	1.435	–	0.9	–
Polyetherether-ketone	PEEK	1.26–1.32	V-0 @ 1.5mm	35	–	Good	–	Fair	<0.01	–
Polyetherimide	PEI	1.27	V-0 @ 0.4mm	47	–	Good	–	Good/Fair	0.5	0.1–0.3
									1.3	0.25

Table 2.15 Average physical properties of plastics

Material	Formula	Density g cm^{-3}	Flammability	Limiting oxygen index %	Optical transmission %	Radiation resistance	Refractive index	Resistance to Ultra violet	Water absorption %	Water absorption equilibrium %	Water absorption – over 24 hours %
Polyethersulfone	PES	1.37	V-0 @ 0.4mm	34–41	–	Good-Fair	1.65	Fair	–	2.2	0.4–1
Polyethylene – Carbon filled	PE	0.96	–	–	–	–	–	Good	–	–	–
Polyethylene – High density	HDPE	0.95	HB	17	–	Fair	1.54	Poor	–	–	<0.01
Polyethylene – Low Density	LDPE	0.92	HB	17	–	Fair	1.51	Poor	–	–	<0.015
Polyethylene – U.H.M.W.	UHMW PE	0.94	HB	17	–	Fair	–	Poor	–	–	<0.01
Polyethylene terephthalate	Polyester, PET, PETP	1.3–1.4	HB	21	–	Good	1.58–1.64	Fair?	–	<0.7	0.1
Polyimide	PI	1.42	V0	53	–	Good	1.66	Poor	–	–	0.2–2.9
Polymethylmethacrylate	PMMA, Acrylic	1.19	HB	17–20	–	Fair	1.49	Good	–	–	0.2
Polymethylpentene	TPX®	0.835	HB	17	–	–	1.463	Poor	–	–	0.01
Polyoxymethylene – Copolymer	Acetal-Copolymer POMC	1.41	HB	15	–	Poor	–	Poor	–	0.6–0.8	0.2–0.25

Material Properties and Selection

Polyoxymethylene – Homopolymer	Acetal – Homo polymer POMH	1.42	HB	15	–	Poor	–	Poor	–	0.6–0.9	0.25
Polyphenylene-oxide	PPO (modified), PPE (modified)	1.06	HB	20	–	Good	–	–	–	–	0.1–0.5
Polyphenylene-oxide (modified), 30% Glass Fiber Reinforced	PPO 30% GFR	1.29	HB	26	–	–	–	–	–	–	0.06–0.33
Polyphenylene-sulfide – 40% Glass Fiber Reinforced	PPS – 40% GFR	1.66	V0	46	–	Good	–	Good	–	–	<0.05
Polyphenylsulfone	PPSu	1.29	V-0	44	–	–	–	–	1.2	0.6	0.35
Polypropylene	PP	0.9	HB	18	–	Fair	1.49	Poor	–	0.03	–
Polystyrene	PS	1.05	HB	19	–	Good	1.59–1.60	Poor	–	–	<0.4
Polystyrene – Conductive	High Impact Conductive Polystyrene	1.04	–	–	–	–	–	–	–	–	–
Polystyrene – Cross-linked	PS – X – Linked	1.05	HB	–	–	Good	1.59	–	–	–	0.02–0.03
Polysulphone	PSu	1.24	HB	30	–	Good	–	Poor	0.40	0.85	–

Table 2.15 Average physical properties of plastics

Material	Formula	Density g cm^{-3}	Flammability	Limiting oxygen index %	Optical transmission %	Radiation resistance	Refractive index	Resistance to Ultraviolet	Water absorption %	Water absorption equilibrium %	Water absorption – over 24 hours %
Polytetrafluoroethylene	PTFE	2.2	V0	95	–	Poor	1.38	Excellent	–	–	0.01
Polytetrafluoroethylene filled with Glass	PTFE 25% GF	2.25	V0	95	–	–	–	Good	–	–	0.15
Polyvinylchloride – Unplasticized	UPVC	1.4	V0	42	–	Fair	1.54	Good	–	–	0.03–0.4
Polyvinylfluoride	PVF	1.37–1.39	V0	35	–	–	1.46	Excellent	–	–	0.05
Polyvinylidenechloride	PVDC	1.63	–	60	–	Fair	–	Poor	–	–	0.1
Polyvinylidenefluoride	PVDF	1.76	V0	44	–	Fair	1.42	Excellent	–	–	0.04
Silicone Elastomer	MQ/VNQ/ PMQ/ P VMQ	1.1–1.5	–	–	–	Poor	–	–	–	–	–
Tetrafluoroethyleneperfluoro(alkoxy vinyl ether) – Copolymer	PFA. Teflon PFA.	2.15	V0	>95%	–	–	1.35	–	<0.03	–	–

Material Properties and Selection

Table 2.16 Hydrostatic design pressures for thermoplastic pipe at different temperatures

Property	Maximum Hydrostatic Design Pressures, N/mm², for life of 10 years						Minimum Temperature of Use[a]
	20°C	40°C	60°C	80°C	100°C	120°C	°C
Low-Density Polyethylene	2.7	1.7					−40
High-Density Polyethylene	4.8	3.9					−20
Polypropylene							
Homopolymer	6.9	4.1	2.2	1.2	1		−1
Copolymer	5.9	3.2	1.7	0.7			−5
CPVC[b]	13.8		6.9	3.5	1.1		−1
High-Impact PVC[b]	6.2	3.1					−20
ABS[b]	11		5.5	1.6			−30
PVF[b]	9	7.7	5.9	4.8	3.4	2.4	

[a] Temperature below which the pipe is very brittle
[b] PVC = polyvinychloride; ABS = acrylonitrile butadiene styrene; CPVC = Chlorinate PVC; PVF = polyvinyl fluoride

Table 2.17 Properties of nylons

Properties	Type Nylon				
	66	66–Glass Filled	8	610	11
Density (g/cm³)	1.14	1.4	1.13	1.09	1.04
Melting point (°C)	260		220	220	190
Moisture Absorption (%)					
Equilibrium at 100% RH	8	5.5	11	305	1.5
Elastic Modulus (N/mm²)					
Dry	3000	7500	2750	1750	1250
Wet	1200	2500	700	1100	900
Tensile Strength (N/mm²)	85	170	80	60	60
Elongation at Break (%)	60	3	100	100	100
Heat Distortion Temperature (°C)	75	245	70	55	50

Table 2.18 Properties of common engineering plastics

Property	Acetal	Polycarbonate	Polyphenylene Oxide	Polysulfone	Nylon 66
Tensile yield stress (N/mm²)	70	60	60	70	70
Elongation at fracture (%)	60	80	25	75	60
Elastic Modulus (dry) (N/mm²)	2800	2500	25000	2500	3000
Heat Deflection Temperature (°C) at 1.8 N/mm² (264psi)	110	135	130	175	75
Water Absorption, 24 hr at 100% RH	0.22	0.15	0.07	0.22	1.5
Equilibrium Immersion	0.8	0.35	0.15		8
Ul Temperature Index (°C)	80	115	105	140	65

Table 2.19 Properties of fiberglass reinforced plastics

Property	Polyester/Glass Mat	Polyester/Woven Glass Cloth
Class (%w/w)	25	60
Specific Gravity	1.5	1.7
Thermal Expansion × 10^{-6} °C	25	12
Modulus of Elasticity (kN/mm²)	7	15
Flexural Strength (N/mm²)	125	345

Material Properties and Selection

Table 2.20 Filler materials and properties imparted to plastics

Filler Materials	Chemical Resistance	Heat Resistance	Electrical Insulation	Impact Strength	Tensile Strength	Dimensional Stability	Stiffness	Hardness	Electrical Conductivity	Thermal Condactivity	Moisture Resistance	Handleability
Alumina Powder									X	X		
Alumina Trihydrate			X				X				X	X
Asbestos	X	X	X	X		X	X	X				
Bronze							X	X	X	X		
Calcium Carbonate		X				X	X	X				X
Calcium Silicate		X				X	X	X				
Carbon Black		X				X	X		X	X		X
Carbon Fiber									X	X		
Cellulose				X	X	X	X	X				
Alpha Cellulose			X		X	X						
Coal (powdered)	X										X	
Cotton (chopped fibers)			X	X	X	X	X	X				
Fibrous Glass	X	X	X	X	X	X	X	X			X	
Graphite	X				X	X	X	X	X	X		
Jute				X			X					
Kaolin	X	X				X	X	X			X	X
Mica	X	X	X			X	X	X			X	
Molybdenum Disufide							X	X			X	X
Nylon (chopped fibers)	X	X	X	X	X	X	X	X				X
Orlon®	X	X	X	X	X	X	X	X		X	X	
Rayon			X	X	X	X	X	X				
Silca, Amorphous			X								X	X
TFE-Fluorocarbon						X	X	X				
Talc		X	X	X			X	X	X		X	X
Wood Flour			X		X	X						

Table 2.21 Chemical resistance of epoxy resin coatings

Material	(%)	5	15	27	38	49	60	71	82
Acetic Acid	1–5	G[a]	C	F[b]	F	F	F	F	P[c]
	5–10	F	F	P	P	P	NR[d]	NR	NR
	10–50	NR	NR	NR	NR	NR	NR	NR	NR
Acetone	1–5	G	G	G	F	F	F	P	P
	10–20	F	F	F	P	P	NR	NR	NR
Alcohols (ethyl)	[e]–	X[f]	X	G	G	F	F	P	P
Alum Sulfate		X	X	X	X	X	X	X	X
Ammonium Chloride		X	X	X	X	X	X	X	X
Ammonium Fluride		X	X	X	X	X	X	X	X
Aromatic Solvent		X	X	X	X	G	G	G	F
Beer		X	X	X	X	X	X	X	X
Black Liquor		X	X	X	X	X	X	X	X
Boric Acid	1–5.0	X	X	X	G	G	G	F	F
Calcium Chloride	1.0–5	X	X	X	X	X	X	G	G
Carbon Tetrachloride		X	X	G	G	G	G	F	F
Chromic Acid	1.0–5	F	F	NR	NR	NR	NR	NR	NR
Citic Acid	1.0–5	F	F	F	G	G	G	G	F
Cooking Oils		X	X	X	X	X	X	X	G
Cooper Salts		X	X	X	X	X	X	X	X
Esters		X	X	X	X	X	X	X	X
Esters (ethyl ether)		X	X	X	X	X	G	G	G
Formaldehyde	1.0–35	X	X	X	X	X	X	X	X
Ferric Chloride		X	X	X	X	G	G	G	F
Ferrous Salts		X	X	X	X	X	X	X	X
Gasoline		X	X	X	X	X	X	X	G
Glycerin		X	X	X	X	X	X	X	X
Hydrochloric Acid	1.0–5	X	X	X	G	G	F	F	F
Hydrofluoric Acid	1.0–5	G	G	G	P	P	NR	NR	NR
Kerosene		X	X	X	X	X	X	X	G
Lactic Acid	1.0–10	X	X	X	G	G	G	F	F
Lead Acetate		X	X	X	X	X	X	X	G
Manganese Salt		X	X	X	X	X	X	X	G

Material Properties and Selection

Table 2.21 (cont.) Chemical resistance of epoxy resin coatings

Material	(%)	5	15	27	38	49	60	71	82
Methyl Ethyl Ketone	1.0–5	G	G	G	F	F	NR	NR	NR
Mineral Spirits		X	X	X	X	G	G	G	F
Naptha		X	X	X	X	X	X	X	X
Nitric Acid	1.0–5	F	F	P	P	P	P	NR	NR
	10.0–20	P	P	P	NR	NR	NR	NR	NR
Oxalic Acid	Saturated	G	G	G	F	F	NR	NR	NR
Phosphoric Acid	20	F	F	P	P	NR	NR	NR	NR
Potassium Hydroxide		X	X	X	X	X	X	X	X
Salt Brine		X	X	X	X	X	X	X	G
Soaps		X	X	X	X	X	X	X	G
Detergents		X	X	X	X	X	X	X	G
Sodium Chromate		X	X	X	X	X	X	X	G
Sodium Dichromate		G	G	G	G	G	G	G	F
Sodium Fruoride		X	X	X	X	X	X	X	X
Sodium Hydroxide	1.0–10	X	X	X	X	X	X	X	G
	50	X	X	X	G	G	G	G	F
Sodium Hypochlorite	3	G	G	F	P	P	NR	NR	NR
Sodium Phosphate		X	X	X	X	X	X	X	X
Sodium Sulfate		X	X	X	X	X	X	X	X
Sodium Sulfite		X	X	X	X	X	X	X	X
Sodium Thiosulfate		X	X	X	X	X	X	X	X
Sulfite Liquor		X	X	X	X	X	X	X	X
Sulfuric Acid	1.0–5.0	X	X	X	G	G	F	F	P
	10–20	X	X	P	P	NR	NR	NR	NR
Vegetable Oils		X	X	X	X	X	X	X	X
Water (frash)		X	X	X	X	X	X	X	X
Water (distilled)		X	X	X	X	X	G	G	G
White Liquor		X	X	X	X	X	X	X	G

Source: Cheremisinoff, N. P., and P. N. Cheremisinoff. Fiberglass–Reinforced Plastics Deskbook (Ann Arbor, MI: Ann Arbor Science Publishers, Inc., 1978).
[a]G=good [b]F=fair [c]P=poor [d]NR= not recommended [e]–=all conditions [f]X= excellent

Table 2.22 Properties of cements

Cement Type	Cure Speed	Viscosity	Trans. %	Shore	Chemical Resistance	Temp Range
TYPE C-59	3hrs @ 70°C	250 CPS	90% to 98%	90	Excellent-Acids	
MIL-A-3920	7 days @ 20°C	To 320 CPS	From 320nm–2.5um		Alkalis-Solvents	–54°C to +100°C
TYPE M-62	1.5 hrs @ 70°C	250 CPS	90% to 98%	90	Excellent-Acids	
MIL-A-3920	4.5 days @ 20°C	320 CPS	From 320nm–2.5um		Alkalis-Solvents	–54°C to +100°C
TYPE F-65	24 HRS @ 20°C	250 CPS	90% to 98%	90	Excellent-Acids	
MIL-A-3920		To 320 CPS	From 320nm–2.5um		Alkalis-Solvents	–54°C to +100°C
TYPE RD 3–74	24 HRS @ 20°C	250 CPS To 320 CPS	90% to 98% From 320nm–2.5um	90	Fair–Some Alkalis-Solvents	–54°C to +100°C
TYPE DC-90	72 HRS @ 20 °C UV Sensitized	250 CPS To 320 CPS	90% to 98% From 320nm–2.5um	90	Excellent-Acids, Alkalis, Solvents	Exceeds –54°C to +100°C
TYPE UV-69	1.5 HRS @ 365nm Element	250 CPS 320 CPS	90% to 98% 320nm–2.5um	90	Excellent-Acids, Alkalis, Solvents	Exceeds –54°C to +100°C

Material Properties and Selection

TYPE UV-74	1.5 HRS @ 365nm	800 CPS	90% TO 98%	90	Excellent-Acids, Alkalis, Solvents	
MIL-A-3920		To 1400 CPS	From 320nm–2.5um		Excellent-Acids, Alkalis, Solvents	Exceeds –54°C to +100°C
TYPE J-91	1 HR @ 365nm	250 CPS To 300 CPS	90% to 98% From 320nm–2um	90	Very Good-Some Alkalis-Solvents	–54°C to +125°C
TYPE VTC-2	1 HR @ 365nm	6000 CPS To 10,000 CPS	90% to 98%	75	Very Good-Some Alkalis and Solvents	–54°C to +125°C
TYPE P-92	1 HR @ 365nm	900 CPS	90% to 98%	35	Very Good-Some Alkalis and Solvents	–30°C to +70°C
TYPE SK-9	1 HR @ 365nm	80 CPS To 100 CPS	90% to 95%	90	Good-Some Alkalis, Acids, Solvents	–30°C to +80°C
TYPE EK-93	8 HRS @ 20°C	>25,000 CP	Opaque Grey	>95	Excellent-Acids, Alkalis, Solvents	–50°C to +90°C

Table 2.23 Recommended materials for pumping different liquids

Liquid	Condition	Specific Gravity	Pump Material
Acetaldehyde	Moisture free	0.98	All iron
Acetaldehyde	Presence of moisture	0.78	Bronze liquid end, 304
Acetone		0.79	All iron, standard fitted
Acetate solvents			Std. ftd. All bronze, all iron
Acid, Acetic	5% Room temp.	1.05	Bronze liquid end, 304, 316
Acetic	20% Room temp.		304, 316
Acetic	50% Room temp.		304, 316
Acetic	50% Boiling		Hastelloy, 316 with caution
Acetic	100% Room temp.		Aluminum, 304, 316, alum bronze
Acetic	100% Boiling		Hastelloy, 316 with caution
Arsenic	Room temp.	2.0–2.5	All iron, 304, 316
Arsenic	90% at 225°F		304 with caution
Benzoic		1.27	304, 316, aluminum
Boric			Bronze liquid end, 304
Butyric	5%–100% Room temp.	0.96	Aluminum, 304, 316,
Carbolic	Concentrated	1.07	All iron, 304, 316
Carbolic in H_2O			All iron, standard fitted
Carbonic	Aqueous solution		Bronze liquid end, alum.,304,316
Chromic			All iron, 304, 316

Table 2.23 (cont.) Recommended materials for pumping different liquids

Liquid	Condition	Specific Gravity	Pump Material
Chromic with H_2SO_4			304, 316
Chromic	50% Boiling		Hastelloy C
Citric			Bronze liquid end, 304, 316
Citric	Concentrated, boiling		316, Hastelloy (all)
Fatty (oleic, palmitic and stearic			Bronze liquid end, 304,316 alum bronze, monel
Formic		1.22	Bronze liquid end, 304, 316
Fruit			Bronze liquid end, 304,316 alum bronze
Gallic	5% Room to boiling		Aluminum, std. ftd. 304, 316
Hydrobromic	Boiling		Hastelloy B
Hydrochloric	5% Unaerated- room temp.	1.19 (38%)	Hastelloy all, high silicon
Hydrochloric	10% Unaerated- room temp.		Hastelloy all, monel, nickel, high silicon
Hydrochloric	All 100°F		Hastelloy all, high silicon, stoneware
Hydrochloric	All 160°F		Hastelloy A and B, high silicon
Hydrochloric	Fumes		Illium, cupro-nickel with caution
Hydrocyanic		0.7	All iron, aluminum, 304,316
Hydrofluoric			Bronze on monel with caution

Table 2.23 (cont.) Recommended materials for pumping different liquids

Liquid	Condition	Specific Gravity	Pump Material
Hydrofluosilicic		1.3	Bronze on monel with caution
Lactic	Room temp.	1.25	Bronze liquid end, 304, aluminum
Mine water			Bronze liquid end, 304
Mixed			Full range from all iron, 304, 316, lead, steel depending on acid conc. and percent water
Naphthenic			Aluminum, 304, 316
Nitric	Concentrated-room temp.	1.5	Aluminum, 304, 316
Nitric	95% Room temp.		Aluminum
Nitric	65% Boiling		304, 316
Nitric	Dilute room temp.		304, 316
Oxalic	5%–10% Room or hot	1.65	Bronze liquid end, 304, 316
Oxalic	10% Boiling		Illium, silicon, bronze, hastelloy
OrthoPhosphoric	Crude	1.87	316
Acid, Phosphoric	Dilute room temp.	1.87	304, 316
Pickling			Depending on conditions
Picric	Concentrated-room temp.	1.76	304, 316, high silicon iron

Table 2.23 (cont.) Recommended materials for pumping different liquids

Liquid	Condition	Specific Gravity	Pump Material
Pyrogallic		1.45	304, 316
Pyroligneous			Bronze liquid end, 304, 316
Sulfuric	77% – Room temp.	1.69–1.84	All iron, 304, 316
Sulfuric	77% – Hot		304, 316
Sulfuric	Very dilute-room temp.		304, 316
Sulfuric	Very dilute-boiling		Hastelloy B
Sulfuric fuming (oleum)		1.92–1.94	Steel
Sulfurous			Bronze, liquid end, 316, alum, bronze, lead
Tannic			Bronze liquid end, 304, 316, monel
Tartaric	Solution		Bronze liquid end, 304
Alcohol, grain ethyl			Bronze liquid end, std. ftd.
Alcohol, wood methyl			Bronze liquid end, std. ftd.
Aluminum sulfate	Solution		High silicon, lead, 304, 316
Ammonium bicarbonate	H_2O Solution		All iron
Ammonium chloride	H_2O Solution		All iron, 316
Ammonium hydroxide	H_2O Solution		All iron
Ammonium nitrate	H_2O Solution		All iron, 304, 316

Table 2.23 (cont.) Recommended materials for pumping different liquids

Liquid	Condition	Specific Gravity	Pump Material
Ammonium phosphate	H_2O Solution		All iron, 304, 316
Ammonium sulphate with H_2SO_4			Bronze liquid end, lead
Ammonium sulphate	H_2O Solution		All iron, 304, 316
Aniline		1.02	All iron
Aniline hydrochloride	H_2O Solution		High silicon iron
Asphalt	Hot	0.98–1.4	All iron
Barium chloride	H_2O Solution		All iron, 316
Barium nitrate	H_2O Solution		All iron, 304, 316
Beer			Bronze liquid end, 304, 316
Beer wort			Bronze liquid end, 304, 316
Beet juice			Bronze liquid end, 304, 316
Benzene		0.88	See benzol
Benzol			All iron, standard fitted
Benzine			See Petroleum ether
Bichloride of mercury			See Mercuric chloride
Bittern			Ni resist
Bitterwasser			Bronze liquid end, 304
Black liquor			All iron, 304, 316, ni resist
Bleach solutions			See respective Hypochlorites

Material Properties and Selection

Table 2.23 (cont.) Recommended materials for pumping different liquids

Liquid	Condition	Specific Gravity	Pump Material
Blue vitriol			See Copper sulfate
Boiler feed water			All iron, bronze, liquid end, standard fitted dependinding on pH
Borax			All iron
Brine, calcium chloride			All iron if pH is 8.5, otherwise bronze liquid end
Brine, calcium and magnesium			Bronze liquid end, ni resist
Brine, calcium and sodium chloride			Bronze liquid end
Brine, sodium chloride	3% salt	1.02–1.20	All iron, bronze liquid end
Brine, sodium chloride	Over 3%		Bronze liquid end, 304, 316
Brine, sea water		1.03	All iron, bronze liquid end
Butane		0.60 (32 °F)	Standard fitted iron fitted
Cachaza			Standard fitted
Cadium plating solution			Chrome-nickel iron, high silicon rubber, stone
Calcium bisulfite		1.06	316
Calcium chlorate			304, 316
Calcium chlorite			See Brine
Calcium hypochlorite			All iron, high silicon iron
Cane juice			Standard fitted bronze liquid end

Table 2.23 (cont.) Recommended materials for pumping different liquids

Liquid	Condition	Specific Gravity	Pump Material
Carbonated water			See Acid, carbonic
Carbon bisulfide		1.26	All iron
Carbonate of soda			See Soda ash
Carbon dioxide			See Acid, carbonic
Carbon tetrachloride	Moisture free	1.5	All iron
Carbon tetrachloride	In presence of moisture		Bronze liquied end
Caustic potash			See Potassium hydroxide
Caustic soda			See Sodium hydroxide
Cellulose acetate			316. high silicon iron
Chlorate of lime			See Calcium chlorate
Chloride of lime			See Calcium hypochlorite
Chlorinated solvents	Moisture free		All iron
Chlorinated solvents	Presence of moisture		Bronze liquid end
Chlorine			Hastelloy, high silicon stoneware, rubber
Chlorobenzene		1.1	Standard fitted 304
Chloroform		1.5	Bronze liquid end, 304, 316, lead
Chrome alum	H_2O Solution		High silicon iron, chrome-nickel iron

Material Properties and Selection

Table 2.23 (cont.) Recommended materials for pumping different liquids

Liquid	Condition	Specific Gravity	Pump Material
Copperas alum			See ferrous sulfate
Copper acetate			316
Copper chloride	H$_2$O Solution		Hastelloy, high silicon, rubber stoneware
Copper nitrate			304, 316
Copper sulfate, blue vitriol			304, 316 high silicon, lead
Clay slip, (paper mill)			All iron, (hardened fittings)
Condensate			Standard fitted
Creosote		1.04–1.10	All iron, std. ftd.
Cresol, meta		1.03	All iron
Cyanide			All iron
Cyanogen	In H$_2$O		All iron
Developing solutions			304
Diethanolamine			All iron
Diethylene glycol			All iron, bronze liquid end, standard fitted
Diphenyl	Moisture free	0.99	All iron, steel
Diphenyl oxide	Moisture free		All iron, steel
Diphenyl	In alcohol		All iron
Distillery wort			Bronze liquid end
Enamel			All iron
Epsom salts			See Magnesium sulfate
Ethyl acetate			All iron 316
Ethyl alcohol (ethanol)			See Alcohol, ethyl

Table 2.23 (cont.) Recommended materials for pumping different liquids

Liquid	Condition	Specific Gravity	Pump Material
Ethylene chloride	Cold	1.28	Lead, high silicon, high chrome, nickel iron
Ferric chloride			Hastelloy, high silicone, rubber, stoneware
Ferric sulfate			304, high silicone
Ferrous chloride			Bronze liquid end, 304
Ferrous sulfate			Bronze liquid end, 304, rubber, stonewear, lead
Formaldehyde		1.08	Bronze liquid end, 304
Fruit juices			Bronze liquid end, 304
Fuel oil			See Oil, fuel
Furfural		1.16	All iron, bronze liquid end, standard fitted
Fuse oil			Bronze liquid end
Gasoline, refined		.68–.75	Standard fitted iron ftd.
Glaubers salt			See sodium sulfate
Glue			Standard fitted
Glue sizing			Bronze liquid end
Glycerin			See Glycerol
Glycerol		1.26	Std. fitted, bronze liquid end
Green liquor			All iron, 304, 316
Heptane		0.69	Standard fitted
Hydrogen peroxide			Aluminum, 304, 316

Table 2.23 (cont.) Recommended materials for pumping different liquids

Liquid	Condition	Specific Gravity	Pump Material
Hydrogen sulfide			Aluminum, 304
Hydrosulfite of soda			See Sodium hydrosulfite
Hyposulfite of soda			See Sodium thiosulfate
Kaolin slip			Bronze liquid end
Kerosene			Std. fitted, bronze liquid end
Lacquer			Std. fitted, all iron, bronze liquid end
Lacquer solvents			Std. fitted, all iron, bronze liquid end
Lard			Std. fitted, all iron
Latex			All iron
Lead (molten)			All iron, steel
Lead acetate	H_2O Solution		304, 316, rubber, stoneware
Lime water			All iron
Lithium chloride	H_2O Solution		All iron, bronze liquid end
Lye			See Sodium hydroxide
Magnesium chloride			Bronze liquid end, lead, high silicone
Magnesium sulfate			Bronze liquid end, all iron
Magma (thick residue)			Bronze liquid end, all iron, 304 ni resist
Magnesium chloride	H_2O Solution		Bronze liquid end, 304, 316

Table 2.23 (cont.) Recommended materials for pumping different liquids

Liquid	Condition	Specific Gravity	Pump Material
Manganous sulfate	H_2O Solution		All iron, bronze liquid end, 304, 316
Mash			Standard fitted, bronze liquid end, 304
Mercuric chloride	Very dilute H_2O solution		304, high silicon iron
Mercuric chloride	Commercial conc. H_2O solution		Hastelloy, high silicon, stoneware
Mercuric sulfate	In H_2SO_4		High silicon iron, stoneware
Mercurous sulfate	In H_2SO_4		High silicon iron, stoneware
Mercury			All iron
Methyl acetate			316
Methyl alcohol			See Alcohol, wood
Methyl chloride		0.52	All iron
Methylene chloride		1.34	All iron, 304
Milk		1.03–1.04	Tinned bronze, 304
Milk of lime			See Lime water
Mine water			See Acid, Mine water
Miscella			All iron
Molasses		0.75	Std. fitted, bronze liquid end
Mustard			Bronze liquid end, 304
Naptha		0.78 – .88	Standard fitted
Naptha, crude		0.92 – .95	Standard fitted
Nickel chloride	Low pH plating solutions		Chrome-nickel iron, high silicon

Material Properties and Selection

Table 2.23 (cont.) Recommended materials for pumping different liquids

Liquid	Condition	Specific Gravity	Pump Material
Nickel sulfate	Low pH plating solutions		Chrome-nickel iron, high silicon
Nicotine sulfate			Chrome-nickel iron, high silicon
Nitre			See potassium nitrate
Nitre cake			See Sodium bisulphate
Nitro-Ethane		1.04	304, 316
Nitro-Methane		1.14	304, 316
Oil, coconut		0.91	All iron, br. ftd. all bronze
Oil, crude (asphalt base)			Standard fitted
Crude (paraffin base)			Standard fitted
Fuel (furnace)			All iron, standard fitted
Fuel (kerosene)			All iron, standard fitted
Lubricating			Standard fitted
Mineral			Standard fitted
Mineral	U. S. P.		304
Palm		0.9	All iron, bronze liquid end, monel
Quenching		0.91	Standard fitted, all iron
Soya bean			All iron, bronze liquid end, monel
Vegetable			All iron, standard fitted
Coal tar			All iron, standard fitted

Table 2.23 (cont.) Recommended materials for pumping different liquids

Liquid	Condition	Specific Gravity	Pump Material
Creosole			All iron, standard fitted
Turpentine		0.87	All iron, standard fitted
Linseed		0.94	All iron, std. ftd.
Rapeseed		0.92	Bronze liquid end, 304, monel
Olive Oil		0.9	All iron, std. ftd.
Paraffin			Standard fitted, all iron
Perchlorethlene		1.62	Std. ftd. iron ftd.
Petroleum			See Oil, crude
Petroleum solvent		0.8	Std. ftd. iron ftd.
Perhydrol			See Hydrogen peroxide
Peroxide of hydrogen			See Hydrogen peroxide
Petroleum ether			All iron, standard fitted
Phenol		1.07	See Acid, carbonic
Photographic developers			See Developing solutions
Pickling acids			See Acids, pickling
Potash			See Potassium carbonate
Potash alum			Bronze liquid end
Potassium bichromate			All iron
Potassium carbonate			All iron

Material Properties and Selection

Table 2.23 (cont.) Recommended materials for pumping different liquids

Liquid	Condition	Specific Gravity	Pump Material
Potassium chloride			Bronze liquid end
Potassium cyanide			All iron
Potassium hydroxide			All iron, 304
Potassium nitrate			All iron
Potassium sulfate			All iron
Propane		0.59 (48 °F)	Std. ftd. all iron
Pyridine sulfate			High chrome-nickel iron, lead
Rectifying pump (distillery)			Bronze liquid end
Rhigolene (oil distillery)			Standard fitted
Rosin (colophony)			All iron
Sal ammoniac			See Ammonium chloride
Salt			See Brines
Salt cake			Bronze liquid end
Sea water			See Water, Sea
Sewage			Standard fitted all iron
Silicate of soda			All iron, standard fitted, bronze liquid end
Silver nitrate			304, 316, high silicon iron
Slop, brewery			Bronze liquid end
Slop, distillery			Bronze liquid end

Table 2.23 (cont.) Recommended materials for pumping different liquids

Liquid	Condition	Specific Gravity	Pump Material
Soap liquor			All iron
Soda ash			See Sodium carbonate
Sodium bicarbonate			All iron, bronze liquid end
Sodium bisulfate			High silicon, lead
Sodium carbonate			All iron
Sodium chloride			See Brine, Sodium
Sodium cyanide			All iron
Sodium hydroxide			All iron
Sodium hydrosulfite			304, 316, lead
Sodium hypochlorite			High silicon, stoneware, lead
Sodium hyposulfite			See Sodium thiosulfate
Sodium nitrate			Ail iron, 304
Sodium phosphate (mono)			Bronze liquid end, 304
Sodium phosphate (di)			Bronze liquid end
Sodium phosphate (tri)			All iron
Sodium phosphate (meto)			304
Sodium silicate (water glass)			All iron
Sodium sulfate			Bronze liquid end all iron
Sodium sulfide			All iron, 304
Sodium sulfite			Bronze liquid end

Table 2.23 (cont.) Recommended materials for pumping different liquids

Liquid	Condition	Specific Gravity	Pump Material
Sodium tetraborate			All iron
Sodium thiosulfate			Bronze liquid end, 304
Stannic chloride			Hastelloy, high silicon, rubber stoneware
Stannous chloride			Hastelloy, high silicon, rubber stoneware
Starch			Standard fitted
Stock, paper			Bronze liquid end of std. fitted, depending on pH value
Strontium nitrate			All iron, 304
Sugar			Bronze liquid end
Sulfite liquor (paper mill)			316, bronze liquid end
Sulfur	In water		All iron, bronze liquid end, ni resist
Sulfur			All iron
Sulfur chloride	Molten		All iron, lead
Syrup	Cold		Bronze liquid end
Tallow		0.9	All iron
Tanning liquor			Bronze liquid end, 304
Tar			All iron
Tar and ammonium	In water		All iron
Tetraethyllead		1.66	Std. fitted, all iron
Toluene (Toluol)		0.87	Std. fitted, all iron

Table 2.23 (cont.) Recommended materials for pumping different liquids

Liquid	Condition	Specific Gravity	Pump Material
Trichloro-ethylene	Moisture free	1.47	All iron
Trichloro-ethylene	Moisture present	1.47	Bronze liquid end
Trisodium phosphate			See Sodium phosphate (tri)
Turpentine			All iron, standard fitted
Urine			Bronze liquid end
Varnish			Standard fitted
Vinegar			Bronze liquid end, 304
Vitriol, blue			See Copper sulfate See
Vitriol, green			Ferrous sulfate See
Vitriol, oil of			Acid, sulfuric
Vitriol, white			See Zinc sulfate
Water, acid mine			See Acid, mine water
Water, distilled			Bronze liquid end, std. fitted
Water, fresh			Std. fitted, bronze liquid end
Water, salt			Bronze liquid end, all iron
Water, sea			Bronze liquid end, all iron
Whiskey			Bronze liquid end, 304
White liquor			Bronze liquid end or std. fitted
Wine			Depending on pH

Material Properties and Selection

Table 2.23 (cont.) Recommended materials for pumping different liquids

Liquid	Condition	Specific Gravity	Pump Material
Wood vinegar			Bronze liquid end, 304
Wort			See Pyroligneous acid
Xylol (xylene)		0.87	Standard fitted, all iron
Yeast			Bronze liquid end, std. fitted
Zinc chloride			316, high silicon iron
Zinc, plating solution			High silicon iron, lead
Zinc sulfate			Bronze liquid end, 304

All Iron Pump-All parts of the pump coming in direct contact with the liquid pumped are to be made of iron or ferrous metal.

Standard Fitted Pump – Iron fitted or bronze fitted includes cast iron casing, steel shaft, either iron or bronze impeller and usually bronze wearing rings and shaft sleeves when used).

Bronze Liquid End – All parts of the pump coming in direct contact with the liquid pumped are to be made of bronze, with stainless steel fastenings.

Source: Cheremisinoff, N. P., Process Engineering Data Book, P.N. and N.P. Cheremisinoff, Technomic Pub., Lancaster, PA., 1995.

3
Property Tables of Various Liquids, Gases, and Fuels

Table 3.1 Properties of gases at standard temperature and pressure (273°K, 1 atm)

Gas	Formula	MW	Density, kg/m³	Freezing Temp, Deg. K	Boiling Temp, Deg. K	Critical Temp, Deg. K	Critical Pressure x 10⁵ Pa	Critical Density kg/m³	Thermal Conductivity (mW/mK)	Dynamic Viscosity µP
Acetylene			1.173	192.44	189.04	308.36	61.39	230.41	21.3	10.22
Air			1.293	59.15	78.67	132.45	37.7	316.56	24.1	181
Ammonia			0.77	83.84	87.32	150.1	112.8	235.29	24.2	10.13
Argon	A	39.95	1.783	84	87	151	48.98	531	16.2	223
n-Butane	C_4H_{10}	58.1	2.672	134.9	272.69	425.22	37.97	227.95	13.77	69.02
Carbon Dioxide	CO_2	44	1.977	216		33	74	460	14.5	147
Carbon Monoxide	CO_2	28.01	1.25	68.19	81.74	132.96	34.99	270.27	24.6	177.5
Chlorine	Cl_2	70.9	3.214	172.16	239.16	417.19	77.1	571.43	8.034	123
Ethane	C_2H_6	30.07	1.356	90.39	184.59	305.06	48.8	203	21.18	93.76
Ethylene	C_2H_2	28.05	1.26	104.04	169.45	282.2	50.4	214.13	17.97	93.76
Helium	He	4.03	0.178	1.89	4.26	5.24	2.3	69.93	146	187
Hydrochloric Acid	HCl	36.5	1.639	159	188	324	83.1	450.45	13.43	134
Hydrogen	H_2	2.02	0.09	14	20.4	33.23	13	31.4	168.4	88

Table 3.1 (cont.) Properties of gases at standard temperature and pressure (273°K, 1 atm)

Gas	Formula	MW	Density, kg/m³	Freezing Temp, Deg. K	Boiling Temp, Deg. K	Critical Temp, Deg. K	Critical Pressure x 10⁵ Pa	Critical Density kg/m³	Thermal Conductivity (mW/mK)	Dynamic Viscosity µP
Hydrogen Sulphide	H_2S	34.1	1.538	187.7	212.8	373.57	89.63	346.02	12.65	11.71
Methane	CH_4	16	0.716	90.71	111.7	190.6	46.04	161.58	30.01	103.3
Methyl Chloride	CH_2Cl	50.5	2.308	175.49	248.97	416.29	66.8	352.86	179.2	100.4
Natural Gas		19.5	0.862							
Neon	Ne		0.902	24.59	27.13	44	26.53	483.79	46.11	297
Nitric Oxide	NO	30	1.389	112.19	121.42	180.19	64.8	518.13	23.69	178
Nitrogen	N_2	28	1.25				33.9	311	24.3	174
Nitrous Oxide	N_2O	44	1.978	182.37	184.71	309.61	72.45	451.67	15.52	134.9
Oxygen	O_2	32	1.429	54	90	154	50.4	430	24.4	200
Propane	C_3H_8	44.1	2.019	85.56	231.15	369.86	42.49	220.36	15.91	75.98
Sulphur Dioxide	SO_2	64.1	2.927	200.04	263.17	430.79	78.84	524.93	8.42	117.9

Property Tables of Various Liquids, Gases, and Fuels 109

Table 3.2 Thermodynamic properties

Gas	Formula	MW	Density (kg/m^3)	R kJ/(kg.K)	Cp kJ/(kg.K)	Cv kJ/(kg.K)	Cp/Cv= γ
Acetylene			1.0925	0.32	1.465	1.127	1.3
Air		29	1.2045	0.287	1.009	0.721	1.4
Ammonia	NH$_4$	17	0.7179	0.49	2.19	1.659	1.32
Argon	A	39.9	1.661	0.208	0.519	0.311	1.67
n-Butane	C$_4$H$_{10}$	58.1	2.4897	0.143	1.654	1.49	1.11
Carbon Dioxide	CO$_2$	44	1.8417	0.189	0.858	0.66	1.3
Carbon Monoxide	CO	28	1.1648	0.297	1.017	0.726	1.4
Chlorine	Cl$_2$	70.9	2.9944	0.117	0.481	0.362	1.33
Ethane	C$_2$H$_6$	30	1.2635	0.277	1.616	1.325	1.22
Ethylene	C$_2$H$_2$	28	1.1744	0.296	1.675	1.373	1.22
Helium	He	4	0.1663	2.078	5.234	3.153	1.66
Hydrochloric Acid	HCl	36.5	1.5273	0.228	0.8	0.567	1.41
Hydrogen	H$_2$	2	0.0837	4.126	14.319	10.155	1.41

Table 3.2 (cont.) Thermodynamic properties

Gas	Formula	MW	Density (kg/m³)	R kJ/(kg.K)	Cp kJ/(kg.K)	Cv kJ/(kg.K)	Cp/Cv=γ
Hydrogen Sulphide	H_2S	34.1	1.4334	0.243	1.017	0.782	1.3
Methane	CH_4	16	0.6673	0.519	2.483	1.881	1.32
Methyl Chloride	CH_2Cl	50.5	2.15	0.165	1.005	0.838	1.2
Natural Gas		19.5	0.8034	0.426	2.345	1.846	1.27
Neon	Ne	20.18	0.84				
Nitric Oxide	NO	30	1.2941	0.277	0.967	0.691	1.4
Nitrogen	N_2	28	1.1648	0.297	1.034	0.733	1.41
Nitrous Oxide	N_2O	44	1.8429	0.189	0.925	0.706	1.31
Oxygen	O_2	32	1.331	0.26	0.909	0.649	1.4
Propane	C_3H_8	44.1	1.8814	0.188	1.645	1.43	1.15
Sulphur Dioxide	SO_2	64.1	2.727	0.129	0.645	0.512	1.26

Properties at Normal Temperature and Pressure (293 K, 1 atm).

Property Tables of Various Liquids, Gases, and Fuels

Table 3.3 Properties of common organic solvents

Solvent	Formula	MW	Boiling Point (°C)	Melting Point (°C)	Density (g/mL)	Solubility in Water (g/100g)	Dielectric Constant	Flash Point (°C)
Acetic Acid	$C_2H_4O_2$	60.05	118	16.6	1.049	Miscible	6.15	39
Acetone	C_3H_6O	58.08	56.2	-94.3	0.786	Miscible	20.7(25)	-18
Acetonitrile	C_2H_3N	41.05	81.6	-46	0.786	Miscible	37.5	6
Benzene	C_6H_6	78.11	80.1	5.5	0.879	0.18	2.28	-11
1-Butanol	$C_4H_{10}O$	74.12	117.6	-89.5	0.81	6.3	17.8	35
2-Butanol	$C_4H_{10}O$	74.12	98	-115	0.808	15	15.8(25)	26
2-Butanone	C_4H_8O	72.11	79.6	-86.3	0.805	25.6	18.5	-7
T-Butyl Alcohol	$C_4H_{10}O$	74.12	82.2	25.5	0.786	Miscible	12.5	11
Carbon Tetrachloride	CCl_4	153.82	76.7	-22.4	1.594	0.08	2.24	–
Chlorobenzene	C_6H_5Cl	112.56	131.7	-45.6	1.1066	0.05	5.69	29
Chloroform	$CHCl_3$	119.38	61.7	-63.7	1.498	0.795	4.81	–
Cyclohexane	C_6H_{12}	84.16	80.7	6.6	0.779	<0.1	2.02	-20
1,2-Dichloroethane	$C_2H_4Cl_2$	98.96	83.5	-35.3	1.245	0.861	10.42	13

Table 3.3 (cont.) Properties of common organic solvents

Solvent	Formula	MW	Boiling Point (°C)	Melting Point (°C)	Density (g/mL)	Solubility in Water (g/100g)	Dielectric Constant	Flash Point (°C)
Diethylene Glycol	$C_4H_{10}O_3$	106.12	245	−10	1.118	10	31.7	143
Diglyme (Diethylene Glycol Dimethyl Ether)	$C_6H_{14}O_3$	134.17	162	−68	0.943	Miscible	7.23	67
Dimethylether	C_2H_6O	46.07	−22	−138.5	NA	NA	NA	−41
Dimethyl-Formamide (DMF)	C_3H_7NO	73.09	153	−61	0.944	Miscible	36.7	58
Dimethyl Sulfoxide (DMSO)								
Dioxane	$C_4H_8O_2$	88.11	101.1	11.8	1.033	Miscible	2.21(25)	12
Ethanol	C_2H_6O	46.07	78.5	−114.1	0.789	Miscible	24.6	13
Ethyl Acetate	$C_4H_8O_2$	88.11	77	−83.6	0.895	8.7	6(25)	−4
Ethylene Glycol	$C_2H_6O_2$	62.07	195	−13	1.115	Miscible	37.7	111
Glycerin	$C_3H_8O_3$	92.09	290	17.8	1.261	Miscible	42.5	160
Heptanes	C_7H_{16}	100.2	98	−90.6	0.684	0.01	1.92	−4
Hexamethyl Phosphoramide	$C_6H_{18}N_3OP$	179.2	232.5	7.2	1.03	Miscible	31.3	105

Table 3.4 Properties of chlorinated solvents

	Methylene Chloride	Perchloroethylene	Trichloroethylene	1,1,1-Trichloroethane
Chemical Formula	CH_2Cl_2	C_2Cl_4	C_2HCl_3	$C_2H_3Cl_3$
Molecular Weight	84.9	165.8	131.4	133.4
Boiling Point °F (°C) @ 760 mm Hg	104 (40)	250 (121)	189 (87)	165 (74)
Freezing Point °F (°C)	−139 (−95)	−9 (−23)	−124 (−87)	−34 (−37)
Specific Gravity (g/cm³) @ 68°F	1.33	1.62	1.46	1.34
Pounds per gallon @ 77°F	10.99	13.47	12.11	11.1
Vapor Density (Air = 1.00)	2.93	5.76	4.53	4.55
Vapor Pressure (mm Hg) @ 77°F	436	18.2	74.3	123
Evaporation Rate @ 77°F				
Ether = 100	71	12	30	37
n-Butyl Acetate = 1	14.5	2.1	4.5	6
Specific Heat (BTU/lb per °F) @ 68°F	0.28	0.21	0.23	0.25
Heat of Vaporization (cal/g) @ boiling pt	78.9	50.1	56.4	56.7

Table 3.4 (cont.) Properties of chlorinated solvents

	Methylene Chloride	Perchloroethylene	Trichloroethylene	1,1,1-Trichloroethane
Viscosity (cps) @ 77°F	0.41	0.75	0.54	0.79
Solubility (g/100 g)				
Water in Solvent	0.17	1.01	0.04	0.05
Solvent in Water	1.7	0.015	0.1	0.07
Surface Tension (dynes/cm) @ 68°F	28.2	32.3	29.5	25.6
Kauri-Butanol (Kb) Value	136	90	129	124
Flash Point				
Tag Open Cup	None	None	None	None
Tag Closed Cup	None	None	None	None
Flammable Limits (% Solvent in Air)				
Lower Limit	13	None	8	8
Upper Limit	23	None	11	13

Table 3.5 Volatility propertied of glycerine water solutions

Parts by Weight of Glycerine in 100 Parts of Aqueous Solution	Boiling Point at 760 mm Hg (°C)(1)	Vapor Pressure of Glycerine Solution at 100°C (mm) Hg[a]
100	290.0	64
99	239.0	87
98	208.0	107
97	188.0	126
96	175.0	144
95	164.0	162
94	156.0	180
93	150.0	198
92	145.0	215
91	141.0	231
90	138.0	247
89	135.0	263
88	132.5	279
87	130.5	295
86	129.0	311
85	127.5	326
84	126.0	340
83	124.5	355
82	123.0	370
81	122.0	384
80	121.0	396
79	120.0	408
78	119.0	419
76	117.4	440

Table 3.5 (cont.) Volatility propertied of glycerine water solutions

Parts by Weight of Glycerine in 100 Parts of Aqueous Solution	Boiling Point at 760 mm Hg (°C)(1)	Vapor Pressure of Glycerine Solution at 100°C (mm) Hg[a]
75	116.7	450
74	116.0	460
73	115.4	470
72	114.8	480
71	114.2	489
70	113.6	496
65	111.3	553
60	109.0	565
55	107.5	593
50	106.0	618
45	105.0	639
40	104.0	657
35	103.4	675
30	102.8	690
25	102.3	704
20	101.8	717
10	100.9	740
0	100.0	760

[a] 1 mm Hg = 0.1333 kPa.

3.1 General Properties of Hydrocarbons

3.1.1 General Information

Hydrocarbons are compounds containing only hydrogen and carbon atoms. Since a hydrocarbon is a chemical combination of hydrogen and carbons, both of which are non-metals, hydrocarbons are covalently bonded. Hydrogen has only one electron in the outer ring and, therefore, will form only one bond by donating one electron to the bond. Carbon, on the other hand, occupies a unique position in the Periodic Table, being halfway to stability with its four electrons in the outer ring. None of these electrons are paired, so carbon uses all of them to form covalent bonds. Carbon's unique structure makes it the basis of organic chemistry.

Carbon not only combines covalently with other non-metals but also with itself. Oxygen also reacts with itself to form O_2; hydrogen reacts with itself to form H_2; nitrogen reacts with itself to form N_2; fluorine reacts with itself to form F_2; and chlorine reacts with itself to form Cl_2. Forming diatomic molecules, however, is the extent of the self-reaction of the elemental gases, while carbon has the ability to combine with itself almost indefinitely. Although the elemental gases form molecules when they combine with themselves, the carbon-to-carbon combination must include another element or elements, generally hydrogen. This combination of carbon with itself (plus hydrogen) forms a larger molecule with every carbon atom that is added to the chain. When the chain is strictly carbon-to-carbon with no branching, the resulting hydrocarbon is referred to as a straight-chain hydrocarbon. Where there are carbon atoms joined to carbon atoms to form side branches off the straight chain, the resulting compound is known as a branched hydrocarbon, or an isomer.

The carbon-to-hydrogen bond is always a single bond. While the resulting bond between carbon and hydrogen is always a single bond, carbon does have the capability to form double and triple bonds between itself and other carbon atoms and/or any other atom that has the ability to form more than one bond. When a hydrocarbon contains only single bonds between carbon atoms, it is known as a saturated hydrocarbon; when there is at least one double or triple bond between two carbon atoms anywhere in the molecule, it is an unsaturated hydrocarbon. When determining the saturation or unsaturation of a hydrocarbon, only the carbon-to-carbon bonds are considered, since the carbon-to-hydrogen bond is always single. Hydrocarbons are among the most useful materials to mankind but are also among the most dangerous in terms of their fire potential.

The simplest analogous series of hydrocarbons is known as the alkanes. In this series, the names of all the compounds end in -ane. The first compound in this series is methane. Methane's molecular formula is CH_4. Methane is a gas and the principal ingredient in the mixture of gases known as natural gas. The next compound in this series is ethane, whose molecular formula is C_2H_6. It is also a gas present in natural gas, although in a much lower percentage than methane. The difference in the molecular formulas of methane and ethane is one carbon and two hydrogen atoms.

Propane is the next hydrocarbon in this series, and its molecular formula is C_3H_8, which is one carbon and two hydrogen atoms different from ethane. Propane is an easily liquefied gas that is used as fuel.

The next hydrocarbon in the series is butane, another easily liquefied gas used as a fuel. Together, butane and propane are known as the LP (liquefied petroleum) gases. Butane's molecular formula is C_4H_{10}, which is CH_2 bigger than propane.

Hence, the series begins with a one-carbon-atom compound, methane, and proceeds to add one carbon atom to the chain for each succeeding compound. Since carbon will form four covalent bonds, it must also add two hydrogen atoms to satisfy those two unpaired electrons and allow carbon to satisfy the octet rule, thus achieving eight electrons in the outer ring. In every hydrocarbon, whether saturated or unsaturated, all atoms must reach stability. There are only two elements involved in a hydrocarbon – hydrogen and carbon. Hydrogen must have two electrons in the outer ring, and carbon must have eight electrons in the outer ring. Since the carbon-hydrogen bond is always single, the rest of the bonds must be carbon-carbon, and these bonds must be single, double, or triple, depending on the compound.

Continuing in the alkane series (also called the paraffin series because the first solid hydrocarbon in the series is paraffin, or candle wax), the next compound is pentane. This name is derived from the Greek word penta, meaning five. As its name implies, it has five carbon atoms, and its molecular formula is C_5H_{12}. From pentane on, the Greek prefix for the numbers five, six, seven, eight, nine, ten, etc., is used to name each alkane by corresponding the Greek prefix to the number of carbon atoms in each molecule. The first four members of the alkane series do not use the Greek prefix method of naming simply because their common names are so universally accepted, thus the names methane, ethane, propane, and butane.

The next six alkanes are named pentane, hexane, heptane, octane, nonane, and decane. Their molecular formulas are C_5H_{12}, C_6H_{14}, C_7H_{16}, C_8H_{18}, C_9H_{20}, and $C_{10}H_{22}$. The alkanes do not stop at the ten-carbon chain however. Since these first ten represent flammable gases and liquids, and most of the derivatives of these compounds comprise the vast majority of hazardous materials encountered, we have no need to go any further in the series. The general formula for the alkanes is C_nH_{2n+2}. The letter n stands for the number of carbon atoms in the molecule. The number of hydrogen atoms then becomes two more than twice the number of carbon atoms. Since there is more than one analogous series of hydrocarbons, one must remember that each series is unique; the alkanes are defined as the analogous series of saturated hydrocarbons with the general formula C_nH_{2n+2}.

3.1.2 Isomers

Within each analogous series of hydrocarbons exist isomers of the compounds within that series. An isomer is defined as a compound with the same molecular formula as another compound but with a different structural formula. In other words, if there is a different way in which the carbon atoms can align themselves in the molecule, a different compound with different properties will exist.

Beginning with the fourth alkane, butane, we find we can draw a structural formula of a compound with four atoms and ten hydrogen atoms in two ways; the first is as the normal butane exists, and the second is as follows with the name isobutane (refer to Table 3.6 for properties).

With isobutane, no matter how you count the carbon atoms in the longest chain, you will always end with three. Notice that the structural formula is different – one carbon atom attached to the other carbon atoms – while in butane (also called normal butane), the largest number of carbon atoms another carbon atom can be attached to is two. This fact does make a difference in certain properties of compounds. The molecular formulas of butane and isobutane are the same and, therefore, so are the molecular weights. However, there is a 38-degree difference in melting points, 20-degree difference in boiling points, and the 310-degree difference in ignition temperatures. The structure of the molecule clearly plays part in the properties of the compounds.

With the five-carbon alkane, pentane, there are three ways to draw the structural formula of this compound with five carbon atoms and twelve hydrogen atoms. The isomers of normal pentane are isopentane and neopentane.

Table 3.6. Typical properties of alkanes[a]

Compound	Formula	Atomic Weight (°F)	Melting Point (°F)	Boiling Point (°F)	Flash Point (°F)	Ignition Temp. (°F)
Methane	CH_4	16	−296.5	−259	gas	999
Ethane	C_2H_6	30	−298	−127	gas	882
Propane	C_3H_8	44	−306	−44	gas	842
Butane	C_4H_{10}	58	−217	31	gas	550
Pentane	C_5H_{12}	72	−201.5	97	<−40	500
Hexane	C_6H_{14}	86	−139.5	156	−7	437
Heptane	C_7H_{16}	100	−131.1	209	25	399
Octane	C_8H_{18}	114	−70.2	258	56	403
Nonane	C_9H_{20}	128	−64.5	303	88	401
Decane	$C_{10}H_{22}$	142	−21.5	345	115	410
Butane	C_4H_{10}	58	−217	31	gas	550
Isobutane	C_4H_{10}	58	−255	11	gas	860
Pentane	C_5H_{12}	72	−201.5	97	<−40	500
Isopentane	C_5H_{12}	72	−256	82	<−60	788
Neopentane	C_5H_{12}	72	2	49	<−20	842

[a]Values are average literature reported.

Note the three identical molecular formulas and weights but significantly different melting, boiling, and flash points and different ignition temperatures. These property differences are referred to as the "structural effect", i.e., differences in the properties of compounds exist for materials having the same molecular formulas but different structural arrangements. This particular structure effect is called the branching effect, and the isomers of all the straight-chain hydrocarbons are called branched hydrocarbons. Another structural effect is produced by the chain formed of consecutively attached carbon atoms.

In noting the increasing length of the carbon chain from methane through decane, the difference in each succeeding alkane is that "unit" made up of one carbon atom and two hydrogen atoms; that "unit" is not a chemical compound itself, but it has a molecular weight of fourteen. Therefore, each succeeding alkane in the analogous series weighs fourteen atomic mass units more than the one before it and fourteen less than the one after it. This weight effect is the reason for the increasing melting and boiling points, the increasing flash points, and the decreasing ignition temperatures. The increasing weights of the compounds also account for the changes from the gaseous state of the first four alkanes, to the liquid state of the next thirteen alkanes, and finally to the solid state of the alkanes, starting with the 17-carbon atom alkane, heptadecane.

Note that the larger a molecule, or the greater the molecular weight, the greater affinity each molecule will have for other molecules, therefore slowing down the molecular movement. The molecules, duly slowed from their frantic movement as gases, become liquids. As the molecules continue to get larger, they are further slowed from their still rapid movement as liquids become solids.

The straight-chain hydrocarbons represent just one group of hydrocarbons, the saturated hydrocarbons known as the alkanes. There are other series of hydrocarbons that are unsaturated; one of those is important in the study of hazardous materials. Additionally, the first hydrocarbon in another series is the only hydrocarbon important in that series. Each of these hydrocarbon series is briefly described below.

3.1.3 Alkenes

The series of unsaturated hydrocarbons that contains just one double bond in the structural formula of each of its members is the analogous series known as the alkenes. Notice that the name of the analogous series is similar to the analogous series of saturated hydrocarbons known as the alkanes, but the structural formula is significantly different. Remembering that the definition of a saturated hydrocarbon is a hydrocarbon with nothing but single bonds in the structural formula and that an unsaturated hydrocarbon is a hydrogen-carbon with at least one multiple bond in the structural formula, then we would expect to find a multiple bond in the structural formulas of the alkenes. The names of all the hydrocarbons in this series end in -ene. The corresponding names for this series of hydrocarbons are similar to the alkanes, with the only difference being the above-mentioned ending. Thus, in the alkene series ethane becomes ethene, propane is propene, butane is butene. The five-carbon straight-chain hydrocarbon in the alkene series is pentene, as opposed to pentane in the alkane series, and so on.

Note that these compounds are covalently bonded compounds containing only hydrogen and carbon. The differences in their structural formulas are apparent; the alkanes have only single bonds in their structural formulas, while the alkenes have only one double bond in their structural formulas. There are different numbers of hydrogen atoms in the two analogous series. This difference is due to the octet rule that carbon must satisfy. Since one pair of carbon atoms shares a double bond, the number of electrons the carbons need (collectively) reduces by two, so there are two fewer hydrogen atoms in the alkene than in the corresponding alkane.

In any hydrocarbon compound, carbon will form four covalent bonds. In saturated hydrocarbons the four bonds will all be single bonds. The definition of an unsaturated hydrocarbon, however, is a hydrocarbon with at least one multiple bond, and the alkenes are an analogous series of unsaturated hydrocarbons containing just one double bond (which is a multiple bond). The double bond must be formed with another carbon atom since hydrogen atoms can form only single bonds, and in a hydrocarbon compounds there are no other elements but hydrogen and carbon. In forming a double bond with another carbon atom, and to satisfy the octet rule, the alkene must form fewer bonds with hydrogen, resulting in less hydrogen in the structural formula of each alkene than in the corresponding alkane.

There are two fewer hydrogen atoms in each of the alkenes than in the alkane with the same number of carbon atoms. This is also shown by the general molecular formula of the alkenes, C_nH_{2n}, as opposed to the general molecular formula of the alkanes, which is C_nH_{2n+2}.

Note that there is no one-carbon alkene corresponding to methane, since hydrogen can never form more than one covalent bond, and there is no other carbon atom in the structural formula. Therefore, the first compound in the alkene series is ethene, while the corresponding two-carbon compound in the alkane series, ethane, is the second compound in the series with methane as the first.

Although the naming of the alkenes is the same as the alkanes with only the ending changed from -ane to -ene, there is a problem with the names of the first three alkenes. The systematic names of hydrocarbons came a long while after the simplest – that is, the shortest chain – of the compounds in each series was known and named. In naming the alkanes, the system of using the Greek names for numbers as prefixes begins with pentane, rather than with methane. That situation occurred because methane, ethane, propane, and butane were known and named long before it was known that there was an almost infinite length to the chain that carbon could form, and that a systematic naming procedure would be needed. Before the new system was adopted, the common names for the shortest-chain compounds had become so entrenched that those names survived unchanged. Therefore, not only are the first four compounds in the alkane series named differently from the rest of the series, the corresponding two-, three-, and four-carbon compounds are not generally known as ethene, propene, and butene. Their common names are ethylene, propylene, and butylene.

As noted earlier, more than one compound may have the same molecular formula (isomers), but a structural formula is unique to one compound. In addition, there are many chemicals that possess more than one chemical name for the same reason mentioned above. The most common organic chemicals are those that have the shortest carbon chains. This fact is also true for their derivatives. The inclusion of a double bond in the structural formula has a profound effect on the properties of a compound. Table 3.7 illustrates those differences through the properties of alkenes. The presence of a double bond and, indeed, a triple bond between two carbon atoms in a hydrocarbon increases the chemical activity of the compound tremendously over its corresponding saturated hydrocarbon. The smaller the molecule, or the shorter the chain, the more pronounced this activity is. A case in point is the unsaturated hydrocarbon ethylene. Disregarding the present differences in combustion properties between it and ethane, ethylene is so chemically active that under the right conditions, polymerization, a much more violent reaction than combustion, occurs. This tendency to polymerize is due to the presence of the double bond and decreases as the molecule gets bigger, or the chain longer. Only the first four or five of the straight-chain hydrocarbons are important in the study of hazardous materials. Few, if any, of the isomers of the alkenes are common.

There are other hydrocarbon compounds that contain multiple bonds. However, discussion here is limited to those compounds containing just one multiple bond in their molecules. This is because the compounds containing just one multiple bond are the most valuable commercially and, therefore, the most common. There is, however, a simple way to recognize when you are dealing with a compound that may contain two double bonds; that is, a name in which the Greek prefix "di-" is used. The compound butadiene is an example. Recognize from the first part of the name ("buta-") that there are four carbon atoms in the chain and that a double bond is present (the ending "-ene"). However, just before the -ene ending is the prefix "di-," meaning two. Therefore, recognize that you are dealing with a four-carbon hydrocarbon with two double bonds.

As in the alkanes, it is possible for carbon atoms to align themselves in different orders to form isomers. Not only is it possible for the carbon atoms to form branches that produce isomers, but it is also possible for the double bond to be situated between different carbon atoms in different compounds. This different position of the double bond also results in different structural formulas, which of course are isomers. Just as in the alkanes, isomers of the alkenes have different properties. The unsaturated hydrocarbons and their derivatives are more active chemically than the saturated hydrocarbons and their derivatives.

Table 3.7 Typical properties of alkenes

Compound	Formula	Molecular Weight	Melting Point (°F)	Boiling Point (°F)	Flash Point (°F)	Ignition Temp. (°F)
Ethylene	C_2H_4	28	−272.2	−155.0	gas	1,009
Propylene	C_3H_6	42	−301.4	−53.9	gas	927
1-Butene	C_4H_8	56	−300.0	21.7	gas	700
2-Butene	C_4H_8	56	−218.2	38.7	gas	615
1-Pentene	C_5H_{10}	70	−265.0	86.0	32	523
2-Pentene	C_5H_{10}	70	−292.0	98.6	32	NA
1-Hexene	C_6H_{12}	84	−219.6	146.4	−15	487
2-Hexene	C_6H_{12}	84	−230.8	154.4	−5	473
1-Heptene	C_7H_{14}	98	−119.2	199.9	28	500
1-Octene	C_8H_{16}	112	−152.3	250.3	70	446

NA = Not Applicable.

3.1.4 Alkynes

The alkynes is another analogous series of unsaturated hydrocarbons that contains just one multiple bond, but instead of being a double bond, it is a triple bond. The names of all the compounds end in -yne. The only compound in this series that is at all common happens to be an extremely hazardous material. It is a highly unstable (to heat, shock, and pressure), highly flammable gas that is the first compound in the series. This two-carbon unsaturated hydrocarbon with a triple bond between its two carbon atoms is called ethyne, and indeed this is its proper name. It is, however, known by its common name, acetylene.

The -ene ending could be confusing, so one must memorize the fact that acetylene is an alkyne rather than an alkene. Its molecular formula is C_2H_2. The fact that it contains this triple bond makes it extremely active chemically, which is what its instability to heat, shock, and pressure means. It takes energy to start a chemical reaction, and heat, shock, and pressure are forms of energy. The fact that the triple bond contains so much energy tied up in the structure means that it will release this energy, which is the input of some slight amount of external energy. When this input energy strikes the molecule of acetylene, the triple bond breaks, releasing the internal energy of the bonds. This produces either great amounts of heat or an explosion, depending on the way in which the external energy was applied.

There are no other alkynes that are of commercial importance, and so acetylene will be the only member of this series that is considered in fire discussions. There are other alkynes, however, along with hydrocarbons that might have one double bond and a triple bond present in the molecule.

3.1.5 Straight-Chain Hydrocarbon Nomenclature

The system for naming the straight-chain hydrocarbons is based on an agreed-upon method of retaining the first three or four common names, then using Greek prefixes that indicate the number of carbon atoms in the chain. The same system is used for isomers, always using the name of the compound that is attached to the chain and the name of the chain.

Recall the first analogous series of hydrocarbons, the alkanes – a series of saturated hydrocarbons, all ending in -ane. For these hydrocarbons and other hydrocarbons to react, a place on the hydrocarbon chain must exist for the reaction to take place. Since all the bonds from carbon to hydrogen are already used, an "opening" on one of the carbon atoms must exist for it to be able to react with something else. This "opening" occurs when one of the hydrogen atoms is removed from its bond with a carbon atom, thus causing that carbon to revert back to a condition of instability with seven electrons in its outer ring, or, as we now state, with one unpaired electron. This one unpaired electron (or half of a covalent bond, or "dangling" bond) wants to react with something, and it will, as soon as another particle that is ready to react is brought near. This chain of carbon atoms (from one carbon to another to another, and so on) with a hydrogen atom missing is a particle that was once a compound, and it is called a radical.

Radicals are created by energy being applied to them in a chemical reaction or in a fire. Remember that a hydrocarbon compound with at least one hydrogen atom removed is no longer a compound but a chemical particle known as a radical. Radicals have names of their own, which are derived from the name of the alkane. When a hydrogen atom is removed from the alkane hydrocarbon, the name is changed from -ane to -yl. Therefore, when a hydrogen atom is removed from the compound methane, the methyl radical is formed. When a hydrogen atom is removed from the compound ethane, the ethyl radical is formed. In the same manner, the propyl radical comes from propane, the butyl radical comes from butane, and so on. Similarly, isobutane will produce the isobutyl radical, and isopentane will produce the isopentyl radical. A list of hydrocarbons and the radicals produced from them when a hydrogen atom is removed is shown in Table 3.8. Note that there are only a few radicals from compounds other than the alcanes. Radicals are referred to as hydrocarbon "backbones". As an example, isobutane is more properly named methyl propane. Another isomer with a different proper name is isopentane, more properly called methyl butane. Neopentane is also named 2,2-dimethyl propane.

The following is a list of rules for proper nomenclature of the isomers and their derivatives:

1. Find the longest continuous chain and name it as if it were an alkane.
2. Name the side branches in the same manner.

Table 3.8 A listing of common radicals

Methane	CH_4	Methyl	$-CH_3$
Ethane	C_2H_6	Ethyl	$-C_2H_5$
Propane	C_3H_8	n-Propyl	$-C_3H_7$
		Isopropyl	$-C_3H_7$
Butane	C_4H_{10}	n-Butyl	$-C_4H_9$
Isobutane	C_4H_{10}	Isobutyl	$-C_4H_9$
		Sec-Butyl	$-C_4H_9$
		Tert-Butyl	$-C_4H_9$
Ethylene	C_2H_4	Vinyl	$-C_2H_3$
Benzene	C_6H_6	Phenyl	$-C_6H_5$

3. Identify the number of the carbon atom on the longest chain to which the branch is attached by counting from the end of the chain nearest to the branch.
4. If there is any confusion as to which carbon atom is meant, put the number in front of the name of the compound, followed by a dash.
5. If there is more than one branch, use the numbers to identify the carbon atom to which they are attached.
6. If the branches are identical, use the prefixes di- for two, tri- for three, tetra- for four, and so on. In this manner, the four isomers of hexane are named 2-methyl pentane, 3-methyl pentane, 2,2,-dimethyl butane, and 2,3-dimethyl butane.

3.1.6 Aromatic Hydrocarbons

The above discussions have concentrated on hydrocarbons, both saturated and unsaturated, with the unsaturated hydrocarbons containing only one multiple bond. The unsaturated hydrocarbons are the alkenes with one double bond and the alkynes with one triple bond. There are other straight-chain hydrocarbons that are unsaturated and contain more than one multiple bond, more than one double bond, or a mixture of double bonds and triple bonds. The combinations and permutations are endless, but there are only a few of the highly unstable materials.

From a commercial standpoint, there is a large body of hydrocarbons that is very important. These hydrocarbons are different in that they are not straight-chain hydrocarbons but have a structural formula that can only be called cyclical. The most common and most important hydrocarbon in this group is benzene. It is the first and simplest of the six-carbon cyclical hydrocarbons referred to as aromatic hydrocarbons.

Benzene's molecular formula is C_6H_6, but it does not behave like hexane, hexene, or any of their isomers. One would expect it to be similar to these other six-carbon hydrocarbons in its properties. There are major differences between benzene and the straight-chain hydrocarbons of the same carbon content. Hexene's ignition temperature is very near to hexane's. The flash point difference is not great, however, there are significant differences in melting points. The explanation for these differences is structure, which in the case of benzene is a cyclical form with alternating double bonds. Initially, it was believed that the alternating double bonds impart very different properties to benzene; however, the fact is that they do not.

Benzene's particular hexagonal structure is found throughout nature in many forms, almost always in a more complicated way and usually connected to many other " benzene rings" to form many exotic compounds. Benzene's derivatives include toluene and xylene. Some typical properties are given in Table 3.9, which illustrates the differences caused by molecular weight and structural formulas. There are other cyclical hydrocarbons, but they do not have the structural formulas of the aromatics unless they are benzene-based. These cyclical hydrocarbons may have three, four, five, or seven carbons in the cyclical structure in addition to the six-carbon ring of the aromatics. None of them have the stability or the chemical properties of the aromatics.

The aromatic hydrocarbons are used mainly as solvents and as feedstock chemicals for chemical processes that produce other valuable chemicals. With regard to cyclical hydrocarbons, the aromatic hydrocarbons are the only compounds discussed. These compounds all have the six-carbon benzene ring as a base, but there are also three-, four-, five-, and seven-carbon rings. These materials will be considered as we examine their occurrences as hazardous materials. After the alkanes, the aromatics are the next most commonly

Table 3.9 Comparison of benzene and some of its derivatives

Compound	Formula	Melting Point (°F)	Boiling Point (°F)	Flash Point (°F)	Ignition Temperature (°F)	Molecular Weight
Benzene	C_6H_6	41.9	176.2	12	1,044	78
Toluene	C_7H_8	–138.1	231.3	40	997	92
o-Xylene	C_8H_{10}	–13.0	291.2	90	867	106
m-Xylene	C_8H_{10}	–53.3	281.9	81	982	106
p-Xylene	C_8H_{10}	–55.8	281.3	81	984	106

shipped chemicals used in commerce. The short-chain olefins (alkenes), such as ethylene and propylene, may be shipped in larger quantities because of their use as monomers, but for sheer numbers of different compounds, the aromatics will surpass even the alkanes in number but not in volume.

3.1.7 Hydrocarbon Derivatives

A hydrocarbon derivative is a compound with a hydrocarbon backbone and a functional group attached to it chemically. A hydrocarbon backbone is defined as a molecular fragment that began as a hydrocarbon compound and has had at least one hydrogen atom removed from the molecule. Such a fragment is also known as a radical. A functional group is defined as an atom or a group of atoms bound together, which impart specific chemical properties to a molecule. A hydrocarbon derivative then is essentially a compound made up of two specific parts; the first part comes from a hydrocarbon, and the second may have many different origins (which includes coming from a hydrocarbon), depending on the chemical makeup of the functional group. The hydrocarbon backbone may come from an alkane, an alkene, an alkyne (any saturated or unsaturated hydrocarbon), or an aromatic hydrocarbon or other cyclical hydrocarbon. Any hydrocarbon compound may form the hydrocarbon backbone portion of the hydrocarbon derivative, as long as it has been converted to a radical by removal of one or more hydrogens in preparation for the reaction. The functional group may have many origins, given how chemists use any chemical compounds as reactants that will produce the desired functional group. The functional groups include the halogens (fluorine, chlorine, bromine, and iodine), the hydroxyl radical, the carbonyl group, oxygen, the carboxyl group, the peroxide radical, the amine radical, and even other hydrocarbon radicals. When these functional groups are chemically attached to hydrocarbon backbones, they form compounds called hydrocarbon derivatives, and each functional group imparts a separate set of chemical and physical properties to the molecule formed by this chemical attachment.

Just as the alkanes and alkenes had general formulas, the carbon derivatives all have general formulas. The hydrocarbon backbone provides a portion of the general formula, and the functional group provides the other part. In each case, the hydrocarbon derivative is represented by the formula R-, and the hydrocarbon backbone has its own specific formula. The term "substituted hydrocarbon" is another name for hydrocarbon derivative, because the functional group is substituted for one or more hydrogen atoms in the chemical reaction.

3.1.8 Halogenated Hydrocarbons

A halogenated hydrocarbon is defined as a derivative of a hydrocarbon in which a halogen atom replaces a hydrogen atom. Since all of the halogens react similarly, and the number of hydrocarbons (including all saturated hydrocarbons, unsaturated hydrocarbons, aromatic hydrocarbons, other cyclical hydrocarbons, and all the isomers of these hydrocarbons) is large, the number of halogenated hydrocarbons can also be very large. The most common hydrocarbon derivatives are those of the first four alkanes, the first three alkenes, and of course the isomers of these hydrocarbons. There are some aromatic hydrocarbon derivatives, but, again, they are of the simplest structure. Whatever the hydrocarbon backbone is, it is represented in the general formula by its formula, which is R-. Therefore, the halogenated hydrocarbons will have formulas such as R-F, R-Cl, R-Br, and R-I for the respective substitution of fluorine, chlorine, bromine, and iodine on to the hydrocarbon backbone. As a rule, the general formula can be written R-X, with the R as the hydrocarbon backbone, the X standing for the halide (any of the halogens), and the "-" is the covalent bond between the hydrocarbon backbone and the halogen. R-X is read as "alkyl halide".

Radicals of the alkanes are referred to as alkyl radicals. There are two other important radicals – the vinyl radical, which is produced when a hydrogen atom is removed from ethylene, and the phenyl radical, which results when a hydrogen atom is removed from benzene. The term halogenated means that a halogen atom has been substituted for a hydrogen atom in a hydrocarbon molecule. The most common halogenated hydrocarbons are the chlorinated hydrocarbons. The simplest chlorinated hydrocarbon is methyl chloride, whose molecular formula is CH_3Cl. The structural formula for methyl chloride shows that one chlorine atom is substituted for one hydrogen atom. Methyl chloride is used as an herbicide, a topical anesthetic, an extractant, a low-temperature solvent, and as a catalyst carrier in low-temperature polymerization. It is a colorless gas that is easily liquified and flammable; it is also toxic in high concentrations. Methyl chloride is the common name for this compound, while its proper name is chloromethane. Proper names are determined by the longest carbon chain in the molecule, and the corresponding hydrocarbon's name is used as the last name of the compound. Any substituted groups are named first, and a number is used to designate the carbon atom that the functional group is attached to, if applicable.

It is possible to substitute more than one chlorine atom for a hydrogen atom on a hydrocarbon molecule. Such substitution is done only when the resulting compound is commercially valuable or is valuable in another chemical process. An example is methylene chloride (the common name for dichloromethane), which is made by substituting two chlorine atoms for two hydrogen atoms on the methane molecule. Its molecular formula is CH_2Cl_2. Methylene chloride is a colorless, volatile liquid with a sharp, ether-like odor. It is listed as a non-flammable liquid, but it will ignite at 1,224°F. It is also narcotic at high concentrations. It is most commonly used as a stripper of paints and other finishes. It is also a good degreaser and solvent extractor and is used in some plastics processing applications.

Substituting a third chlorine on the methane molecule results in the compound whose proper name is trichloromethane. Tri- is for three, chloro- is for chlorine, and methane is for the hydrocarbon's name for the one-carbon chain). It is more commonly known as chloroform. Its molecular formula is $CHCl_3$. Chloroform is a heavy, colorless, volatile liquid with a sweet taste and characteristic odor. It is classified as non-flammable, but it will burn if exposed to high temperatures for long periods of time. It is narcotic by inhalation and toxic in high concentrations. It is an insecticide and a fumigant and is very useful in the manufacture of refrigerants. The total chlorination of methane results in a compound whose proper name is tetrachloromethane (tetra- for four), but its common name is carbon tetrachloride (or carbon tet). This is a fire-extinguishing agent that is no longer used since it has been classified as a carcinogen. It is still present though, and its uses include refrigerants, metal degreasing, and chlorination of organic compounds. Its molecular formula is CCl_4. It is possible to form analogues of methyl chloride (methyl fluoride, methyl bromide, and methyl iodide), methylene chloride (substitute fluoride, bromide, and iodide in this name also), chloroform (fluoroform, bromoform, and iodoform), and carbon tetrachloride (tetrafluoride, tetrabromide, and tetraiodide). Each of these halogenated hydrocarbons has commercial value.

What was true for one hydrocarbon compound is true for most hydrocarbon compounds, particularly straight-chain hydrocarbons; that is, you may substitute a functional group at each of the bonds where a hydrogen atom is now connected to the carbon atom. Where four hydrogen atoms exist in methane, there are six hydrogen atoms in ethane. You recall that the difference in make-up from one compound to the next in an analogous series is the "unit" made up of one carbon and two hydrogens. Therefore, it is possible to substitute six functional groups on to the ethane molecule. You should also be aware that the functional groups that would be substituted for the hydrogens need not be the same; that is, you may substitute chlorine at one bond, fluorine at another, the hydroxyl radical at a third, an amine radical at a fourth, and so on. Substituting one chlorine atom for a hydrogen atom in ethane produces ethyl chloride, a colorless, easily liquefiable gas with an ether-like odor and a burning taste, which is highly flammable and moderately toxic in high concentrations. It is used to make tetraethyl lead and other organic chemicals. Ethyl chloride is an excellent solvent and analytical reagent, as well as an anesthetic. Its molecular formula is C_2H_5Cl. Although we are using chlorine as the functional group, it may be any of the other halogens. In addition, we are giving the common names, while the proper names may be used on the labels and shipping papers. Ethyl chloride's proper name is chloroethane.

Substituting another chlorine produces ethylene dichloride, whose proper name is 1,2-dichloroethane. In this case, an isomer is possible, which would be the chlorinated hydrocarbon where both chlorines attached themselves to the same carbon atom, whereby 1,1-dichloroethane is formed. These compounds have slightly different properties and different demands in the marketplace. As further chlorination of ethane occurs, we would have to use the proper name to designate which compound is being made. One of the analogues of ethylene dichloride is ethylene dibromide, a toxic material that is most efficient and popular as a grain fumigant, but it is known to be a carcinogen in test animals. There are many uses for the halogenated hydrocarbons. Many of them are flammable, and most are combustible. Some halogenated hydrocarbons are classified as neither, and a few are excellent fire-extinguishing agents (the Halons®), but they will all decompose into smaller, more harmful molecular fragments when exposed to high temperatures for long periods of time.

3.1.9 Alcohols

The compounds formed when a hydroxyl group (-OH) is substituted for a hydrogen atom are called alcohols. They have the general formula R-OH. The hydroxyl radical looks exactly like the hydroxide ion, but it is not an ion. Where the hydroxide ion fits the definition of a complex ion – a chemical combination of two or more atoms that have collectively lost or (as in this case) gained one or more electrons – the hydroxide radical is a molecular fragment produced by separating the -OH from another compound and has no electrical charge. It does have an unpaired electron waiting to pair up with another particle having its own unpaired electron. The alcohols, as a group, are flammable liquids in the short-chain range, combustible liquids as the chain grows longer, and solids that will burn if exposed to high temperatures as the chain continues to become longer. As in the case of the halogenated hydrocarbons, the most useful alcohol compounds are of the short-carbon-chain variety. Also just as in the case of the halogenated hydrocarbons, the simplest alcohol is made from the simplest hydrocarbon, methane. Its name is methyl alcohol and its molecular formula is CH_3OH.

Nature produces a tremendous amount of methyl alcohol, simply by the fermentation of wood, grass, and other materials made to some degree of cellulose. In fact, methyl alcohol is known as wood alcohol and also such names as wood spirits and methanol, which is its proper name. The proper names of all alcohols end in -ol. Methyl alcohol is a colorless liquid with a characteristic alcohol odor. It has a flash point of 54°F and is highly toxic. It has too many commercial uses to list here, but among them are uses as a denaturant for ethyl alcohol (the addition of the toxic chemical methyl alcohol to ethyl alcohol in order to form denatured alcohol), antifreezes, gasoline additives, and solvents. No further substitution of hydroxyl radicals is performed on methyl alcohol.

The most widely known alcohol is ethyl alcohol, simply because it is the alcohol in alcoholic drinks. It is also known as grain alcohol, or by its proper name, ethanol. Ethyl alcohol is a colorless, volatile liquid with a characteristic odor and a pungent taste. It has a flash point of 55°F, is classified as a depressant drug, and is toxic when ingested in large quantities. Its molecular formula is C_2H_5OH. In addition to its presence in alcoholic beverages, ethyl alcohol has many industrial and medical uses, such as a solvent in many manufacturing processes as antifreeze, antiseptics, and cosmetics.

The substitution of one hydroxyl radical for a hydrogen atom in propane produces propyl alcohol, or propanol, which has several uses. Its molecular formula is C_3H_7OH. Propyl alcohol has a flash point of 77°F and, like all the alcohols, burns with a pale blue flame. More commonly known is the isomer of propyl alcohol, isopropyl alcohol. Since it is an isomer, it has the same molecular formula as propyl alcohol but a different structural formula. Isopropyl alcohol has a flash point of 53°F. Its ignition temperature is 850°F, while propyl alcohol's ignition temperature is 700°F, another effect of the different structure. Isopropyl alcohol, or 2-propanol (its proper name), is used in the manufacture of many different chemicals, but is best known as rubbing alcohol.

The above-mentioned alcohols are by far the most common. Butyl alcohol is not as commonly used as the first four in the series, but it is used. Secondary butyl alcohol and tertiary butyl alcohol, named because of the type of carbon atom in the molecule to which the hydroxyl radical is attached, must be mentioned because they are flammable liquids, while isobutyl alcohol has a flash point of 100°F. All of the alcohols of the first four carbon atoms in the alkanes, therefore, are extremely hazardous because of their combustion characteristics.

Whenever a hydrocarbon backbone has two hydroxyl radicals attached to it, it becomes a special type of alcohol known as a glycol. The simplest of the glycols, and the most important, is ethylene glycol, whose molecular formula is $C_2H_4(OH)_2$. The molecular formula can also be written CH_2OHCH_2OH and may be printed as such on some labels. Ethylene glycol is a colorless, thick liquid with a sweet taste, is toxic by ingestion and by inhalation, and among its many uses is a permanent antifreeze and coolant for automobiles. It is a combustible liquid with a flash point of 240°F.

The only other glycol that is fairly common is propylene glycol which has a molecular formula of $C_3H_6(OH)_2$. It is a combustible liquid with a flash point of 210°F, and its major use is in organic synthesis, particularly of polyester resins and cellophane.

The last group of substituted hydrocarbons produced by adding hydroxyl radicals to the hydrocarbon backbone are the compounds made when three hydroxyl radicals are substituted; these are known as glycerols. The name of the simplest of this type of compound is just glycerol. Its molecular formula is $C_3H_5(OH)_3$. Glycerol is a colorless, thick, syrupy liquid with a sweet taste. It has a flash point of 320°F and is used to make such diverse products as candy and explosives, plus many more. Other glycerols are made, but most of them are not classified as hazardous materials.

3.1.10 Ethers

The ethers are a group of compounds with the general formula R-O-R'. The R, of course, stands for any hydrocarbon backbone, and the R' also stands for any hydrocarbon backbone, but the designation R' is used to indicate that the second hydrocarbon backbone may be different from the first. In other words, both the hydrocarbon backbones in the formula may be the same, but the apostrophe is used to indicate that it may also be different. R-O-R as the general formula for the ethers is also correct. The fact that there are two hydrocarbon backbones on either side of an oxygen atom means that two hydrocarbon names will be used.

The simplest of the ethers would be ether that has the simplest hydrocarbon backbones attached; those backbones are the radicals of the simplest hydrocarbon, methane. Therefore, the simplest of the ethers is dimethyl ether, whose formula is CH_3OCH_3. Dimethyl is used because there are two methyl radicals, and "di-" is the prefix for two. This compound could also be called methyl methyl ether, or just plain methyl ether, but it is better known as dimethyl ether. It is an easily liquefied gas that is extremely flammable, it has a relatively low ignition temperature of 66°F, and it is used as a solvent, a refrigerant, a propellant for sprays, and a polymerization stabilizer.

The next simplest ether is the ether with the simplest alkane as one of the hydrocarbon backbones and the next alkane, which is methyl ethyl ether. Its molecular formula is $CH_3OC_2H_5$. It is a colorless gas with the characteristic ether odor. It has a flash point of 31°F and an ignition temperature of only 374°F. This property, of course, makes it an extreme fire and explosion hazard.

The next simplest ether is actually the one most commonly referred to as "ether". It is diethyl ether, whose molecular formula is $C_2H_5OC_2H_5$, sometimes written as $(C_2H_5)_2O$. This ether is the compound that was widely used as an anesthetic in many hospitals. One of the hazards of all ethers, and particularly diethyl ether because of its widespread use, is that once ethers have been exposed to air, they possess the unique capability of adding an oxygen atom to their structure and converting into a dangerously unstable and explosive organic peroxide. The peroxide-forming hazard aside, diethyl ether has a flash point of –56°F and an ignition temperature of 356°F. It is a colorless, volatile liquid with the characteristic ether odor. In addition to its use as an anesthetic, it is useful in the synthesis of many other chemicals, but it is an extremely hazardous material.

Another important ether is vinyl ether, a colorless liquid with the characteristic ether odor. Its molecular formula is $C_2H_3OC_2H_3$. Vinyl ether has a flash point of –22°F and an ignition temperature of 680°F. It is highly toxic by inhalation and is used in medicine and the polymerization of certain plastics.

3.1.11 Ketones

The ketones are a group of compounds with the general formula R-C-R'. The -C- functional group is known as the carbonyl group or carbonyl radical; it appears in many different classes of hydrocarbon derivatives. There are only a few important ketones, and they are all extremely hazardous.

The first is the simplest, again with two methyl radicals, one on either side of the carbonyl group. Its molecular formula is CH_3COCH_3. Its proper name is propanone; propa- is used because of the relationship to the three-carbon alkane, propane, and -one is used because it is a ketone. It could logically be called dimethyl ketone, but it is universally known by its common name, acetone. Acetone is a colorless, volatile liquid with a sweet odor, has a flash point of 15°F and an ignition temperature of 1,000°F, is narcotic in high concentrations, and could be fatal by inhalation or ingestion. It is widely used in manufacturing many chemicals and is extremely popular as a solvent.

The next most common ketone is methyl ethyl ketone, commonly referred to as MEK. Its molecular formula is $CH_3COC_2H_5$. MEK has a flash point of 24°F and an ignition temperature of 960°F. It is a colorless liquid with a characteristic ketone odor. It is as widely used as acetone and is almost as hazardous.

3.1.12 Aldehydes

The aldehydes are a group of compounds with the general formula R-CHO. The aldehyde functional group is always written -CHO, even though this does not represent the aldehyde's structural formula. It is written in this way so that the aldehydes will not be confused with R-OH, the general formula of the alcohols. The simplest of the aldehydes is formaldehyde, whose molecular formula is HCHO. The second hydrocarbon backbone of the ketone is replaced by a hydrogen atom. Formaldehyde is a gas that is extremely soluble in water; it is often sold commercially as a 50 percent solution of the gas in water. The gas itself has an ignition temperature of 806°F and a strong, pungent odor, and it is flammable and toxic by inhalation. Inhalation at low concentrations over long periods of time has produced illness in many people. Besides its use as an embalming fluid, formaldehyde is used in the production of many plastics and numerous other chemicals. The next aldehyde is acetaldehyde, a colorless liquid with a pungent taste and a fruity odor. Its molecular formula is CH_3CHO. It has a flash point of −40°F, an ignition temperature of 340°F, and is toxic by inhalation. Acetaldehyde is used in the manufacture of many other chemicals. Other important aldehydes are propionaldehyde, butyraldehyde, and acrolein.

3.1.13 Peroxides

The peroxides are a group of compounds with the general formula R-O-O-R'. All peroxides are hazardous materials, but the organic peroxides may be the most hazardous of all.

3.1.14 Esters

The esters are a group of compounds with the general formula R-C-O-O-R'. They are not generally classified as hazardous materials, except for the acrylates, which are monomers and highly flammable. Few of the rest of the class are flammable. There are some esters that are hazardous.

3.1.15 Amines

The amines are a group of compounds with the general formula R-NH$_2$, and all the common amines are hazardous. As a class, the amines pose more than one hazard being flammable, toxic, and, in some cases, corrosive. The amines are an analogous series of compounds and follow the naming pattern of the alkyl halides and the alcohols; that is, the simplest amine is methyl amine with the molecular formula of CH$_3$NH$_2$. Methyl amine is a colorless gas with an ammonia-like odor and an ignition temperature of 806°F. It is a tissue irritant and toxic, and it is used as an intermediate in the manufacture of many chemicals. Ethyl amine is next in the series, followed by propyl amine, isopropyl amine, butyl amine and its isomers, and so on.

3.2 Fuel Properties

This section contains data on common fuel properties. The following tables of information are provided in this section:

- Table 3.10 Approximate heating values of various common fuels.
- Table 3.11 Heating values of various fuel gases.
- Table 3.12 Wood heating value by species.
- Table 3.13 Fuel economy cost data.
- Table 3.14 Ultimate analysis of coal.
- Table 3.15 Typical syngas compositions.
- Table 3.16 Calorific value and compositions of syngas.
- Table 3.17 Glass transition and decomposition temperature of some materials.
- Table 3.18 Typical biomass compositions.

Property Tables of Various Liquids, Gases, and Fuels

The following definitions are for terms used in the tables provided:

Gross (or high, upper) Heating Value - The gross or high heating value is the amount of heat produced by the complete combustion of a unit quantity of fuel. The gross heating value is obtained when all products of the combustion are cooled down to the temperature before the combustion and the water vapor formed during combustion is condensed.

Net (or lower) Heating Value - The net or lower heating value is obtained by subtracting the latent heat of vaporization of the water vapor formed by the combustion from the gross or higher heating value.

The following are common units used for heating value:

- 1 Btu/lb = 2,326.1 J/kg = 0.55556 kcal/kg
- 1 J/kg = 0.00043 Btu/lb = 2.39 × 10^{-4} kcal/kg
- 1 kcal/kg = 1.80 Btu/lb = 4,187 J/kg

Other useful units also include the following:

- MJ = million joules or about 948 BTUs
- BTU = the amount of heat necessary to raise 1 pound of water 1 degree Fahrenheit
- 1 MJ/kg = about 430 BTU/pound
- kg = kilogram, about 2.2 pounds
- m^3 = cubic meter, about 35.3 cubic feet
- cord = 128 cubic feet, about 3.6 cubic meters

A brief description of common types of fuels is provided at the end of this section.

3.2.1 Crude Oil

Crude oil is used both directly as a fuel and as a feedstuff for the petrochemical factories to produce commercial fuels, synthetic rubbers, plastics, and additional chemicals. Oil refineries were originally placed near the oil fields partly because natural gas, which could not then be economically transported long distances, was available to fuel the highly energy-intensive refining process. But since 1950, crude oil has been transported by tankers and oleoducts to local refineries for strategic reasons.

Table 3.10 Approximate heating values of various common fuels

Natural Gas	1,030 Btu/cu ft	100,000 Btu/therm
Propane	2,500 Btu/cu ft	92,500 Btu/gal
Methane	1,000 Btu/cu ft	
Landfill Gas	500 Btu/cu ft	
Butane	3,200 Btu/cu ft	130,000 Btu/gal
Methanol		57,000 Btu/gal
Ethanol		76,000 Btu/gal
Fuel Oil		
Kerosene	135,000 Btu/gal	
#2	138,500 Btu/gal	
#4	145,000 Btu/gal	
#6	153,000 Btu/gal	
Waste Oil	125,000 Btu/gal	
Biodiesel – Waste Vegetable Oil	120,000 Btu/gal	
Gasoline	125,000 Btu/gal	
Wood		
Softwood	2–3,000 lb/cord	10–15,000,000 Btu/cord
Hardwood	4–5,000 lb/cord	18–24,000,000 Btu/cord
Sawdust – Green	10–13 lb/cu ft	8–10,000,000 Btu/ton
Sawdust – Kiln Dry	8–10 lb/cu ft	14–18,000,000 Btu/ton
Chips – 45% Moisture	10–30 lb/cu ft	7,600,000 Btu/ton
Hogged	10–30 lb/cu ft	16–20,000,000 Btu/ton
Bark	10–20 lb/cu ft	9–10,500,000 Btu/ton
Wood Pellets – 10% Moisture	40–50 lb/cu ft	16,000,000 Btu/ton
Hard Coal (Anthracite)	13,000 Btu/lb	26,000,000 Btu/ton

Table 3.10 (cont.) Approximate heating values of various common fuels

Soft Coal (Bituminous)	12,000 Btu/lb	24,000,000 Btu/ton
Corn – Shelled	7,800–8,500 Btu/lb	15–17,000,000 Btu/ton
Corn – Cobs	8,000–8,300 Btu/lb	16–17,000,000 Btu/ton
Rubber – Pelletized	16,000 Btu/lb	32–34,000,000 Btu/ton
Plastic	18–20,000 Btu/lb	

Table 3.11 Heating values of various fuel gases

Gas	Gross Heating Value		Net Heating Value	
	(Btu/ft^3)	(Btu/lb)	(Btu/ft^3)	(Btu/lb)
Acetylene (Ethyne) – C_2H_2	1,498	21,569	1,447	20,837
Blast Furnace Gas	92	1,178	92	1,178
Blue Water Gas		6,550		
Butane – C_4H_{10}	3,225	21,640	2,977	19,976
Butylene (Butene)	3,077	20,780	2,876	19,420
Carbon Monoxide – CO	323	4,368	323	4,368
Carburetted Water Gas	550	11,440	508	10,566
Coal gas	149	16,500		
Coke Oven Gas	574	17,048	514	15,266
Digester Gas (Sewage or Biogas)	690	11,316	621	10,184
Ethane – C_2H_6	1,783	22,198	1,630	20,295
Hydrogen (H_2)	325	61,084	275	51,628
Landfill Gas	476			
Methane – CH_4	1,011	23,811	910	21,433

Table 3.11 (cont.) Heating values of various fuel gases

Gas	Gross Heating Value		Net Heating Value	
	(Btu/ft^3)	(Btu/lb)	(Btu/ft^3)	(Btu/lb)
Natural Gas (Typical)	950	19,500	850	17,500
	–	–	–	–
	1,150	22,500	1,050	22,000
Producer Gas		2,470		
Propane – C$_3$H$_8$	2,572	21,500	2,365	19,770
Propene (Pro-pylene) – C$_3$H$_6$	2,332	20,990	2,181	19,630
Sasol	500	14,550	443	13,016
Water Gas (Bituminous)	261	4,881	239	4,469

Table 3.12 Wood heating value by species

Species	Million BTUs/cord
Ash	16.5–20.0
Aspen	10.3–12.5
Beech, American	17.3–21.8
Birch, Yellow	17.3–21.3
Douglas Fir, Heartwood	13.0–18.0
Elm, American	14.3–17.2
Hickory, Shagbark	20.7–24.6
Maple, Red	15.0–18.6
Maple, Sugar	18.4–21.3
Oak, Red	17.9–21.3
Oak, White	19.2–22.7
Pine, Eastern White	12.1–13.3
Pine, Southern Yellow	14.2–20.5

Source: http://synerjy.com/SITE%20PAGES/Fuel%20Heat%20Values%20Page.htm#solids_liq_gas

Property Tables of Various Liquids, Gases, and Fuels

Table 3.13 Fuel economy cost data

Type of Fuel	Softwood (kiln dried ~13% Moisture)	Hardwood (kiln dried ~8% Moisture)	Wood Pellets	Natural Gas	Electricity	Firewood (Seasoned ~20% Moisture)	Switchgrass (Ovendried)	Bituminous Coal	Shelled Corn (~15% Moisture)	No. 2 Fuel Oil	No. 6 Fuel Oil	Propane
Gross Heating Value	15,824,000 (Btu/ton)	15,996,000 (Btu/ton)	16,400,000 (Btu/ton)	1,025,000 (Btu/1000 ft3)	3,412 (Btu/kWh)	20,000,000 (Btu/cord)	15,500,000 (Btu/ton)	30,600,000 (Btu/ton)	392,000 (Btu/bu)	138,800 (Btu/gal)	150,000 (Btu/gal)	91,300 (Btu/gal)
Efficiency, %	78	79	83	80	98	77	80	85	80	83	83	79
Net Heating Value	12,300,000 (Btu/ton)	12,600,000 (Btu/ton)	13,600,000 (Btu/ton)	820,000 (Btu/1000 ft3)	3,340 Btu/kWh	15,300,000 (Btu/cord)	12,400,000 (Btu/ton)	26,000,000 (Btu/ton)	314,000 (Btu/bu)	115,000 (Btu/gal)	124,000 (Btu/gal)	71,900 (Btu/gal)
$/Million Btu	$/ton	$/ton	$/ton	$/1000 ft3	$/kWh	$/cord	$/ton	$/ton	$/bu	$/gal	$/gal	$/gal
1.0	12.30	12.62	13.61	0.82	0.003	15.35	12.40	26.01	0.31	0.11	0.12	0.07
1.5	18.45	18.94	20.42	1.23	0.005	23.02	18.60	39.02	0.47	0.17	0.19	0.11
2.0	24.60	25.25	27.22	1.64	0.007	30.70	24.80	52.02	0.63	0.23	0.25	0.14
2.5	30.75	31.56	34.03	2.05	0.008	38.37	31.00	65.03	0.78	0.29	0.31	0.18
3.0	36.90	37.87	40.84	2.46	0.010	46.05	37.20	78.03	0.94	0.34	0.37	0.22
3.5	43.05	44.18	47.64	2.87	0.012	53.72	43.40	91.04	1.10	0.40	0.43	0.25
4.0	49.20	50.50	54.45	3.28	0.013	61.39	49.60	104	1.25	0.46	0.50	0.29

Table 3.13 (cont.) Fuel economy cost data

Type of Fuel	Softwood (kiln dried ~13% Moisture)	Hardwood (kiln dried ~8% Moisture)	Wood Pellets	Natural Gas	Electricity	Firewood (Seasoned ~20% Moisture)	Switchgrass (Ovendried)	Bituminous Coal	Shelled Corn (~15% Moisture)	No. 2 Fuel Oil	No. 6 Fuel Oil	Propane
4.5	55.35	56.81	61.25	3.69	0.015	69.07	55.80	117	1.41	0.52	0.56	0.32
5.0	61.50	63.12	68.06	4.10	0.017	76.74	62.00	130	1.57	0.57	0.62	0.36
5.5	67.65	69.43	74.87	4.51	0.018	84.42	68.20	143	1.72	0.63	0.68	0.40
6.0	73.80	75.74	81.67	4.92	0.020	92.09	74.40	156	1.88	0.69	0.74	0.43
6.5	79.94	82.06	88.48	5.33	0.022	99.77	80.60	169	2.04	0.74	0.80	0.47
7.0	86.09	88.37	95.28	5.74	0.023	107	86.80	182	2.20	0.80	0.87	0.50
7.5	92	95	102	6.15	0.025	115	93.00	195	2.35	0.86	0.93	0.54
8.0	98	101	109	6.56	0.027	123	99.20	208	2.51	0.92	0.99	0.57
8.5	105	107	116	6.97	0.028	130	105	221	2.67	0.97	1.05	0.61
9.0	111	114	123	7.38	0.030	138	112	234	2.82	1.03	1.11	0.65
9.5	117	120	129	7.79	0.032	146	118	247	2.98	1.09	1.18	0.68

Property Tables of Various Liquids, Gases, and Fuels

10.0	123	126	136	8.20	0.033	153	124	260	3.14	1.15	1.24	0.72
11.0	135	139	150	9.02	0.037	169	136	286	3.45	1.26	1.36	0.79
13.0	160	164	177	10.66	0.043	200	161	338	4.08	1.49	1.61	0.93
14.0	172	177	191	11.48	0.047	215	174	364	4.39	1.60	1.73	1.01
15.0	184	189	204	12.30	0.050	230	186	390	4.70	1.72	1.86	1.08
16.0	197	202	218	13.12	0.054	246	198	416	5.02	1.83	1.98	1.15
17.0	209	215	231	13.94	0.057	261	211	442	5.33	1.95	2.10	1.22
18.0	221	227	245	14.76	0.060	276	223	468	5.64	2.06	2.23	1.29
19.0	234	240	259	15.58	0.064	292	236	494	5.96	2.18	2.35	1.37
20.0	246	252	272	16.40	0.067	307	248	520	6.27	2.29	2.48	1.44
30.0	369	379	408	24.60	0.100	460	372	780	9.41	3.44	3.71	2.16
40.0	492	505	544	32.80	0.134	614	496	1040	12.54	4.58	4.95	2.87
50.0	615	631	681	41.00	0.167	767	620	1301	15.68	5.73	6.19	3.59
60.0	738	757	817	49.20	0.201	921	744	1561	18.82	6.87	7.43	4.31

(source: http://www.fpl.fs.fed.us/documnts/techline/fuel-value-calculator.pdf)

Table 3.14 Ultimate analysis of coal

Rank	As Received, Btu/Lb	Percent by Mass					
		O	H	C	N	S	Ash
Anthracite	12,700	5.0	2.9	80.0	0.9	0.7	10.5
Semianthracite	13,600	5.0	3.9	80.4	1.1	1.1	8.5
Low-Volatile Bituminous	14,350	5.0	4.7	81.7	q.4	1.2	6.0
High-Volatile Bituminous A	13,800	9.3	5.3	75.9	1.5	1.5	6.5
High-Volatile Bituminous B	12,500	13.8	5.5	67.8	1.4	3.0	8.5
High-Volatile Bituminous C	11,000	20.6	5.8	59.6	1.1	3.5	9.4
Subbituminous B	9,000	29.5	6.2	52.5	1.0	1.0	9.8
Lignite	6,900	44.0	6.9	40.1	0.7	1.0	7.3

Table 3.15 Typical syngas compositions

	Project						
	PSI Wabash	Tampa Polk	El Dorado	Shell Pernis	Sierra Pacific	IBIL	Schwarze Pumpe
Fuel	Coal	Coal	Pet Coke/ Waste Oil	Vacuum Residue	Coal	Lignite	*
H	24.8	27.0	35.4	34.4	14.5	12.7	61.9
CO	39.5	35.6	45.0	35.1	23.5	15.3	26.2
CH_4	1.5	0.1	0.0	0.3	1.3	3.4	6.9
CO_2	9.3	12.6	17.1	30.0	5.6	11.1	2.8
N_2+Air	2.3	6.8	2.1	0.2	49.3	46.0	1.8
H_2O	22.7	18.7	0.4	–	5.7	11.5	–
LHV, KJ/M^3	8350	7960	9535	8235	5000	4530	12500
T_{fuel}, °C	300	371	121	98	538	549	38
Oxidant	O_2	O_2	O_2	O_2	Air	Air	O_2

*Lignite/oil slurry with waste plastic & waste oil.

PROPERTY TABLES OF VARIOUS LIQUIDS, GASES, AND FUELS

Table 3.16 Calorific value and compositions of syngas

Developer	Heating Value MJ/m³	Heating Value Btu/ft³	Syngas Composition (Wt. %) H₂	CO	CO₂	CH₄	C₂⁺	S as H₂S	H₂O	N₂	Other	CO/H₂
COAL GASIFICATION TECHNOLOGIES												
Entrained Flow Geometry												
1. Hitachi EAGLE	10.46	280										
2. Shell Coal Gasification Process (SCGP)	8.235	221	34.4	35.1	1–5	0.3			–	0.2		2.0–2.6
3. Mitsubishi Heavy Industries (MHI)	4.187	112										
4. Texaco	10–12	268–321										
5. Babcock Borsig Power (Noell)												
6. E-Gas (Destec)	10.340 327	277	34.4	45.3	15.8	1.9				1.9		

Table 3.16 (cont.) Calorific value and compositions of syngas

Developer	Heating Value		Syngas Composition (Wt. %)								
	MJ/m³	Btu/ft³	H_2	CO	CO_2	CH_4	C_2^+	S as H_2S	H_2O	N_2	Other	CO/H_2
7. Prenflo	est. <10	est. <268										
Fluidized Bed Geometry												
8. Integrated Drying Gasification Combined Cycle (IDGCC)									Hi			
9. Air Blown Gasification Cycle (ABGC)	3.6	96										
10. BHEL (Indian Institute of Tech.)												
11. High Temperature Winkler (HTW)												

Property Tables of Various Liquids, Gases, and Fuels

12. Kellog Rust Westinghouse (KRW)								
13. Transport Reactor Gasifier	4	107						
Moving Bed								
14. BHEL pilot plant								
15. Lurgi Dry Ash process								
16. Schwarze Pump Complex (Germany)								
17. British Gas/Lurgi (BGL)								
18. Wellman Process Engr.	5.53	148	6.9	29.5	6.1	22.2	35	

Table 3.16 (cont.) Calorific value and compositions of syngas

Developer	Heating Value		Syngas Composition (Wt. %)								
	MJ/m³	Btu/ft³	H_2	CO	CO_2	CH_4	C_2^+	S as H_2S	H_2O	N_2	Other	CO/H_2

BIOMASS GASIFICATION TECHNOLOGIES

19. Volund (Ansaldo)	2.6–5.0	70–134	4.4	11.6	14.7	4				64		
20. Union Carbide Purox Process	13.7	367	23.43	39.06	24.41	5.47	4.93	0.05			2.65	
21. Sofresid-Caliqua	Low		NA	NA	NA	NA	NA	NA	NA			

Bubbling Fluid Bed

22. Gas Techn. Inst.	12.97	350	25.3	16	39.4	17.8	1.5	–	–	0	–	
23. MTCI	16.24	438	43.3	9.22	28.1	4.73	9.03		5.57		0.08	
24. Citicorp Ind. Credit	6.9	186	12.67	15.5	15.88	5.72	2.27			48		
25. Energy Products of Idaho	5.6	150	5.8	17.5	15.8	4.65	2.58	0	0	52	0.8	

26. ASCAB/ Stein Industrie	5.52	155	19.87	25.3	40	0	0		13	
27. Tampella Power Inc.	5	140	11.3	13.5	12.9	4.8		17.7	40	
28. BECON Iowa State	4.5	126	4.1	23.9	12.8	3.1			56	0.2
29. BCL/ FERCO	18.7	500	14.9	46.5	14.6	17.8	6.2			
30. TPS-Thermal Process-Studsvik	5.5	147	7–9	9–13	12–14	6–9	–	–	47–52	0.5–1.0
31. Lurgi Energy	5.8	155	20.2	19.6	13.5	3.8			43	0.1
32. Aerimpianti	5	134	7–9	9–13	12–14	6–9		10–14	47–52	0.5–1.0
33. Foster Wheeler	7.5	201	15–16	21–22	10–11	5–6			46–47	
34. Sydkraft AB	5.8	121	9.5–12	16–19	14.4–17.5	5.8–7.5			48–52	

Source: J. Rezaiyan and N. Cheremisinoff, Gasification Technologies, Marcel Dekker Pub., NY, NY, 2005.

Table 3.17 Glass transition and decomposition temperature of some materials

Material	Density, kg/m^3	Softening T_g	Decomposition T_d	Minimum Ignition T_{flas}	Auto-ignition $T_{self-ign}$	Combustion HHV [MJ/kg]
Bakelite (Phenol-Formaldehyde, PF)	1300	NA	–	–	–	–
Cellulose (90% in Cotton)	1600	NA	280°C	300°C	400°C	17
Coal	1400	NA	200°C	–	–	28
Methacrylate (PMMA)	1180	85°C	180°C	300°C	450°C	26
Nylon (Polyamide, PA)	1140	220°C	300°C	450°C	500°C	32
Polyacrylonitrile (PAN)		–	250°C	450°C	550°C	–
Polyester	1380	260°C	–	–	–	18
Polyethylene (PE)	930	100°C	350°C	350°C	370°C	46.5
Polypropylene (PP)	910	170°C	350°C	350°C	370°C	46
Polystyrene (PS)	1040	100°C	300°C	350°C	500°C	42
Polyurethane (PU)	1100	NA	–	300°C	400°C	–
PVC (polyvinyl chloride)	1400	75°C	200°C	350°C	450°C	20
Teflon (PTFE)	2250	330°C	500°C	550°C	600°C	–
Wood	500	NA	200°C	300°C	400°C	25

Table 3.18 Typical biomass compositions (wt. %)

	Sugar Cane Bagasse	Pine Sawdust	Almond Shells	Grape Stalks
Moisture, %	7	9	12	8
C	46	45	41	41
H	5	5	5	6
O	40	39	39	40
N	0	0	1	0
S	0	0	0	0
Ash	1	1	3	5
PCI kJ/kg	16 200	16 400	16 000	16 70

Most data are highly variable with crude-oil field. The following are typical ranges:

- Density – Typically 900 kg/m^3 (from 700 kg/m^3 to 1000 kg/m^3 at 20°C; floats on water); linear temperature variation fit; the density of spilled oil will also increase with time as the more volatile and less dense components are lost, so that after considerable evaporation, the density of some crude oils may increase enough for the oils to sink below the water surface.
- Freezing and boiling points – At 100 kPa solid below 210 K, solid-liquid between 210 K and 280 K, liquid-vapour above 280 K, and starts to decompose at 900 K.
- Viscosity – 5 × 10^{-6} ~ 20 × 10^{-6} m^2/s at 20°C; exponential temperature variation fit; pour point= 5 ~ 15°C.
- Vapour pressure – 5 ~ 20 kPa at 20°C (40 ~ 80 kPa at 38°C); vapors are heavier than air (2 to 3 times); the characteristic time for evaporation of crude oil spills at sea is 1 day (25% vol. evaporated).
- Composition – Each crude-oil field has a different composition, that can be established by a combination of gas-chromatography, fluorescence-spectroscopy, and infrared-spectroscopy techniques, and that may be used,

for instance, in forensic analysis of oil spills at sea (Even after refining, crude-oil derivatives may be associated to their source field). Saturated hydrocarbons' content is around 60%wt, aromatics 30%wt, resins 5%wt. Sulfur content is 0.5 ~ 2%wt. Heavy metals <100 ppm. Crude-oil vapors are mainly short-chain hydrocarbons (only about 10% in volume have more than 4 carbons).

- Flash-point and autoignition temperature – Some 230 K and 700 K approximately.
- Ignition limits – Lower 0.5 ~ 1%, upper 7 ~ 15%.
- Solubility – <0.4%wt, due mainly to volatile compounds.
- Surface tension – 0.029 N/m with its vapors, 0.023 N/m with water.

3.2.2 Gasoline

- Types – In EU: Eurosuper-95, Eurosuper-98 (both lead-free); In the USA: Regular (97 RON) and Premium (95 RON).
- Density – 750 kg/m^3 (from 720 kg/m^3 to 760 kg/m^3 at 20°C); thermal expansion coefficient = 900 × [inline-formula]10^{-6} K^{-1}. (Automatic temperature compensation for volume metered fuels is mandatory in some countries).
- Boiling and solidification points – Not well defined because they are mixtures, e.g., when heating a previously subcooled sample at constant standard pressure, some 10% in weight of gasoline is in the vapour state at 300 K, and some 90% when at 440 K.
- Viscosity – 0.5 × 10^{-6} m^2/s at 20°C.
- Vapour pressure – 50 ~ 90 kPa at 20°C, typically 70 kPa at 20°C.
- Heating value – Average Eurosuper values are HHV = 45.7 MJ/kg, LHV = 42.9 MJ/kg.
- Theoretical air/fuel ratio – A = 14.5 kg air by kg fuel.
- Octane number (RON) – 92 ~ 98; this is a measure of autoignition resistance in a spark-ignition engine, being the volume percentage of iso-octane in an iso-octane/n-heptane mixture having the same anti-knocking

characteristic when tested in a variable-compression-ratio engine.
- Cetane number – 5 ~ 20, meaning that gasoline has a relative large time-lag between injection in hot air and autoignition, although this is irrelevant in typical gasoline applications (spark ignition).
- Composition – Gasoline composition has changed in parallel with SI-engine development. Lead tetraethyl, $Pb(C_2H_5)_4$, a colourless oily insoluble liquid, was used as an additive from 1950 to 1995 in some 0.1 grams of lead per litre to prevent knocking; sulfur was removed at that time because it inhibited the octane-enhancing effect of the tetraethyl lead. Average molar mass is $M = 0.099$ kg/mol, and ultimate analysis is 87%C and 13%H (corresponds roughly to $C_{7.2}H_{12.6}$).

3.2.3 Bioethanol and ETBE

Bioethanol is bio-fuel substitute of gasoline, i.e., it is ethanol obtained from biomass, not from fossil fuels, and is used as a gasoline blend.

Pure bioethanol (E100-fuel) is the most produced biofuel, mainly in Brazil and the USA. More widespread practice has been to add up to 20% to gasoline by volume (E20-fuel or gasohol) to avoid the need of engine modifications. Nearly pure bioethanol is used for new "versatile fuel vehicles." (E80-fuel only has 20% gasoline, mainly as a denaturaliser.) Anhydrous ethanol (<0.6% water) is required for gasoline mixtures, whereas for use-alone up to 10% water can be accepted.

ETBE (ethanol tertiary butyl ether, $C_6H_{14}O$, density = 760 kg/m³, LHV = 36 MJ/kg) is a better ingredient than bioethanol because it is not so volatile, not so corrosive, and has less affinity for water. ETBE-15 fuel is a blend of gasoline with 15% in volume of ETBE. ETBE is obtained by catalytic reaction of bioethanol with isobutene (45%/55% in weight), noting that isobutene comes from petroleum. The other gasoline-substitute ether, MTBE (methanol tertiary butyl ether, $(CH_3)_3\text{-CO-CH}_3$), is a full petroleum derivate (65% isobutene, 35% methanol).

Bioethanol is preferentially made from cellulosic biomass materials instead of from more expensive traditional feedstock such as starch crops. Obtaining it from sugar-feedstock is even more expensive. In Japan, a bacteria has been bred which produces ethanol from paper or rice-straw without any pre-treatment. Steps processes in ethanol production are:

- Milling – The feedstock passes through hammer mills, which grind it into a fine meal.
- Saccharification – The meal is mixed with water and an enzyme (alpha-amylase) and kept at 95°C to reduce bacteria levels and to get a pulpy state. The mash is cooled and a secondary enzyme (gluco-amylase) is added to convert the liquefied starch into fermentable sugars (dextrose).
- Fermentation – Yeast is added to the mash to ferment the sugars into ethanol and carbon dioxide (CO_2), a byproduct sold to the carbonate-beverage industry. Using a continuous process, the fermenting mash is allowed to flow, or cascade, through several fermenters until the mash is fully fermented and leaves the final tank. In a batch fermentation process, the mash stays in one fermenter for about 48 hours before the distillation process is started.
- Distillation – The fermented mash contains about 10% ethanol, as well as all the non-fermentable solids from the feedstock and the yeast cells. The mash is pumped to the continuous-flow, multicolumn distillation system where the alcohol is removed from the solids and the water. The alcohol leaves the top of the final column at 96% strength, and the residue from the base of the column is further processed into a high protein-content nutrient used for livestock feed.
- Dehydration – To get rid of the water in the azeotrope, most ethanol plants use a molecular sieve to capture the remaining water and get anhydrous ethanol (>99.8%wt pure).
- Denaturing – Fuel ethanol is denatured with a small amount (2%–5%) of some product such as gasoline to make it unfit for human consumption.

3.2.4 Diesel Oil, Kerosene, Jet A1, and Biodiesel

Kerosene is similar to diesel but with a wider-fraction distillation (see Petroleum fuels). Jet A1 fuel is just kerosene with special

Property Tables of Various Liquids, Gases, and Fuels

additives (<1%) used in most jetliners. (JP-4 fuel, approximately C_9H_{20}, was used before, as in military aircraft). Biodiesel is a biomass-derived fuel and a safer, cleaner, renewable, non-toxic, and biodegradable direct substitute of diesel oil in CI engines but is more expensive.

Types – Diesel-oil. In EU: type A for road vehicles, B for industries (agriculture, fishing; same properties as type A, but red-coloured for different taxation), C for heating (not for engines; blue-coloured). In USA: No. 1 Distillate (Kerosene), and No. 2 Distillate (Diesel).

- Density – 830 kg/m³ (780 ~ 860 kg/m³ at 40°C).
- Thermal expansion coefficient – 800 × 10⁻⁶ K⁻¹.
- Boiling and solidification points – Not well defined because they are mixtures; in general, these fuels remain liquid down to –30°C (some antifreeze additives may be added to guaranty that).
- Viscosity – 3 × 10⁻⁶ m²/s (2.0 × [inline-formula]10⁻⁶ 4.0 × 10⁻⁶ m²/s at 40°C) for diesel; 4.0 × 10⁻⁶ ~ 6.0 × 10⁻⁶ m²/s for biodiesel; 8.0 × 10⁻⁶ m²/s for Jet A1 at 20°C.
- Vapor pressure – 10 × 20 kPa at 38°C for diesel (and JP-4); 0.7 ~ 1 kPa at 38°C for kerosene and Jet-A1.
- Cetane number – 45 (between 40 Symbol (T1) 55); 60 Symbol (T1) 65 for biodiesel; this is a measure of a fuel's ignition delay, the time period between the start of injection and start of combustion (ignition) of the fuel, with larger cetane numbers having lower ignition delays; this is only of interest in compression-ignition engines, and only valid for light distillate fuels. (Because of the test engine for heavy fuel oil, a different burning-quality index is used, calculated from the fuel density and viscosity).
- Flash point – 50°C typical (40°C minimum); in the range 310 Symbol (T1) 340 K (370 Symbol (T1) 430 K for biodiesel).
- Heating value – HHV = 46 MJ/kg, LHV = 43 MJ/kg (HHV = 40 MJ/kg for biodiesel).
- Composition – 87%wt C and 13%wt H (they can be approximated to $C_{11}H_{21}$, or $C_{12}H_{23}$, or $C_{13}H_{26}$, or $C_{14}H_{30}$); saturated hydrocarbons 66%vol, aromatics 30%, olefins 4%; must have <0.2% sulfur; 77%wt C, 12%wt H, 11%wt O, 0.01%wt S for biodiesel.

3.2.5 Fuel Oil

Types – There are two basic types of fuel oil: distillate fuel oil (lighter, thinner, better for cold-start) and residual fuel oil (heavier, thicker, more powerful, better lubrication). Often, some distillate is added to residual fuel oil to get a desired viscosity. They are only used for industrial and marine applications because, although fuel oil is cheaper than diesel oil, it is more difficult to handle; it must be settled, pre-heated and filtered, and leave a sludge at the bottom of the tanks. Notice that sometimes, particularly in the USA, the term "fuel oil" also includes diesel and kerosene.

- Density – 900 ~ 1010 kg/m^3; varies with composition and temperature.
- Viscosity – widely variable with composition; 1000 × 10^{-6} m^2/s at 20°C (4000 × [inline-formula]10^{-6} m^2/s at 10°C, (10 ~ 30) × 10^{-6} m^2/s at 100°C); varies with composition and temperature; must be heated for handling; it is usually required to have <500 × 10^{-6} m^2/s for pumping and <15 × 10^{-6} m^2/s for injectors.
- Pour point – In the range 5 ~ 10°C; fuel oils are usually graded by their viscosity at 50°C (ISO-8217). A typical marine heavy-fuel-oil (HFO), graded IF-300 (Intermediate Fuel) has 300 × 10^{-6} m^2/s at 50°C (300 cSt), 25 × 10^{-6} m^2/s at 100°C; density = 990 kg/m^3 at 15°C, HHV = 43 MJ/kg, and the flash-point at 60 ~ 80°C.
- Vapor pressure – 0.1 ~ 1 kPa at 20°C.
- Composition – Distillate fuel oils are similar to diesel oil. Residual fuel oil consists of semi-liquid phase with dispersed solid or semi-solid particles (asphaltenes, minerals and other leftovers from the oil source, metallic particles from the refinery equipment, and some dumped chemical wastes), plus some 0.5% water. Residual fuel oil leaves a carbonaceous residue in the tanks and may have up to 5% of sulfur. Residual fueloil-C (or bunker-C) has a composition of 88%wt C, 10%wt H, 1%wt S, 0.5%wt H$_2$O, 0.1%wt ash.

3.2.6 Natural Gas, Biogas, LPG and Methane Hydrates

Natural gas is a flammable gaseous mixture, composed mainly of methane, 70–99% CH_4 (70% Lybia, 99% in Alaska), 1–13% C_2H_6, 0–2% C_3H_8, and minor concentrations of H_2O, CO, CO_2, N_2, He, etc. It is found on many underground cavities, either as free deposits (e.g. Indonesia, Algeria, New Zealand) or linked to petroleum fields (e.g. Saudi Arabia, Nigeria, Alaska). Since the mid 20th century it has been traded by large continental gasoducts (up to 2 m in diameter, with sensors and control valves every 25 km and pumping stations every 100 km) and LNG-ships (Liquefied Natural Gas carriers typically of 140 000 m³ in capacity, 250 000 m³ for new ones). The liquefaction of natural gas requires the removal of the non-hydrocarbon components of natural gas such as water, carbon dioxide, and hydrogen sulfide to prevent solid plugs and corrosion.

Biogas is a flammable gaseous mixture, composed mainly of methane and carbon dioxide obtained by anaerobic fermentation of condensed biomass (manure or sewage). The production may range from 20–70 m³ of biogas per cubic metre of manure, lasting 10 ~ 30 days within a digestor (depending on the temperature, that is 20 ~ 40°C), which is where biomass is first hydrolysed by some bacteria in absence of oxygen, yielding monomers that are made to ferment by other bacteria, yielding alcohol that later turns to acetic acid and finally decomposes to methane plus carbon dioxide, the later step being the controlling stage.

LPG (liquefied petroleum gas) are petroleum derivative mixtures (gaseous at ambient temperature, but handled as liquids at their vapour pressure, 200–900 kPa) mainly constituted by propane, n-butane, iso-butane, propylene, and butylenes, with composition varying widely from nearly 100% propane in cold countries to only 20 ~ 30% propane in hot countries (e.g. 100% in UK, 50% in the Netherlands, 35% in France, 30% in Spain, 20% in Greece). In Spain, the traditional bottle for domestic use (UD-125) holds 12.5 kg of commercial butane (56% n-butane, 25% propane, 17% iso-butane, 2% pentane, 0,1 g/kg H_2O and 1 mg/kg mercaptans, with $T_b = -0.5°C$ and $\rho_L = 580$ kg/m³ at 20°C). The new aluminium bottle holds 6 kg (13 kg total, 290 mm diameter and 376 mm height). For higher rates or cold ambient, propane bottles works better. For vehicles EN-589-1993 applies. LPG is also marketed in small expandable containers for laboratory use (containing some 50–300 g of LPG, 190 g is the most common) and portable "camping gas" bottles (containing some 2–4 kg of LPG, 2.8 kg is most common) and has a rough molar composition of 40% propane and 60% butanes (n-butane and iso-butane).

All gaseous fuels are odorless (except those containing traces of H_2S), and odor markers (sulfur-containing chemicals as thiols or mercaptans) are introduced for safety.

Pure methane, propane, and butane can be easily found from local chemicals suppliers, if the commercial mixtures traded (natural gas, commercial propane, and commercial butane) are not good for some laboratory work. For small lab demonstrations they may also be obtained in situ; e.g., methane may be easily produced by means of $Al_4C_3(s)+6H_2O(l) = 3CH_4(g) + 2Al_2O_3(s)$, or by heating a 50/50 mix of anhydrous sodium acetate and sodium hydroxide, $NaOH(s) + NaC_2H_3O_2(s) = CH_4(g)+Na_2CO_3(s)$, as did his discoverer, the American Mathews, in 1899.

Methane hydrates are solid icy-balls of some centimetres in size found trapped under high pressure (>30 MPa) and in chilling temperatures (0–5°C) such as in plant-covered moist places like the continental sediments on the sea floor and permafrost soil on high-latitude lands. They might be the major source of natural gas in the future; presently they are a nuisance in high-pressure gasoducts, where they may block valves. Strictly speaking, they are not hydrates (chemical compounds of a definite formula) but clathrates, i.e., an unstable network (they tend to the liquid state) of host polar molecules like water, characterized by H-bonds and regular open cavities and stabilised to a solid state by incorporating small guest non-polar molecules of appropriate size to which they are not bonded; only van-der-Waals forces act to stabilise the network. Besides methane, carbon dioxide, hydrogen sulphide, and larger hydrocarbons such as ethane and propane can stabilize the water lattices and form "hydrates." Smaller molecules like nitrogen, oxygen, or hydrogen are much more difficult to stabilize in water.

Methane hydrates (approx. $CH_4 \cdot 6H_2O$) fizzle and evaporate quickly when depressurized, yielding some 150 times its volume of methane. Since this methane comes from very large-time biomass decomposition, the problem of global warming remains. It yields CO_2 on burning, and released CH_4 losses are worse being 20 times more relative greenhouse effect that CO_2. Hydrates soils are prone to accidental landslides, particularly during exploitation, which constitutes a high risk to extraction platforms.

3.2.7 Hydrogen

In the long term, hydrogen-energy appears as the final solution to face the energy-environment dilemma of scarcity and pollution, not only for the much-pursued nuclear-fusion power stations (using hydrogen isotopes), but also for the use of hydrogen as an intermediate energy carrier (like electricity) cleanly produced from water and solar energy and cleanly converted back to water to drive fuel cells engines and clean combustors.

Pure hydrogen (H_2) is an artificial product on Earth (1 ppm in the atmosphere), but makes up nearly 100% of Jupiter's atmosphere and 90% of all atoms in the Universe (nearly 3/4 of its total mass). On Earth, it is found combined in water, living matter, and fossil matter. It was discovered in 1766 by Cavendish (used in 1520 by Paracelsus as inflammable air) and named by Lavoisier in 1781. The first massive production was in 1782 by Jacques Charles (Fe(s) + 2HCl(aq) = FCl_2(aq)+H_2(g)) to inflate a balloon. Present use is mainly for chemical synthesis (e.g. ammonia), metallurgy, ceramics, and for the hydrogenation of fats. It is also used as a cryogenic fuel in rockets, in cryogenic research, and as a fuel-cell fuel.

Production at large (world production in 2000 was $30 \cdot 10^9$ kg) is based on fossil feedstock. The major share of world production (some 2/3 and increasing) is by natural gas reforming: CH_4+aH_2O = $(2+a)H_2$+bCO_2+cCO at 1150 K with Co-Ni catalysts, but >2000 ppm-CO is left and PEFC-type fuel cells required <20 ppm-CO. Instead of fully purifying the H_2, it is easier to purify it until <1000 ppm-CO and add oxygen to get rid of the CO at the catalyst. But if more O_2 is used, it reacts with H_2 at the catalyst, producing just heat. In MCFC & SOFC the CO is an additional fuel. Natural gas reforming is presently the best method to produce hydrogen while renewable sources are being developed, but availability of natural gas is in question if new major sources, such as seabed clathrates, are not made available (and in that case with CO_2 sequestration).

Some 30% of world production was based on coal reforming (declining rapidly): $C+H_2O = H_2+CO$ at 1300 K and $CO+H_2O = H_2 + CO_2$ with FeO-CrO_2-ThO_2 catalyst.

A small percentile of world production has been based on water electrolysis from cheap hydroelectric energy in Canada and Scandinavia: $H_2O = H_2+(1/2)O_2$, with $h_e = 0.65$. It is the purest H_2. As fossil fuels are being exhausted, water electrolysers seem to be the most popular hydrogen sources in the future. Electrolysers with liquid potash lye produce hydrogen cheaper than other kinds of electrolysers, e.g., proton-exchange-membrane, PEM, electrolysers.

Production is at intermediate locations (for transportation or for stationary applications) by reforming. (See 'Reforming' details below).

Reforming a fuel is producing another fuel from it. Several reforming processes exist, such as partial oxidation (PO) and sometimes catalytic (PCO). It is the exothermic reaction with deficient oxygen; it is the simplest reforming process but gives rise to very high temperatures (2000 K) and pollutants (NO_x, NH_3, HCN) without appropriate catalysts. For methanol reforming, $CH_3OH(vap)$ + [inline-formula][inline-formula][inline-formula]O_2 = $2H_2$ + CO_2 + 667 kJ/mol (in reality yields 40% $H_2(g)$ instead of 67%), a Pd catalyst is used. For natural-gas reforming, $CH_4(g)$ + [inline-formula][inline-formula][inline-formula]O_2 = $2H_2$+CO+36 kJ/mol (in reality yielding 1.3 mol_{H_2}/mol_{CH_4} instead of 2), a Pt or Ni catalysts on alumina are used, working at 1200 K. (A big problem is the formation of soot, 2CO = C+CO_2, that clogs the catalyst).

Steam reforming (SR) – The endothermic reaction with water vapor; it is the most widely used reforming process and the one that yields more hydrogen, but it is a very complex one because of the required external heating (only used for production >500 kg/day).

Methanol reforming – methanol has 12.5%wt of hydrogen.

Ethanol reforming – Ethanol has 13%wt of hydrogen; may be the best H_2 production method in the long term and not electrolysis with renewable-energy.

Gasoline reforming – Gasoline has some 16%wt of hydrogen; diesel has a little less and is not used. Gasoline may be thermally decomposed at >800°C (or best at 300°C with Ni-catalyst), $CuHv+uH_2O$ = $(u+v/2)H_2$+uCO, syngas=synthesis gas, endothermic but good energy rate (78%), but contaminates a lot (and desulfurisation is required); more $H_2O(g)$ may yield H_2+CO_2 (40%/60%); the more aromatics in gasoline the worse the reforming, which is why diesel is bad.

NG (natural gas, methane) or LPG (propane+butane) reforming – Methane has 25%wt of hydrogen; water has only 11%; however, gaseous fuels are not good for storage and transportation. Natural gas is most used today in low-temperature fuel cells (PEFC and PAFC); it is first desulfurized (from previously added safety odorants), then steam-reformed (yields 10% CO), afterwards shift-converted (reduces CO to 1%), and (only for PEFC) finally selectively oxidized (to <10 ppm CO).

Coal reforming – Not good because of pollution and high temperature work.

Methanol reforming – $CH_3OH(vap)+H_2O(vap) = 3H_2+CO_2-49$ kJ/mol (in reality yields 70% $H_2(g)$ instead of 75%, and consumes up to 20% of HHV, at 250°C, with Ni plus a final pass through Pt to further oxidise CO to CO_2, $CO+H_2O = CO_2+H_2+41$ kJ/mol; or with Cu/ZnO); although it works for PEFC, it is best suited to PAFC because of the CO_2, and PAFC have the advantage that the vapor is produced with the by-product heat.

Natural-gas reforming – $CH_4(g)+H_2O = 3H_2+CO-206$ kJ/mol is carried out at 900 K to 1200 K in a gas furnace, with Ni-catalyst on alumina plus a final pass through Pt to further oxidise CO to CO_2; although low pressure favors the reaction, it is not used in practice.

Autothermic reforming (AR) – It is just a combined PO+SR process that it is adiabatic overall; this is presently the most economic method of H_2 production.

Thermal decomposition (TD) – It is the endothermic cracking at high temperature; for methanol, $CH_3OH(vap) = 2H_2+CO-95$ kJ/mol (it is not used alone but adding water, i.e. SR); for methane, $CH_4(g)=2H_2(g)+C(s)-75$ kJ/mol, using a Ni-catalyst on silica, that must be regenerated with oxygen from time to time to get rid of the carbon deposited; notice that no CO is involved.

Carbon-dioxide reforming – Not used much; for methane, $CH_4(g)+CO_2(g) = 2H_2+2CO-248$ kJ/mol.

Membrane reactor – It is not reforming itself but a post-processing stage to reforming; after SR (or PO or AR) the gas flow is exposed to a selective membrane that yields 90% H_2.

4

General Guidelines on Fire Protection, Evacuation, First Responder, and Emergency Planning

4.1 Flammability Properties

4.1.1.1 General Information

The majority of information organized in this section has been taken from the *Handbook of Industrial Toxicology and Hazardous Materials* for which permission has been granted.[1]

Two main categories of liquids are flammable and combustible and are determined mainly by the liquid's flash point. Both categories of liquids will burn, but it is into which of these two categories the liquid belongs that determines its relative fire hazard. Flammable liquids are generally considered the more hazardous of the two categories mainly because they release ignitable vapors.

A flammable material is any liquid having a flash point below 100°F. The U.S. National Fire Protection Agency (NFPA) expands this definition by including the stipulation that the vapor cannot exceed 40 psi (pounds per square inch) at a liquid temperature of 100°F, with the theory being that such liquids are capable of releasing vapor at a rate sufficient to be ignitable. Since this aspect of the definition relating to vapor pressure has little fire-ground application, it is often ignored. However, it is important to note that if the heat from a fire raises the liquid temperature to a temperature above the liquid's flash point, it will automatically increase the vapor pressure inside a closed container. Any other source of sufficient heat will produce the same result.

---

[1]Cheremisinoff, N. P., Handbook of Industrial Toxicology and Hazardous Materials, Marcel Dekker Publishers, New York, New York, 1999.

Within the combustible liquid category are those materials with a flash point above 100°F. Combustible liquids are considered less hazardous than flammable liquids because of their higher flash points. However, this statement can be misleading since there are circumstances when it is not a valid assumption. It is possible for certain combustible liquids to be at their flash point when a hot summer sun has been striking their metal container for some time. Additionally, during the transportation of some combustible products, the product is either preheated or a heat source is maintained to make the product more fluid than it would be at atmospheric temperatures. One reason this is done is to facilitate transportation or pumping, i.e., to aid with the movement of a material that is very viscous, such as asphalt or tar. Also, some materials classified as combustible solids will be heated to their melting point. Naphthalene is one example of this treatment. Naphthalene might be heated to a temperature above its melting point, which is about 176°F. Despite its fairly high ignition temperature (almost 980°F), it would not be unreasonable to surmise that a spill of liquid naphthalene could present a serious fire hazard. Fortunately, with naphthalene, quick action with adequate amounts of water applied as spray streams should cool and solidify it, thus greatly minimizing the fire risk.

It is important to note that a combustible liquid at or above its flash point will behave in the same manner that a flammable liquid would in a similar emergency. As an example, No. 2 fuel oil when heated to a temperature of 150°F can be expected to act or react in the same way gasoline would at 50°F. In most instances, however, to reach this elevated temperature will require the introduction of an external heat source. Some common examples of combustible petroleum liquids are given in Table 4.1.

Table 4.1 Examples of petroleum liquids that are combustible

Product	Flash Point (°F)
Kerosene	100+
Fuel oils	100 – 140
Diesel oil	130
Lubricating oil	300
Asphalt	400

General Guidelines on Fire Protection

It is important to note that the extinguishing techniques, controlling actions, or fire-prevention activities implemented can differ greatly depending upon which of the two categories the liquid falls in. To have the ability to categorize a liquid correctly when it is not identified so, it is only necessary to know its flash point. By definition, the flash point of a liquid determines whether a liquid is flammable or combustible.

The categories of liquids are further subdivided into classes according to the flash point plus the boiling point of certain liquids. These divisions are summarized in Table 4.2, which shows that flammable liquids fall into Class 1, and combustible liquids fall into Classes 2 and 3. The products that are at the low end (100°F) of the Class 2 combustible-liquid group might be thought of as borderline cases. These could act much like flammable liquids if atmospheric temperatures were in the same range. It is not a common industry practice to identify either stationary or portable (mobile) liquid containers by the class of liquid it contains. The usual practice is to label either "Flammable" or "Combustible" and include the required United Nations hazard placard.

Flash point is the temperature a liquid must be at before it will provide the fuel vapor required for a fire to ignite. A more technical definition for flash point is the lowest temperature a liquid may be at and still have the capability of liberating flammable vapors at a

Table 4.2 Classes of flammable and combustible liquids

Class	Flash Point (°F)	Boiling Point (°F)
1	Below 100	—
1A	Below 73	Below 100
1B	Below 73	At or above 100
1C	73 – 99	—
2	100 – 139	—
3	140 or above	Below 100
3A	140 – 199	At or above 100
3B	200 or above	—

sufficient rate that, when united with the proper amounts of air, the air-fuel mixture will flash if a source of ignition is presented. The amounts of vapor being released at the exact flash-point temperature will not sustain the fire, and after flashing across the liquid surface the flame will go out. It must be remembered that the liquid is releasing vapors at the flash point temperature, and as with other ordinary burnable materials, it is the vapors that burn. The burning process for both ordinary combustible solids and liquids requires the material to be vaporized. It may also be in the form of a very fine mist, which will be instantly vaporized if a source of heat is introduced. It is not the actual solid or the liquid that is burning but the vapors being emitted from it. For this reason, when we speak of a fuel we are referring to the liberated vapor. It is an accepted phenomenon, assuming sufficient amounts of air to be present, that the greater the volume of released vapor the larger the fire will be.

The technical literature sometimes refers to the "fire point," which in most instances is just a few degrees above the flash point temperature, and is the temperature the liquid must be before the released vapor is in sufficient quantity to continue to burn, once ignited. However, because a flash fire will normally ignite any Class "A" combustible present in the path of the flash, it is reasonable to accept the flash point as being the critical liquid temperature in assessing a fire hazard. Any of the other combustibles ignited by the flash fire, that is, wood, paper, cloth, etc., once burning, could then provide the additional heat necessary to bring the liquid to its fire point.

A crucial objective upon arrival of the first responding fire forces is to determine if the liquid present is a product that is vaporizing at the time or not, and what condition may be present that is capable of providing the required heat to cause the liquid to reach its flash point. This information would have a direct influence on the selection of control and/or extinguishing activities. An emergency involving a petroleum liquid, which is equal to or above flash point, means that a fuel source consisting of flammable vapors will be present. This, in turn, means the responding firefighting forces will be faced with either a highly hazardous vapor cloud condition or with a fire if ignition has occurred before arriving at the scene. Conversely, if it is a liquid at a temperature below its flash point, then fuel would not be immediately available to burl.

A source of air, or more specifically oxygen, must be present. A reduction in the amount of available air to below ideal quantities causes the fire to diminish. Moreover, reduce the fuel quantity available and the fire will also diminish in size. Almost all extinguishing techniques developed are methods of denying the fire one or both of these requirements. By cooling a material below its flash point, vapor production is halted, thus removing the fuel from the fire. When utilizing a smothering-type extinguishing agent, the principle involved consists of altering the air-fuel mixture. When the vapor is no longer in its explosive range, the fire dies either due to insufficient fuel or a lack of oxygen.

The flash point tells us the conditions under which we can expect the fuel vapor to be created, but it is the explosive range that tells us that a certain mixture of fuel vapor and air is required for the vapor to become ignitable. The terms flammable limit and combustible limit are also used to describe the explosive range. These three terms have identical meanings and can be used interchangeably. This information is reported as the lower explosive limits (LEL) and the upper explosive limits (UEL). The values that are reported for the LEL and UEL are given as a percentage of the total volume of the air-fuel mixture. The area between the LEL and the UEL is what is known as the explosive range. The figures given for the amount of fuel vapor required to place a substance within its explosive range are shown as a percentage of the total air-fuel mixture. To compute how much air is required to achieve this mixture, subtract the listed percentage from 100 percent. The remainder will be the amount of air needed. Even though it is only the oxygen contained in the air that the fire consumes, flammable ranges are shown as air-fuel ratios because it is the air that is so readily available. Any air-fuel mixture in which the vapor is above the UEL, or any air-fuel mixture in which the vapor is below the LEL, will not burn. Using gasoline as an example, the explosive range can be computed as follows:

	LEL(%)	UEL (%)
Gasoline Vapor	1.5	7.6
Air	98.5	92.4
Total Volume	100	100

This example helps to illustrate that large volumes of air are required to burn gasoline vapors. The explosive ranges for the different grades of gasoline, or even those of most other petroleum liquids, are such that average explosive-range figures that are suitable for use by the fire fighter would be the LEL at 1 percent and the UEL at about 7 percent. The vapor content of a contaminated atmosphere may be determined through the use of a combustible gas-detecting instrument referred to as an explosimeter.

If a fire involving a petroleum liquid does occur, an extinguishing technique that may be appropriate is the altering of the air-fuel mixture. One technique utilized will necessitate the use of an extinguishing agent, such as foam, with the capability of restricting the air from uniting with the vapor. Another technique is to prevent the liquid from having the ability to generate vapor. Usually this is a cooling action and is accomplished with water spray streams. In both cases, extinguishment is accomplished as a result of altering the air-fuel mixture to a point below the LEL for the specific liquid.

Flammability, the tendency of a material to burn, can only be subjectively defined. Many materials that we normally do not consider flammable will burn given high enough temperatures. Neither can flammability be gauged by the heat content of materials. Fuel oil has a higher heat content than many materials considered more flammable because of their lower flash point. In fact, flash point has become the standard for gauging flammability.

4.1.1.2 *Flammability Designation*

The most common systems for designating flammability are the Department of Transportation's (DOT) definitions, the National Fire Protection Association's (NFPA) system, and the Environmental Protection Agency's (EPA) Resource Conservation and Recovery Act's (RCRA) definition of ignitable wastes, all of which use flash point in their schemes. The NFPA diamond, which comprises the backbone of the NFPA Hazard Signal System, uses a four-quadrant diamond to display the hazards of a material. The top quadrant (red quadrant) contains flammability information in the form of numbers ranging from zero to four. Materials designated as zero will not burn. Materials designated as four rapidly or completely vaporize at atmospheric pressure and ambient temperature and will burn readily (flash point <73°F and boiling point <100°F). The NFPA defines a flammable liquid as one having a

General Guidelines on Fire Protection

flash point of 200°F or lower and divides these liquids into five categories:

1. Class IA – Liquids with flash points below 73°F and boiling points below 100°F. An example of a Class 1A flammable liquid is n-pentane (NFPA Diamond: 4).
2. Class IB – Liquids with flash points below 73°F and boiling points at or above 100°F. Examples of Class IB flammable liquids are benzene, gasoline, and acetone (NFPA Diamond: 3).
3. Class IC – Liquids with flash points at or above 73°F and below 100°F. Examples of Class IC flammable liquids are turpentine and n-butyl acetate (NFPA Diamond: 3).
4. Class II – Liquids with flash points at or above 100°F but below 140°F. Examples of Class II flammable liquids are kerosene and camphor oil (NFPA Diamond: 2).
5. Class III – Liquids with flash points at or above 140°F but below 200°F. Examples of Class III liquids are creosote oils, phenol, and naphthalene. Liquids in this category are generally termed combustible rather than flammable (NFPA Diamond: 2). The DOT system designates those materials with a flash point of 100°F or less as flammable, those between 100°F and 200°F as combustible, and those with a flash point of greater than 200°F as nonflammable. EPA designates those wastes with a flash point of less than 140°F as ignitable hazardous wastes. To facilitate the comparison of these systems, they are presented graphically in Figure 4.1.

The elements required for combustion are few – a substrate, oxygen, and a source of ignition. The substrate, or flammable material, occurs in many classes of compounds but most often is organic. Generally, compounds within a given class exhibit increasing heat contents with increasing molecular weights (MW).

These designations serve as useful guides in storage, transport, and spill response. However, they do have limitations. Since these designations are somewhat arbitrary, it is useful to understand the basic concepts of flammability.

Other properties specific to the substrate that are important in determining flammable hazards are the auto-ignition temperature,

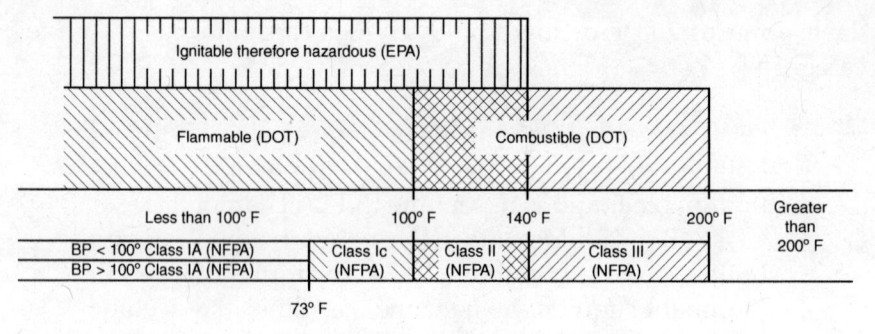

Figure 4.1 Illustrative classification of flash point designators.

boiling point, vapor pressure, and vapor density. Auto-ignition temperature, the temperature at which a material will spontaneously ignite, is more important in preventing fire from spreading, e.g., knowing what fire protection is necessary to keep temperatures below the ignition point, but can also be important in spill or material handling situations. For example, gasoline has been known to spontaneously ignite when spilled onto an overheated engine or manifold. The boiling point and vapor pressure of a material are important, not only because vapors are more easily ignited than liquids, but also because vapors are more readily transportable than liquids. They may disperse or, when heavier than air, flow to a source of ignition and flash back. Vapors with densities greater than one do not tend to disperse but rather settle into sumps, basements, depressions in the ground, or other low areas, thus representing active explosion hazards.

Oxygen, the second requirement for combustion, is generally not limiting. Oxygen in the air is sufficient to support combustion of most materials within certain limits. These limitations are compound specific and are called the explosive limits in air. The upper and lower explosive limits (UEL and LEL) of several common materials are given in Table 4.3.

The source of ignition may be physical, such as a spark, electrical arc, small flame, cigarette, welding operation, or a hot piece of equipment, or it may be chemical in nature, such as an exothermic reaction. In any case, when working with or storing flammables, controlling the source of ignition is often the easiest and safest way to avoid fires or explosions.

Once a fire has started, control of the fire can be accomplished in several ways: through water systems by reducing the temperature,

General Guidelines on Fire Protection

Table 4.3 Explosive limits of hazardous materials

Compound	LEL %	UEL %	Flash Point °F	Vapor Density
Acetone	2.15	13	−4	2.0
Acetylene	2.50	100	Gas	0.9
Ammonia	16	25	Gas	0.6
Benzene	1.30	7.1	12	7.8
Carbon Monoxide	12.4	74	Gas	1.0
Gasoline	1.4	7.6	−45	3 – 4
Hexane	1.1	7.5	−7	3.0
Toluene	1.2	7.1	40	3.1
Vinyl Chloride	3.6	33	Gas	2.2
p-xylene	1.0	6.0	90	3.7

through carbon dioxide or foam systems by limiting oxygen, or through removal of the substrate by shutting off valves or other controls. This chapter provides detailed discussion on the theories of fire and specific information on hydrocarbons, as well as chemical specific fire characteristics.

4.1.2 Ignition Temperature

Consider an emergency situation where there is a spill of gasoline. We may immediately conclude that two of the requirements for a fire exist. First, the gasoline, which would be at a temperature above its flash point, will be releasing flammable vapors; thus, a source of fuel will be present. Moreover, there is ample air available to unite with the fuel, thus there is the potential for the mixture to be in its explosive range. The only remaining requirement needed to have a fire is a source of heat at or above the ignition temperature of gasoline. Technically speaking, all flammable vapors have an exact minimum temperature that has the capability of igniting the specific air-vapor mixture in question. This characteristic is referred to as the ignition temperature and could range from as low as 300°F for the vapor from certain naphthas to over 900°F for asphaltic material vapor. Gasoline vapor is about halfway between at 600°F. A rule of thumb for the ignition temperature of petroleum-liquid vapors is 500°F. This figure may appear low for several of the hydrocarbon vapors, but it is higher than that of most ordinary combustibles and is close enough to the actual ignition temperatures of the products most frequently present at emergency scenes to give a suitable margin of safety.

In emergency situations, it is best to take conservative approaches by assuming that all heat sources are of a temperature above the ignition temperature of whatever liquid may be present. This approach is not an overreaction when it is realized that almost all the normally encountered spark or heat sources are well above the ignition temperature of whatever petroleum liquid might be present. Among the more common sources of ignition would be smoking materials of any kind (cigarettes, cigars, etc.), motor vehicles, equipment powered by internal combustion engines, electrically operated tools or equipment, as well as open-flame devices such as torches and flares. The removal of any and all potential ignition sources from the area must be instituted immediately and methodically. The operation of any motor vehicle, including diesel-powered vehicles, must not be permitted within the immediate vicinity of either a leak or spill of a flammable liquid. The probability of a spark from one of the many possible sources on a motor vehicle is always present. Also, under no circumstances should motor vehicles be allowed to drive through a spill of a petroleum product.

Ignition sources are not necessarily an external source of heat; it could be the temperature of the liquid itself. Refineries and chemical plants frequently operate processing equipment that contains a liquid above its respective ignition temperature. Under normal operating conditions, when the involved liquid is totally contained within the equipment, no problems are presented because the container or piping is completely filled with either liquid or vapor. If full and totally enclosed, it means there can be no air present; thus, an explosive or ignitable mixture cannot be formed. If the enclosed liquid, which in certain stages of its processing may be above the required ignition temperature, should be released to the atmosphere, there is a possibility that a vapor-air mixture could be formed, and hence ignition could occur. This type of ignition is referred to as auto-ignition. Auto-ignition is defined as the self-ignition of the vapors emitted by a liquid heated above its ignition temperature that, when escaping into the atmosphere, enter into their explosive range. Some typical ignition temperatures for various petroleum liquids are 600°F for gasoline, 550°F for naphtha and petroleum ethers, 410°F for kerosene, and 725°F for methanol.

From the above discussions, the important elements that are responsible for a fire are:

- Fuel in the form of a vapor that is emitted when a liquid is at or above its flash point temperature.
- Air that must combine with the vapor in the correct amounts to place the mixture in the explosive range.
- Heat, which must be at least as hot as the ignition temperature, must be introduced.

4.1.3 Flammability Limits

Petroleum liquids have certain characteristics that can exert an influence on the behavior of the liquid and or vapor that is causing the problem. For this reason, these features may have a bearing on the choice of control practices or extinguishing agents under consideration. These characteristics include the weight of the vapor, the weight of the liquid, and whether the liquid will mix readily with water. The specific properties of importance are vapor density, specific gravity, and water solubility. Before discussing these important physical properties, let's first examine the data in Table 4.4, which lists the flammability limits of some common gases and liquids. Two general conclusions can be drawn. First, the lower the material's LEL, obviously the more hazardous. Also note that there are some materials that have wide explosive ranges. This aspect is also significant from a fire standpoint. As an example, comparing hydrogen sulfide to benzene, although the LEL for H_2S is more than 3 times greater, its explosive range is 7 times wider. This would suggest that H_2S is an extremely hazardous material even though its LEL is relatively high. In fact, H_2S fires are generally so dangerous that the usual practice is to contain and allow burning to go to completion rather than to fight the fire.

4.1.4 Vapor Density

Vapor density is a measure of the relative weight of vapor compared to the weight of air. Published data on the characteristics of petroleum products usually include the vapor density. The value of

Table 4.4 Limits of flammability of gases and vapors, % in air

Gas or Vapor	LEL	UEL
Hydrogen	4.00	75.0
Carbon Monoxide	12.5	74.0
Ammonia	15.5	26.60
Hydrogen Sulfide	4.30	45.50
Carbon Disulfide	1.25	44.0
Methane	5.30	14.0
Ethane	3.00	12.5
Propane	2.20	9.5
Butane	1.90	8.5
Iso-butane	1.80	8.4
Pentane	1.50	7.80
Iso-pentane	1.40	7.6
Hexane	1.20	7.5
Heptane	1.20	6.7
Octane	1.00	3.20
Nonane	0.83	2.90
Decane	0.67	2.60
Dodecane	0.60	
Tetradecane	0.50	
Ethylene	3.1	32.0
Propylene	2.4	10.3
Butadiene	2.00	11.50
Butylene	1.98	9.65
Amylene	1.65	7.70
Acetylene	2.50	81.00

Table 4.4 (cont.) Limits of flammability of gases and vapors, % in air

Gas or Vapor	LEL	UEL
Allylene	1.74	
Benzene	1.4	7.1
Toluene	1.27	6.75
Styrene	1.10	6.10
o-Xylene	1.00	6.00
Naphthalene	0.90	
Antluacene	0.63	
Cyclo-propane	2.40	10.4
Cyclo-hexene	1.22	4.81
Cyclo-hexane	1.30	8.0
Methyl Cyclo-hexane	1.20	
Gasoline-regular	1.40	7.50
Gasoline-73 octane	1.50	7.40
Gasoline-92 octane	1.50	7.60
Gasoline-100 octane	1.45	7.50
Naphtha	1.10	6.00

unity has been arbitrarily assigned as the weight of air. Hence any vapor that is reported to have a density of greater than 1 is heavier than air, and any vapor with a density of less than 1 is lighter than air. Vapors weighing more than 1 will usually flow like water, and those weighing less will drift readily off into the surrounding atmosphere. Even heavier-than-air flammable petroleum-liquid vapor can be carried along with very slight air currents. It may spread long distances before becoming so diluted with enough air as to place it below the lower explosive limit (LEL), at which time it would become incapable of being ignited. There are catastrophic incidences that have occurred whereby ignitable air-vapor mixtures have been detected as far as one-half mile from the vapor source.

For this reason, while responding to a spill or leak, we must consider environmental and topographical features of the surroundings, such as wind direction, the slope of the ground, and any natural or artificial barriers that may channel the liquid or vapors. It is critical in a non-fire incident, such as a spill or leak, to determine the type of petroleum liquid present and its source. Information about the material's vapor density enables us to make reasonable predictions as to the possible behavior of the emitting vapor. These factors may influence the route of approach, the positioning of firefighting apparatus and personnel, the need for and the route of evacuation, and the boundaries of the potential problem area. It is essential that no apparatus or other motor vehicles or personnel be located in the path that a vapor cloud will most likely follow.

As a rule of thumb one should approach a hydrocarbon spill (non-fire situation) under the assumption that the liquid is vaporizing (the vapors will be invisible) and that the liberated vapors are heavier than air unless proven otherwise. The expected conduct of a heavier-than-air vapor is for it to drop and spread at or below ground level much as a liquid would. The big difference is that a liquid will be visible and its boundaries well defined. One can expect that the invisible heavier-than-air vapor will settle and collect in low spots such as ditches, basements, sewers, etc. As the vapor travels, it will be mixing with the air, thus some portions of the cloud may be too rich to burn, other sections too lean, and still others well within the explosive range. Some typical vapor densities for petroleum products are 3 to 4 for gasoline, 2.5 for naphtha, and 1.1 for methanol. For comparison, the vapor density for hydrogen gas is 0.1.

4.1.5 Specific Gravity

The property of specific gravity indicates a petroleum liquid's weight relative to the weight of an equal volume of water. The specific gravity of water is assigned the value of unity as a reference point. Hence, other liquids are evaluated relative to water. Those lighter than water have a value less than unity, whereas those that are heavier have a value that is greater than unity. In general, petroleum products are lighter than water. As a result, they can be expected to float on and spread over the water's surface. The exceptions to this are thick, viscous materials such as road tars and heavy "bunker" fuel.

With their low specific gravity, most petroleum liquids, if spilled onto a pool of water, have a tendency to spread quickly across the water's surface. Unless the fluid contacts an obstacle, the oil will continue to spread until it is of microscopic thickness. For this reason, a relatively small amount of oil floating on water is capable of covering a large area of the surface. In a spill situation, knowledge of the liquid's specific gravity can help determine which one of several tactics will be implemented to mitigate the spread of contamination or to eliminate a fire hazard. Knowing the specific gravity will help determine the following:

- Will be possible to use only water, or must a different agent be applied for purposes of smothering the fire?
- How great is the probability that the burning liquid could result in the involvement of exposures because of it floating on any water that is applied?
- Can the displacement of the fuel by water be considered an effective technique to control a leak from a container?

Some specific gravities of common petroleum liquids are 1 to 1.1 for asphalt, 0.8 for gasoline, and 0.6 for naphtha.

4.1.6 Water Solubility

Another important property is water solubility, which may be described as the ability of a liquid to mix with water. Since most petroleum products are lighter than water, and even if they are well mixed with water, they will separate into a layer of water and a layer of the hydrocarbon. Exceptions to this are polar solvents such as methanol and other alcohols. These types of materials will readily mix with water and can even become diluted by it.

Information on solubility is important in an emergency situation because a petroleum fire will require the application of a regular-type foam or an alcohol-resistant-type foam for extinguishment or for vapor-suppression purposes. For use on a water-soluble liquid, a good-quality alcohol-resistant foam is generally applied.

4.1.7 Responding to Fires

In an emergency situation involving a flammable liquid, the product can be expected to behave as follows:

- When accidentally released from its container it almost always results in a fire response.

- If a fire does occur, flammable liquids prove to be virtually impossible to extinguish by cooling with water.
- If the liquid is contained, the confined space will consist of a vapor-rich mixture.
- After extinguishment, there is still the strong possibility for a reflash owing to the continued production of vapor.

It is important to remember that during an emergency the escaping flammable liquids are low-flash products and, as a result, are releasing vapor at the usual atmospheric temperatures. These materials are, therefore, very susceptible to ignition. Because of this, they are generally encountered as an event requiring fire-control procedures. Also, for the same reasons and the frequent need for large quantities of chemical extinguishing agents, they can present difficult extinguishment problems. The fact that the temperature at which many flammable liquids release vapor is well below the temperature of the water that is being used for fire-control purposes means that extinguishment by the cooling method is not feasible. This does not mean water cannot be used. In fact, the use of water serves an important function in spite of its limitations. However, to obtain extinguishment other tactics utilizing a different agent and techniques are needed. Exactly what agent and what technique will be dictated by the size of the fire, the type of storage container or processing equipment involved, and the firefighting resources that are available. When dealing with a low flash point flammable liquid, the probability of a reflash occurring after initial extinguishment is achieved is high.

The probability of an ignitable air-fuel mixture existing inside a closed storage container of a flammable liquid is nominally low. Flammable liquids have the capacity of generating vapor below the commonly encountered atmospheric temperatures; thus, the space between the tank top and the liquid surface will most usually contain a vapor-rich atmosphere, i.e., conditions will be above the UEL. As the vapor being liberated drives the air from the container, the vapor-rich mixture being above the UEL cannot be ignited. Moreover, if a fire should occur outside the tank or vessel, it will not propagate a flame back into the tank. The major exceptions would be those instances where the product had a flash point temperature about the same as the prevailing atmospheric temperature. Another example would be when a tank containing a low vapor pressure product with a flash point in the same range as the prevailing atmospheric temperature is suddenly cooled. A thunderstorm accompanied by a downpour of rain could cause the tank to breathe in air if the liquid is cooled below the temperature at which it is capable of emitting vapor (the boiling point).

In contrast to flammable liquids, an emergency situation that involves a combustible liquid will have a much different behavior. The expected behavior of a combustible product would be for the liquid to present no significant vapor problem, a fire to be readily extinguished by cooling the liquid with water, and the atmosphere above the liquid level to be below the LEL of any confined product.

Most combustible liquids do not present a vapor problem if accidentally released into the atmosphere. The probability of a fire, therefore, is considerably less than it would be if the spill were of a flammable material. If, however, the combustible liquid is at a temperature higher than its flash point, then it can be expected to behave in the same manner as a flammable liquid. One major difference between the two in a fire situation is that the potential exists for cooling the combustible liquid below its flash point by the proper application of water, generally applied in the form of water spray. In the event that the liquid is burning, and if the fire forces are successful in achieving the required reduction in liquid temperature, then vapor production will cease and the fire will be extinguished because of a lack of vapor fuel. Unless this reduction in liquid temperature can be brought about, the fire will necessitate the same control considerations as a low-flash liquid fire would.

With a fire in a storage tank containing a combustible liquid, normally the application of a foam blanket is the only practical method to achieve extinguishment. This is normally done when the entire exposed surface is burning and is necessitated by the fact that the sheer size of the fire makes it very difficult to apply water spray in the amounts and at the locations needed. Also, the volume of oil that has become heated means that the large quantities of water needed to cool the liquid would introduce the possibility of overflowing the tank. Another problem is that the water requirements for both protecting exposures and attempting extinguishment might be far greater than what is available. These factors alone would cause the emergency forces to treat all refined product tank fires alike and, regardless of the flash point of the liquid, initiate the required steps to apply a foam blanket to the liquid surface. In addition there is always concern for any reaction the water may cause when contacting an oil or hydrocarbon heated above the boiling point of water (212°F). In general, because a spill is generally shallow, spilled or splashed combustible liquids do not present the type of problems that a large storage tank does. It is reasonable to expect to be able to extinguish shallow pools or surface spills with water spray.

In the case of a storage tank with a combustible product, the atmosphere inside the container will normally consist of air. On occasion, there will be detectable odors such as that associated with fuel oil. These odors, which are a good indicator of the presence of a combustible product, are not considered fuel vapor. During a fire situation, flame impingement or radiant heat on a container could cause the liquid to become heated to the temperature at which it would emit vapors. Should this occur, and the vapors being generated then start to mix with the air already in the tank, at some instant the space in the container would then contain a mixture that was within the explosive range of the product. If ignition occurred, a forceful internal explosion could result.

The case of crude oil is somewhat unique compared to fires with refined petroleum products. Burning crude oil has the capability of developing a "heat wave." Crude oil has a composition of different fractions of petroleum products. In a manner similar to a refinery operation that distills, or heats crude oil to separate it into the various usable products, such as gasoline and asphalt, a fire accomplishes the same effect. As crude oil burns it releases the fractions that have lower flash points first, and these are burned. The heavier fractions sink down into the heated oil. This movement of light fractions going up to the fire and heavier, heated fractions going down into the crude produces the phenomenon known as a heat wave. Crude oil has the same basic characteristics as other flammable liquids. There are different grades of crude oil produced from various geographic locations throughout the world, with some crude oils having more heavy-asphalt type material than others. Some have greater quantities of sulfur, creating the problem of poisonous hydrogen sulfide gas generation during a fire, and others have more light, gaseous fractions; however, a common characteristic among them is that all will have varying amounts of impurities and some entrained water.

When liquids of this type burn, creating a heat wave which is comprised of the higher boiling point components plus whatever impurities may be present in the product, radiant heat from the flame heats the liquid surface, the light products boil-off, thus creating the vapor that is burning. The remaining hot, heavier materials transfer their heat down into the oil. As it is formed, this heat wave, or layer of heated crude oil components, can reach temperatures as high as 600°F and spread downward at a rate of from 12 to 18 inches per hour faster than the burn-off rate of the crude oil. This would mean that with a crude oil burn-off of 1 foot per hour, at the end of two hours the heat wave would be somewhere between 24 and 36 inches thick. Once this heat wave is created, the chances of extinguishing a crude-oil tank fire, unless it is of small size, are poor, and any water or foam applied could result in a "slopover" of burning oil.

Crude oil normally contains some entrained water and/or an emulsion layer of water and oil. In addition, crude-oil storage tanks will have some accumulations of water on the uneven tank bottoms. In a fire, when a heat wave is formed and comes into contact with any water, a steam explosion will occur, thus agitating the hot oil above it with great force. The evolution of the steam explosion can be attributed to the reaction of water to high temperatures. When water is heated to its boiling point of 212°F, water vapor or steam is generated. Steam that is produced expands approximately 1700 times in volume over the volume of the water that boils away. If a heat wave of a temperature well above 212°F contacts any water entrained in the oil or some of the bottom water, which is usually in larger quantities, the instantaneous generation of steam will act like a piston, causing the oil to be flung upward with considerable violence and force. This reaction is so strong that it causes the oil to overflow the tank shell. This sudden eruption is known as a boilover. When the hot oil and steam reaction takes place, the oil is made frothy, which in turn further increases its volume. The reaction resulting from the heat wave contacting entrained water can be expected to be of lesser activity than from contact with bottom water. The reason for this difference is that the quantities of water converted to steam in a given spot are usually less.

Another phenomenon associated with a crude oil fire is slopover. Basically, the same principles that are responsible for a boilover are the cause of a slopover. The fundamental difference is that in a slopover the reaction is from water that has entered the tank since the start of the fire. Usually this introduction is the result of firefighting activities. A slopover will occur at some point after the heat wave has been formed. Either the water from the hose streams or, after the bubbles collapse, the water in the foam will sink into the oil, contacting the heat wave where it is converted to steam, and the agitation of the liquid surface spills some amount of oil over the tank rim.

The proper extinguishing agents suitable for petroleum liquid emergencies must be capable of performing the identical functions as those agents used in combating structural-type fires. There will be times when circumstances dictate the use of a cooling agent, whereas at other times it will be a smothering agent, and on some occasions both agents will be necessary. There are a variety of agents capable of accomplishing each of these objectives, as well as being appropriate to combat Class B-type fires. Agents suitable for use on Class B-type fires include halon, carbon dioxide (COD), dry chemicals, foam, and aqueous film forming foam. A description of the major types of firefighting agents is given below.

4.1.8 Firefighting Agents

4.1.8.1 Water

For chemical fires water can be used for the dissipation of vapors, for the cooling of exposed equipment, for the protection of personnel, for control purposes, and for actual extinguishment. For cooling down exposed equipment such as pipelines, pinups, or valves, it is recommended that water be applied at the minimum flow rate of 1 gallon per minute per 10 square feet of exposed surface area. Some general guidelines to consider when applying water are as follows:

- All areas of any piping, containers, etc., that are exposed to the fire's heat or flame should be kept wet during the course of the fire.
- The use of water streams to push and move the burning liquid away from exposed equipment is recommended, provided that it can be done safely.
- The rate of flow from any hose stream should not be less than 100 gal/min regardless of its purpose.
- The use of spray streams is recommended whenever possible.
- Back-up lines for lines in active service should be provided and they should be at least the same capacity as the attack line.
- Any equipment being protected is cool enough if water applied to it no longer turns to steam.

With storage tanks or processing equipment exposed to fire or radiant heat, the cooling of any metal above the liquid level inside the vessel is critical. Metal surfaces that have a constant film of water flowing over them will not reach a surface temperature above the boiling point of water. This temperature is well below that which would subject the metal to loss of integrity because of softening.

Water flows employed to cool exposed vertical storage tanks can be calculated using a requirement one hose line (flowing 200 gal/min) per 10 feet of tank diameter. Assuming an average tank height of 50 feet, this would give a water flow capability in excess of the recommended rate. On the fire ground, wind conditions, personnel deficiencies (fatigue, lack of experience or training, etc.), strewn feathering, and so forth have historically resulted in not all the water that is flowing actually doing its intended function. The rule of thumb of one line flowing a minimum of 200 gal/min for each 10 feet of tank diameter provides a suitable safety margin to overcome the loss of water not reaching its target. This flow is only required on the side or sides of the tank being heated; therefore, if a 100-foot-diameter tank is receiving heat on just one-half of its circumference, it would require five hose streams of 200 gal/min each applied to the heated area for cooling purposes. It must be anticipated that these minimum flows will need to be maintained for a time period of at least 60 minutes. Tank truck incidents in which the fire burns for several hours are not unusual. It should be appreciated that a relatively minor fire on a tank truck or rail car could require in excess of 20,000 gallons of water for control and/or extinguishment. It is imperative that as early into the event as possible, an accurate assessment of water flow requirements should be made and flow rates adjusted accordingly. First responders must be constantly alert for any indication of an increase in the internal pressure of a container. Such an increase or any visible outward distortion of the tank shell would be an indication that additional water flows are required. These warning signs, which would be an indication of increased internal pressures, could justify the immediate use of unmanned monitors or hose holders, larger size nozzles, and the pulling back of all personnel to a safe location.

In preparing emergency response plans for petroleum liquid spills or fires, it should be taken into consideration that the required water rates could be needed for long periods of time. Provisions for an uninterrupted supply at a suitable volume must be built into the plans. The rates stipulated in the foregoing do not include amounts of water that may be needed for the protection of firefighting personnel who are involved in activities such as rescue work or valve closing and block off operations. If these or other water-consuming activities are required, additional water must be provided. Of equal importance to the amount of water being used is that water be used in the right place. In general, the application of water should have one or more of the following goals as its objective:

- Cool the shell of any container that is being subjected to high heat levels. This is most effectively accomplished by applying the water to the uppermost portions of the container and allowing it to cascade down the sides.
- Cool any piece of closed-in equipment containing a liquid or gas and exposed to high heat levels. This is most effectively accomplished by applying the water spray over the entire area being heated.
- Protect any part of a container, piping, or item of processing equipment receiving direct time contact. This is most effectively accomplished with a very narrow spray pattern or even a straight stream directed at the point of flame contact.
- Cool steel supports of any container or pipe rack that may be subjected to high heat levels. This is most effectively accomplished by the application of narrow spray streams to the highest part of the support being heated and permitting the water to run down the vertical length of the support.

4.1.8.2 *Foam*

The application of a foam blanket is the only means available to the fire forces for the extinguishment of large petroleum storage tank fires. The foam blanket extinguishes by preventing vapor rising from the liquid surface from uniting with the surrounding air and forming a flammable mixture. Although the water in a foam does provide some incidental cooling action, this is considered more important for cooling heated metal parts, thus reducing the possibility of re-ignition, than as an extinguishing factor. A good-quality foam blanket of several inches in thickness has also been proved effective as a vapor suppressant on low flash-point liquids. Foam may also be used to suppress vapor, hence the layer of foam will be instrumental both in preventing ignition and reducing the contamination of the surrounding atmosphere. Since foam is still water, even if in a different form than usually used, it may conduct electricity. Consequently, its use on live electrical equipment is not recommended.

There are basically two methods of foam application to fires. The first involves the application of chemical foam, which is generated from the reaction of a powder with water. This type of foam has been replaced largely by a technique that involves the formation of foam bubbles when a foaming agent and water are expanded by the mechanical introduction of air. This type, which is not a chemical reaction, is referred to as mechanical foam. Another name for the same material is air-foam. There are a variety of foam concentrates designed to fit different hazards. These include regular protein-based foams, fluoroprotein foams, aqueous film forming types, alcohol-resistant types, as well as foams that are compatible with dry chemical powders and those that will not freeze at below-zero temperatures. Of the many types, the most suitable for general all-around petroleum use would be either a good-quality fluoroprotein or a good-quality aqueous film forming foam (AFFF). Foam liquids are also available in a wide range of concentrates, from 1 percent to 10 percent. The 3 percent and 6 percent protein and fluoroprotein types are usually employed as low-expansion agents with an expansion ratio of about 8 to 1. That is, for each 100 gallons of foam solution (water/concentrate mix) to which air is properly introduced, it will then develop approximately 800 gallons of finished foam. Foam concentrates of other than 3 percent or 6 percent generally are either high-expansion (as high as 1,000 to 1) or alcohol-resistant types.

In foam applications, the manufacturer will provide a percentage rating of a concentrate, which identifies the required quantity of concentrate to add to the water to achieve a correct solution mixture. For each 100 gallons of solution flowing, a concentrate rated as 3 percent would mean 3 gallons of concentrate per 97 gallons of water, whereas a 6 percent concentrate would mix with 94 gallons of water. This readily explains why only half as much space is required to store or transport the amount of 3 percent concentrate needed to generate a given quantity of foam than would be needed for a 6 percent concentrate to make the same volume of foam.

Once the application of foam is initiated, it must be applied as gently as possible in order to develop a good vapor-tight blanket on the liquid surface. Any agitation of the foam blanket or the burning liquid surface will serve to prolong the operation and to waste foam supplies. Water streams cannot be directed into the foam blanket or across the foam streams because the water will dilute and break down the foam. To be assured that all metal surfaces are cool enough and a good, thick (4 inches or more) blanket of foam has been applied, continue application for a minimum of five minutes after all visible fire is extinguished.

One of the agents considered suitable for the extinguishment of petroleum-liquid fires is aqueous film forming foam. This is a liquid concentrate that contains a fluorocarbon surfactant to help float and spread the film across the petroleum-liquid surface and is commonly referred to as "A Triple F." AFFF concentrates of 1, 3, or 6 percent are available, all with about an eight to one expansion ratio. This material is one of the mechanical-type foaming agents. The same kind of air-aspirating nozzles and proportioners that are used for protein-based foams are usable with AFFF concentrates. The primary advantage of AFFF over other foaming agents is its ability to form a thin aqueous film that travels ahead of the usual foam bubbles. This film has the ability to flow rapidly across the burning liquid surface, thus extinguishing the fire by excluding the air as it moves across the surface. The regular foam bubbles formed and flowing behind the film have good securing qualities, which serve to prevent reflashing from occurring. As with all types of foams, care must be exercised so that the foam blanket, once formed, is not disturbed. Water streams should not be directed into the foam blanket or onto the same target a foam stream is aimed at. Water will dilute the foam below the needed concentration, and the force of the stream will destroy the foam's blanketing effect simultaneously. The blanket must be maintained until all flames are extinguished, all heated metal surfaces are cooled, and other sources of ignition are removed from the vicinity.

4.1.8.3 Alcohol-Resistant Foams

Foams that are suitable for water-soluble polar solvents are formulated to produce a bubble that is stable in those fuels and tends to mix and unite with water. Fuels of this type dissolve the water contained in regular foam very rapidly, resulting in the collapse of the bubbles. The breakdown is so fast and complete with regular protein or fluoroprotein-based foams that, unless the rate at which the foam is being applied is well above the recommended rate, the blanket will not form at all. Alcohol-type foam concentrates are most commonly available at strengths of 3 percent, 6 percent, or 10 percent. Because of the possibility of breakdown, regular foams are not considered suitable for polar solvent-type fires. The exception would be a fire in a container of fairly small diameter or a shallow spill, either of which would allow for the possibility of applying foam at sufficient rates to the point of overwhelming the fire.

4.1.8.4 *High Expansion Foams*

High-expansion foam includes foaming agents with the expansion ratio of 20:1 to as high as 1,000:1 between the solution and the foam bubbles. This agent has been found suitable when combating certain types of Class A and Class B fires. Originally developed to help fight fires inside mines, it is most effective when used in confined areas. Extinguishment is accomplished both by the smothering action of the foam blanket and the cooling action obtained from the water as the bubbles break down. Light, fluffy bubbles break apart and are easily dispersed by even relatively moderate wind currents. Bubbles formed at ratios greater than about 400:1 are most likely to be adversely affected by regular air movement as well as the thermal updrafts created by the fire. In an effort to overcome the susceptibility of the bubbles to wind currents, medium expansion foams have been introduced, which have expansion ratios ranging from about 20:1 to 200:1. High-expansion foam concentrates require special foam generators for both proportioning the liquid with water and aspirating the mixture. Many high-expansion foam-dispensing devices have a discharge range of only a few feet; thus, they must be operated fairly close to the area being blanketed.

4.1.8.5 *Other Extinguishing Agents*

Other extinguishing agents that are suitable for use on fires involving petroleum liquids include dry chemical powders, carbon dioxide gas, and halon gases. Each of these agents, while being capable of extinguishing Class B fires, usually is available in either hand-held extinguishers or the larger wheeled or trailer-mounted portable units.

In some petroleum refineries or chemical plants, an on-site fire brigade equipped with an apparatus capable of dispensing large volumes of dry chemical or a vehicle with a large-capacity carbon dioxide (CO_2) cylinder is common practice.

4.1.8.6 Carbon Dioxide

Carbon Dioxide (CO_2) is used in fighting electrical fires. It is non-conductive and, therefore, the safest to use in terms of electrical safety. It also offers the least likelihood of damaging equipment. However, if the discharge horn of a CO_2 extinguisher is allowed to accidentally touch an energized circuit, the horn may transmit a shock to the person handling the extinguisher.

The very qualities that cause CO_2 to be a valuable extinguishing agent also make it dangerous to life. When it replaces oxygen in the air to the extent that combustion cannot be sustained, respiration also cannot be sustained. Exposure of a person to an atmosphere of high concentration of CO_2 will cause suffocation.

4.1.9 Electrical Fire Prevention

To prevent electrical fires, pay special attention to the following areas:

- Against short circuit – Electrical short circuit has three situations: grounding short circuit, short circuit between the lines, and completely short circuit.
    - To prevent short circuits:
        - Install electrical equipment according to the circuit voltage, current strength and the use of nature, and correct wiring. In acidic, hot, or humid places, it is necessary to use acid with anti-corrosion, high temperature, and moisture-resistant wires.
        - Power tools should be well isolated to prevent shorts. Cords should be inspected for damage.
        - Power master switch should be installed for the use of the current strength of the insurance unit.
        - Conduct periodic current operation to eliminate hidden perils.
- Prevent overloads – Prevent circuit overload by ensuring that all electrical equipment are in strict accordance with the electrical safety codes, matching the corresponding

General Guidelines on Fire Protection 191

wires and the correct installation, and are not allowed to randomly indiscriminate access. Check to see that the appropriate use of load is induced on wires. Make sure that electrical equipment do not exceed load limitations and that production levels and needs have separation, or control use. To prevent single-phase three-phase motor running, it is necessary to install three-phase switching, single-phase power distribution boards running lights. The circuit master switch should be installed in traffic safety and wire line fusing easy for the insurance.

- Anti-contact resistance – When a conductor with another wire or wires and switches, protective devices, meters, and electrical connections are exposed, they form a resistance called contact resistance. If the contact resistance is too large, then when the current is passed it will result in heating to the point where the wire insulation layers will catch on fire, metal wire fuse, and sparks are generated. If there is any source of fuel in close proximity then the potential for a fire exists. To avoid this, all connections should be secured and layers should be thoroughly clean of grease, dirt, and oxide films. Also, apply 6–102 mm cross-sectional area of wire, the welding method of connecting. 102 mm cross-section above the wire should be connected by wiring tablets. Regular line connections should be made, and any connections found to be tap loose or hot should be immediately corrected.
- Lighting fires – Lamp fire such as bulbs, including incandescent, iodine-tungsten lamp, and high-pressure mercury lamps, generate high temperature surfaces. The greater the power along with continuous service, the higher temperatures will be experienced. Thermal radiation from hot surfaces can be a source of spontaneous combustion if fuels and combustible materials are stored nearby. Lamps should have solid metal enclosures; they should not be closer than 30 cm from cloth, paper or clothing, and should be isolated from flammable materials. Ballasts can be the cause for fluorescent fires. Attention should be given to the installation of ventilation in order to provide cooling to the fluorescent and to prevent leakage. Ballasts should be

installed from the bottom up, not down. During installation care should be given to prevent asphalt seals from spilling and resulting in burning on the ballast. Ballasts may be installed with high temperature alarms that automatically cut off the power supply when temperature becomes excessive or smoke is detected.

Check electrical equipment regularly for sparks or arcing. These conditions can trigger spontaneous combustion of flammable gases or cause dust explosions. To prevent electrical equipment from contributing to fires: (1) Conduct regular inspections of insulation resistance and monitor the quality of insulation layers. (2) Inspect and correct naked wires and metal contacts to prevent short circuit. (3) Install explosion-proof seals or isolated lighting fixtures, switches and protective devices.

4.1.10 Firefighting Guidance

4.1.10.1 Types

The following are types of firefighting:

- **Structural firefighting** is fighting fires involving buildings and other structures. Most city firefighting is structural firefighting. Structural firefighting is what most people think of when they think of what their local fire department does.
- **Wildland firefighting** is fighting forest fires, brush or bush fires, and fires in other undeveloped areas.
- **Proximity firefighting** is fighting fires in situations where the fire produces a very high level of heat, such as aircraft and some chemical fires. Proximity firefighting is not the same thing as structural firefighting and in some cases, but not all, takes place outdoors.
- **Entry firefighting** is a highly specialized form of firefighting involving the actual direct entry of firefighters into a fire with a very high level of heat. It is not the same thing as proximity firefighting, although it is used in many of the same situations, such as flammable liquid fires.

There are three elements necessary for a fire to burn: oxygen, fuel, and a source of heat. Remove any one of the three and the fire can no longer burn. All firefighting tactics are based on removing one or more of the three elements in the fire triangle. Firefighting involves a range of methods and equipment for fighting fires, depending on the circumstances and availability.

4.1.10.2 Firefighting Agents and Extinguishers

Water: The standard method is to pour water onto the fire. This reduces the heat, hopefully to the point that the fire can no longer burn. However, water is not suitable for liquid fuel fires, because it is likely to spread the fuel and make the fire larger, nor for electrical fires because of the risk of electric shock.

Foam: Foam is put onto fires fuelled by liquid fuel. It works by preventing oxygen getting to the fire but does not spread the fuel around.

Powder: Powder is used on electrical fires, as it does not conduct electricity. Like foam, it works by cutting off the oxygen supply to the fire.

Gas: Non-combustible gases are sometimes used in fire-suppression systems installed in buildings. They work by being heavier than air, thus displacing the oxygen. However, because the oxygen is displaced from the building, humans are unable to survive for long either, so gas systems are only used in special circumstances.

Fire beaters and rakes: Fire beaters and rakes are low-tech ways to fight small fire in scrub. They work by spreading the fire out, thus reducing the concentration of heat, and by removing the source of fuel from the fire.

Extinguishers: Fire extinguishers are canisters of water, foam, or powder that can be carried by one person to the fire. They are of limited capacity but can be used to stop a fire before it becomes large.

Vehicles: Various vehicles are used by firefighters, including trucks that pump water from a nearby supply onto a fire and tankers that carry water to the site of the fire when no local supply is available.

Fire breaks: For fires in forested areas, bulldozers or other heavy equipment may be used to cut a clear path through the forest in front of an advancing fire, or such fire breaks may be permanently maintained in forested areas. Fire breaks provide an area with little fuel content for the purpose of the fire stopping when it reaches the break. However, strong winds might carry sparks from the fire across the firebreak to further forest on the other side. Even so, it may slow down the fire enough for fire crews to deal with the rest.

Back-burning: Instead of constructing fire breaks, an area ahead of an advancing fire might be set alight and burnt in a controlled manner to remove the fuel from that area, thus impeding the advance of the main fire.

For small fires, a fire extinguisher can be used. Remember P.A.S.S. (Pull - Aim - Squeeze - Sweep). Fire extinguishers are rated for the types of fires they are effective against. Class A is used for ordinary flammable solids, such as wood and paper; Class B is used for flammable liquids, such as grease, oil, and gasoline; Class C is used for electrical fires; and Class D is used for flammable metals.

Fire extinguishers are divided into four categories based on different types of fires. Each fire extinguisher also has a numerical rating that serves as a guide for the amount of fire the extinguisher can handle. The higher the number, the more firefighting power the extinguisher has. The following is a quick guide to help choose the right type of extinguisher.

- **Class A** extinguishers are for ordinary combustible materials such as paper, wood, cardboard, and most plastics. The numerical rating on these types of extinguishers indicates the amount of water it holds and the amount of fire it can extinguish.
- **Class B** fires involve flammable or combustible liquids such as gasoline, kerosene, grease and oil. The numerical rating for Class B extinguishers indicates the approximate number of square feet of fire it can extinguish.
- **Class C** fires involve electrical equipment, such as appliances, wiring, circuit breakers and outlets. Never use water to extinguish Class C fires - the risk of electrical shock is far too great! Class C extinguishers do not have a numerical rating. The C classification means the extinguishing agent is non-conductive.
- **Class D** fire extinguishers are commonly found in a chemical laboratory. They are for fires that involve combustible metals, such as magnesium, titanium,

potassium, and sodium. These types of extinguishers also have no numerical rating, nor are they given a multi-purpose rating; they are designed for Class D fires only.

Some fires may involve a combination of these classifications:

- **Water extinguishers** or APW extinguishers (air-pressurized water) are suitable for **Class A fires only**. Never use a water extinguisher on grease fires, electrical fires, or Class D fires - the flames will spread and make the fire bigger! Water extinguishers are filled with water and pressurized with oxygen. Again, water extinguishers can be very dangerous in the wrong type of situation. Only fight the fire if you're certain it contains ordinary combustible materials only.
- **Dry chemical extinguishers** come in a variety of types and are suitable for a combination of **Class A, B, and C fires**. These are filled with foam or powder and pressurized with nitrogen.
  - **BC** – This is the regular type of dry chemical extinguisher. It is filled with sodium bicarbonate or potassium bicarbonate. The BC variety leaves a mildly corrosive residue that must be cleaned immediately to prevent any damage to materials.
  - **ABC** – This is the multipurpose dry chemical extinguisher. The ABC type is filled with mono-ammonium phosphate, a yellow powder that leaves a sticky residue that may be damaging to electrical appliances such as computers.

Dry chemical extinguishers have an advantage over CO_2 extinguishers since they leave a non-flammable substance on the extinguished material, reducing the likelihood of re-ignition.

- **Carbon Dioxide (CO_2) extinguishers** are used for **Class B and C fires**. CO_2 extinguishers contain carbon dioxide, a non-flammable gas, and are highly pressurized. The pressure is so great that it is not uncommon for bits of dry ice to shoot out the nozzle. They don't work very well on Class A fires because they may not be able to displace enough oxygen to put the fire out, causing it to re-ignite.

CO_2 extinguishers have an advantage over dry chemical extinguishers since they don't leave a harmful residue - a good choice for an electrical fire on a computer or other favorite electronic device such as a stereo or TV.

These are only the common types of fire extinguishers. There are many others to choose from. Base your selection on the classification and the extinguisher's compatibility with the items you wish to protect.

4.1.10.3 Vehicles

The "fire truck" and "fire engine" are two different vehicles, and the terms should not be used interchangeably. In general, fire engines are outfitted to pump water at a fire. Trucks do not pump water and are outfitted with ladders and other equipment for use in ventilation of the building, rescue, and related activity. In many jurisdictions it is common to have a ladder (or truck) company and an engine company stationed together in the same firehouse.

4.1.10.4 Firefighting Gear

Basic firefighting gear includes the following:

- Bunker gear – refers to the set of protective clothing worn by firefighters. This is made of a modern fire-resistant material, such as **Nomex**, and has replaced the old rubber coat which was traditionally worn. Nomex is a registered trademark for flame resistant meta-aramid material developed in the early 1960's by DuPont. The original use was for parachutes in the space program. A Nomex hood is a common piece of firefighting equipment. It is placed on the head on top of a firefighter's facemask. The hood protects the portions of the head not covered by the helmet and facemask from the intense heat of the fire.
- Self-contained breathing apparatus (SCBA) – worn in any IDLH (immediately dangerous to life or health) situation, which includes entry into burning buildings, where hot air and smoke inhalation present a danger, or

any situation involving hazardous materials exposure. It consists of a breathing mask and a tank of air worn on the back.
- Helmet, boots, and gloves.
- Tools – traditional hand tools such as the pike, halligan, pulaski, axe, hydrant wrench, and power tools such as saws.
- Hoses and nozzles.

"Two in, two out" refers to the rule that firefighters never enter a building or other IDLH situation alone but go in as a pair. Two go in and two come out; also, the two in can refer to the two who enter the building while a backup team of two remains outside the building ready to enter if needed.

To minimize the risk of electrocution, electrical shock, and electricity-related burns while fighting wildland fires, NIOSH recommends that fire departments and firefighters take the following precautions [IFSTA 1998a,b; NWCG 1998; NFPA 1997; 29 CFR* 1910.332(b); 29 CFR 1910.335(b); Brunacini 1985]:

Fire departments should do the following:

- Keep firefighters a minimum distance away from downed power lines until the line is de-energized. This minimum distance should equal the span between two poles.
- Ensure that the Incident Commander conveys strategic decisions related to power line location to all suppression crews on the fireground and continually reevaluates fire conditions.
- Establish, implement, and enforce standard operating procedures (SOPs) that address the safety of firefighters when they work near downed power lines or energized electrical equipment. For example, assign one of the fireground personnel to serve as a spotter to ensure that the location of the downed line is communicated to all fireground personnel.
- Do not apply solid-stream water applications on or around energized, downed power lines or equipment.
- Ensure that protective shields, barriers, or alerting techniques are used to protect firefighters from electrical hazards and energized areas. For example, rope off the energized area.

- Train firefighters in safety-related work practices when working around electrical energy. For example, treat all downed power lines as energized and make firefighters aware of hazards related to ground gradients.
- Ensure that firefighters are equipped with the proper, personal, protective equipment (Nomex® clothing compliant with NFPA standard 1500 [NFPA 1997], leather boots, leather gloves, etc.) and that it is maintained in good condition.
- Ensure that rubber gloves, dielectric overshoes, and tools (insulated sticks and cable cutters) for handling energized equipment are used by properly trained and qualified personnel.

Firefighters should do the following:

- Assume all power lines are energized and call the power provider to de-energize the line(s).
- Wear appropriate, personal, protective equipment for the task at hand – Nomex® clothing compliant with NFPA standard 1500, rubber gloves, and dielectric overshoes and tools (insulated sticks and cable cutters).
- Do not stand or work in areas of dense smoke. Dense smoke can obscure energized electrical lines or equipment and can become charged and conduct electrical current.

4.1.11 Specialized Rescue Procedures

There are many common types of rescues such as building search, victim removal, and extrication from motor vehicles. On the other side, there are certain specialized rescues such as water rescues, ice rescues, structural collapse rescues, and elevator/escalator rescues. These specialized rescues are generally low volume calls, depending on coverage area, and thus firefighters should be trained for them to better assess the situation.

Structural collapse, while not a common incident, may occur for any number of reasons: weakening from age or fire, environmental causes (earthquake, tornado, hurricane, flooding, rain, or snow buildup on roofs), or an explosion (accidental or intentional). Structural collapses can create numerous voids where victims could be trapped.

When arriving on scene the number of potential victims should be acquired. Certain hazards to look for in a structural collapse are a secondary collapse, live electrical wires, and gas leaks. The electricity and gas should be shut-off. Structural collapses come in three different types: pancake collapse, lean-to collapse, and v-type collapse.

A pancake collapse is characterized by both supporting walls failing or by the anchoring system failing and the supported roof or upper floor falling parallel to the floor below. Small voids where victims can be found are created by debris.

A lean-to collapse occurs when only one side of the supporting walls or floor anchoring system fails. One side of the collapsed roof is attached to the remaining wall or anchoring system. The lean-to collapse creates a significant void near the remaining wall.

A v-type collapse can happen when there is a large load in the center of the floor or roof above. The roof may be overloaded from a buildup of snow and/or have been weakened by fire, rot, termites, and improper removal of support beams. Both sides of the supporting wall are still standing but the center of the floor or roof above is compromised and thus collapsed in the center. The v-type collapse usually leaves a void on each side of the supporting walls.

Knowing where the voids are can help a firefighter locate survivors of the structural collapse. After the firefighter has identified the type of collapse and where the voids may be, they need to make a safe entranceway. The firefighter needs to have a basic knowledge of cribbing, shoring, and tunneling, all of which are useful in collapse operations. Cribbing is the use of various sizes of lumber arranged in systematic stacks to support an unstable load. Shoring is the use of wood to support and/or strengthen weakened structures such as roofs, floors, and walls. This avoids a secondary collapse during rescue. Tunneling could be necessary to reach survivors if there is no other way to reach the victim. When tunneling there should be a set destination and a good indication that there is a victim in that void.

4.1.12 First Responder to Electrical Fire Incidents

This information is essential for anyone who might encounter emergency situations where energized electrical equipment creates a hazard. The language is intentionally non-technical in an effort to provide easily understandable information for those without an in-depth knowledge of electricity. Because of the wide variety of emergency situations first responders (i.e., law enforcement officers, firefighters, ambulance attendants, etc.) might encounter, it is not possible to cover every situation.

You must always maintain proper respect for downed wires even though some may appear harmless. Electrical equipment requires the same respect, awareness, and caution you would accord a firearm – always "consider it loaded." In situations where no emergency exists and human life is not in any immediate danger, wait for the local utility personnel to secure the area; they have the knowledge and equipment to complete the job safely.

Law enforcement officers, firefighters, or ambulance attendants are usually first on the scene when overhead wires are down, usually as a result of storms, damaged utility poles, or fallen branches. They need to be aware of the hazards and procedures involved in dealing with emergencies resulting from fallen energized wires.

Electricity seeks the easiest path to ground itself and "does not care" how it gets there. If you or your equipment create that path, you will be placing yourself and possibly others in a life threatening situation. In some situations fallen wires snap and twist, sending out lethal sparks as they strike the ground. At other times the wires lie quietly, producing no sparks or warning signals, as quiet as a rattlesnake and potentially as dangerous. The first rule is to consider any fallen or broken wire extremely dangerous and not to approach within eight feet of it.

Next, notify the local utility and have trained personnel sent to the scene. Have an ambulance or rescue unit dispatched if necessary. Remember, do not attempt to handle wires yourself unless you are properly trained and equipped.

Set out flares, and halt or reroute traffic. Keep all spectators a safe distance (at least 100 feet) from the scene. Electric power emergencies often occur when it is raining; wet ground increases the hazard. After dark, light the scene as well as you can. Direct your spotlight on the broken or fallen wires. *Remember that metal or cable guardrails, steel wire fences, and telephone lines may be energized by a fallen wire and may carry the current a mile or more from the point of contact.*

An ice storm, windstorm, tornado, forest fire, or flood may bring down power lines by the hundreds. Under those circumstances electric companies customarily borrow skilled professionals from one another to augment their own work forces. First Responders have their own jobs to perform at such times, usually as part of a task force,

GENERAL GUIDELINES ON FIRE PROTECTION 201

which lessens the need for individual decision-making. But every first responder should be prepared for when he or she faces an electric power emergency alone and must make decisions about people, power, and the hazards involved. Remember that electricity from a power line, like lightening from a thundercloud, seeks to reach the ground, so it is imperative when working with fallen wires not to let yourself or others create a circuit between a wire and the ground.

In a typical power emergency, a car strikes a utility pole and a snapped power line falls on it. Advise the car's occupants that they should stay in the car. Call the local power company, but remember to not come in contact with either the car or its occupants. If the car catches fire, instruct the occupants to leap, not step, from the car. To step out would put them in the circuit between the wire, the energized car, and the ground with deadly results. If firefighters are on the scene, they may be best able to handle the situation; most full-time firefighters are trained to deal with electric power emergencies and will have the proper equipment to do so. If you must extinguish a car fire without the aid of firefighters, use only dry chemical or CO_2 extinguishers. If the car's occupants are injured and cannot leap to safety, you may be able to use your vehicle to push them out of contact with the wire. If you do this, it is critical to look around the vehicle before leaving your car; there may be another fallen wire behind you or a wire hooked to your bumper. If there is, or you suspect there is, leap from your vehicle.

Once a victim has been removed from the electric hazard, immediately check vital signs. If the victim has no pulse and is not breathing, begin cardio-pulmonary resuscitation immediately and any other appropriate first-aid treatment until he or she is placed in the ambulance.

In any rescue attempt it is essential that you protect yourself. It is a truism that dead heroes rescue no one. Do not, under any circumstance, rely on rubber boots, raincoats, rubber gloves, or ordinary wire cutters for protection. Above all, do not touch or allow your clothing to touch a wire, a victim, or a vehicle that is possibly energized.

4.1.13 Evacuation Planning

Each facility should have a written plan for the orderly evacuation of each building at a facility. The plan should establish the necessary procedures for fire emergencies, bomb threats, etc. Each employee should be familiar with the plan. The following are recommended essential elements to include in the plan.

4.1.13.1 Designated Roles and Responsibilities

There should be a Building Coordinator for each building who is the designated Responsible Individual. The Building Coordinator is responsible for seeing that the plan is implemented and will appoint an adequate number of Floor Marshals, assuring everyone is familiar with this plan, and acting as a liaison with the HSO and First Responders such as the Fire Department.

For the Floor Marshals, at least one individual per floor should be designated and there should be designated back-ups. Floor Marshals will assist in the implementation of the plan by knowing and communicating evacuation routes to occupants during emergency evacuation and report the status of the evacuation to the Building Coordinator.

4.1.13.2 Preparation & Planning for Emergencies

Pre-planning for emergencies is a crucial element of the plan. The following steps should be taken in planning for emergency evacuation of each building:

1. All exits are labeled and operable.
2. Evacuation route diagrams have been approved by the HSO or Safety Division and are posted on all floors and at all exits, elevator lobbies, training/conference rooms, and major building junctions.
3. Occupants do not block exits, hoses, extinguishers, corridors or stairs by storage or rearrangement of furniture or equipment. Good housekeeping is everyone's responsibility.
4. All Floor Marshals have been trained in their specific duties and all building occupants have been instructed in what to do in case of an emergency evacuation.
5. Fire evacuation drills are held at least annually in this building and are critiqued and documented. Prior to holding a fire evacuation drill where the alarm is to be triggered, the Electric Shop and the University's Fire Marshal are notified.
6. Appendix A contains instructions that are posted and/or used in instructing students and visitors using this building's facilities.

4.1.14 Evacuation Procedure

When a fire is discovered, the following should occur:

1. Anyone who receives information or observes an emergency situation should immediately call a central designated number. The Plan should identify the Emergency Number to call, which most often is the Fire Department, Police Department, or Civil Defense.
2. In the incident building, occupants will be notified of emergencies by a fire alarm, paging system, or word of mouth.

The following are general evacuation instructions to include in the plan:

1. Know at least two exits from the building.
2. Be familiar with the evacuation routes posted on the diagram of your floor.
3. To report a fire or emergency, call the Emergency Number. Give your name, room number and the floor that locates the fire. State exactly what is burning, or what is smoking or what smells like a fire to you. Then notify the Building Coordinator, or other designated person, and activate the building notification system.
4. When notified to evacuate, do so in a calm and orderly fashion: Walk, don't run; keep conversation level down; take your valuables and outer garments; close all doors behind you; use the stairs, not the elevators; and assist others in need of assistance.
5. Go to the designated assembly area or as instructed during the notification. Upon exiting the building, move at least 150 feet from the building to allow others to also safely exit the building. (Specify designated assembly areas or indicate areas on evacuation diagram).
6. Persons with disabilities who may have impaired mobility should establish a buddy system to help ensure that any needed assistance will be available to them in an emergency.

4.1.15 General

During evacuation Floor Marshals should assure that every person on his/her floor has been notified and that evacuation routes are clear. If possible, the Floor Marshal will check that all doors are closed and will be the last one out. Upon leaving the floor, the Floor Marshal will report the status of the floor evacuation to the Building Coordinator.

Regarding persons with disabilities (mobility, hearing, sight), each individual who requires assistance to evacuate is responsible for pre-arranging with someone else in their immediate work area to assist them. Anyone knowing of a person with a disability or injury who was not able to evacuate will report this immediately to a Floor Marshal, the Building Coordinator, or the Incident Commander.

4.1.16 Template for Emergency Evacuation Plan

The following are general guidelines for implementation:

1. Assignment of Responsibility – Administrative responsibility for evacuation of each building must be clearly defined. The HSO shall designate an individual with a thorough understanding and appropriate authority as the Building Coordinator under the plan. The Building Coordinator shall appoint Floor Marshals as appropriate and ensure that they are adequately instructed in their duties and responsibilities. There must be adequate alternates to assume responsibilities in the absence of the Building Coordinator or Floor Marshals. These designations should also be made in the pre-planning stage.
2. Coordination – In buildings that are under the control of more than one manager or HSO, there should be one evacuation plan for that building that has been coordinated with all key persons.
3. Notification – The Building Coordinator must assure that building occupants know whom to call in case of an emergency and the proper sequence of notification. If it is a fire emergency, call a designated three-digit number (in the United States, 911 for example), then

notify the Building Coordinator and activate the building notification system. Similar emergency incidents may be covered under the same plan. If it is a tornado, the Building Coordinator or other designated person will notify occupants through the Floor Marshal of safe places of haven. If it is a bomb threat, call a centralized Emergency Notification Number identified in the Plan, then notify the Building Coordinator, but do not activate the building notification system. Responsible parties will review the specific circumstances and mandate evacuation when deemed necessary or will otherwise advise Management when evacuation is not necessary. Occupants must be aware of whom to notify in the event the Building Coordinator is absent. The Building Coordinator must assure that there is an effective method to notify occupants of an emergency. Notification may be by means of an alarm system, public address system, telephone fan out system, or oral communication, although this last method is not advisable for work areas with ten or more persons.

4. Preparation and Planning – Proper preparation and planning for emergencies is essential in order for evacuation to be effective and efficient. Emergency route diagrams can be produced using the building "key plans." These should be made accessible to all employees. Building/room diagrams should also be made available through the Facilities Management Information System (FMIS). Use the key plan to highlight the main corridor exit ways and stairways and to identify the exits. You can also use this diagram to indicate where fire extinguishers and firefighting equipment are located by using color coding or a symbol. The diagram can also be used to indicate designated assembly areas. Post the diagrams at elevator lobbies, junctions in buildings where directions are routinely posted, information bulletin boards, training/conference rooms, or other strategic locations throughout the building. The Fire Marshall and the HSO must review all emergency route diagrams. Make sure all exits are labeled, operable, and not blocked and there is proper illumination of pathways and exit signs. Floor Marshals

should be fully familiar with the building evacuation plan and their role. Make sure there are alternates to cover during absences. Decide on a designated meeting place outside the building where Floor Marshals can quickly report to the Building Coordinator and where occupants can be accounted for. Communicate the program to all staff in each building.
5. Fire Evacuation Drills – All building occupants must be familiar with what they should do during an evacuation. The most effective method of familiarizing them is to hold a fire drill at least annually. Holding a fire drill has other advantages, as well. It will provide you with an opportunity to evaluate your notification and evacuation procedures, and it will give you an opportunity to test your fire alarm system and make occupants aware of the sound.

Steps in Conducting a Fire Drill:

1. If you have an alarm system, you must first contact the administrator or alarm company to make appropriate arrangements.
2. Contact the Fire Marshal before holding your drill so that a representative can be on site to assist in critiquing the evacuation.
3. Although the fire drills should be unannounced, you may need to give advanced notice to key personnel in your building.

5
Chemical Data

The following tables of information are contained in this section for more than 400 chemicals:

- Table 5.1 Physical Properties Data
- Table 5.2 Chemical Synonyms

Table 5.1 Physical properties data

Chemical Name	Physical State	Boiling Point, °C	Freezing Point, °C	Specific Gravity
Acetaldehyde	Liquid	20.4	–123	0.780
Acetic Acid	Liquid	117.9	16.7	1.051
Acetic Anhydride	Liquid	139	–74.1	1.080
Acetone	Liquid	56.1	–94.7	0.971
Acetone Cyanohydrin	Liquid	Decomposes	–21	0.925
Acetonitrile	Liquid	81.6	–45.7	0.787
Acetyl Bromide	Liquid	76	–96.5	1.660
Acetyl Chloride	Liquid	51	–112	1.104
Acetyl Peroxide	Liquid	Decomposes	–8	1.200
Acrydine	Solid	346	110	1.200
Acrolein	Liquid	53	–87	0.843
Acrylamide	Liquid	Data not available	84	1.050
Acrylic Acid	Liquid	141.3	12.3	1.050
Acrylonitrile	Liquid	77.4	–83.6	0.808
Aldrin	Solid	Not pertinent	104	1.600
Allyl Alcohol	Liquid	96.9	–129	0.852
Allyl Chloroformate	Liquid	45	–80	1.139
Allyl Chlorosilane	Liquid	116	Not pertinent	1.215

207

Table 5.1 (cont.) Physical properties data

Chemical Name	Physical State	Boiling Point, °C	Freezing Point, °C	Specific Gravity
Aluminum Chloride	Solid	Not pertinent	Not pertinent	2.440
Aluminum Nitrate	Solid	Not pertinent	73	>1
Ammonium Bifluoride	Solid	239.5	125.6	1.500
Ammonium Carbonate	Solid	Not pertinent	Not pertinent	1.500
Ammonium Dichromate	Solid	Not pertinent	Not pertinent	2.150
Ammonium Fluoride	Solid	Not pertinent	Not pertinent	1.320
Ammonium Hydroxide	Liquid	Not pertinent	Not pertinent	0.890
Ammonium Lactate	Solid or Liquid	Not pertinent	Not pertinent	1.200
Ammonium Nitrate	Solid	Not pertinent	169.9	1.720
Ammonium Nitrate-Sulfate Mixture	Solid	Not pertinent	Not pertinent	1.800
Ammonium Oxalate	Solid	Not pertinent	Not pertinent	1.500
Ammonium Perchlorate	Solid	Not pertinent	Not pertinent	1.950
Ammonium Perchlorate	Solid	Not pertinent	Not pertinent	1.980
Ammonium Silicofluoride	Solid	Not pertinent	Not pertinent	2.000
Amyl Acetate	Liquid	146	<–100	0.876
Aniline	Liquid	184.2	–6.1	1.022
Anisoyl Chloride	Liquid	262	22	1.260
Anthracene	Solid	341.2	216.5	1.240
Antimony Pentachloride	Liquid	175	3	2.354
Antimony Pentafluoride	Liquid	143	7	2.340
Animony Potassium Tartrate	Solid	Not pertinent	Not pertinent	2.600
Antimony Trichloride	Solid	223	73	3.140

Chemical Data

Table 5.1 (cont.) Physical properties data

Chemical Name	Physical State	Boiling Point, °C	Freezing Point, °C	Specific Gravity
Antimony Trifluoride	Solid	Not pertinent	292	4.380
Anitmony Trioxide	Solid	Not pertinent	Not pertinent	5.200
Arsenic Acid	Solid	Not pertinent	Not pertinent	2.200
Arsenic Disulfide	Solid	565	307	3.500
Arsenic Trichloride	Liquid	130.2	−13	2.156
Arsenic Trioxide	Solid	457	315	3.700
Arsenic Trisulfide	Solid	Not pertinent	300	3.430
Asphalf	Liquid	Not pertinent	Not pertinent	1.000
Atrazine	Solid	Decomposes	175	1.200
Azinphosmethyl	Solid	Decomposes	73	1.400
Barium Carbonate	Solid	Not pertinent	Not pertinent	4.300
Barium Chlorate	Solid	Not pertinent	414	3.180
Barium Nitrate	Solid	Decomposes	592	3.240
Barium Perchlorate	Solid	Decomposes	505	Not pertinent
Barium Permanganate	Solid	Decomposes	Not pertinent	3.770
Barium Peroxide	Solid	Decomposes	450	4.960
Benzaldehyde	Liquid	179	Not pertinent	1.046
Benzene	Liquid	80.1	5.5	0.879
Benzene Hexachloride	Solid	Not pertinent	Not pertinent	1.891
Benzene Phosphorous Dichloride	Liquid	221	−51	1.140
Benzene Phosphorous Thiodichloride	Liquid	270	−24	1.378
Benzoic Acid	Solid	249.2	122.3	1.316
Benzonitrile	Liquid	191	−12.8	1.010
Benzophenone	Solid or Liquid	305.5	47.9	1.085
Benzoyl Chloride	Liquid	197.3	−0.6	1.211
Benzyl Alcohol	Liquid	205	−15.3	1.050
Benzylamine	Liquid	184.5	−46	0.980

Table 5.1 (cont.) Physical properties data

Chemical Name	Physical State	Boiling Point, °C	Freezing Point, °C	Specific Gravity
Benzyl Bromide	Liquid	198	−3.9	1.441
Benzyl Chloride	Liquid	179.4	−39.2	1.100
Beryllium Fluoride	Solid	Not pertinent	Not pertinent	1.990
Beryllium Metallic	Solid	Not pertinent	Not pertinent	1.850
Beryllium Nitrate	Solid	Not pertinent	Not pertinent	1.560
Beryllium Oxide	Solid	Not pertinent	Not pertinent	3.000
Beryllium Sulfate	Solid	Not pertinent	Not pertinent	1.710
Bismuth Oxychloride	Solid	Not pertinent	Not pertinent	7.700
Bisphenol A	Solid	Not pertinent	Not pertinent	1.195
Boric Acid	Solid	Not pertinent	Not pertinent	1.510
Boron Trichloride	Liquid	12.4	−1.7	1.350
Bromine	Liquid	58.8	−7.2	3.120
Bromine Trifluoride	Liquid	125.8	8.8	2.810
Bromobenzene	Liquid	156	−30.6	1.490
Butadeane, Inhibited	Gas	−4.4	−108.9	0.621
Butane	Gas	−0.48	−138	0.600
N-Butyl Acetate	Liquid	126	−73.5	0.875
Sec-Butyl Acetate	Liquid	112	−73.5	0.872
Iso-Butyl Acrylate	Liquid	137.9	−61.1	0.889
N-Butyl Acrylate	Liquid	148.8	−64	0.899
N-Butyl Alcohol	Liquid	117.7	−89.3	0.810
Sec-Butyl Alcohol	Liquid	99.5	−114.7	0.807
Tert-Butyl Alcohol	Liquid	82.6	25.7	0.780
N-Butylamine	Liquid	77.4	−49	0.741
Sec-Butylamine	Liquid	63	−104	0.721
Tert-Butylamine	Liquid	45	Not pertinent	0.696
Butylene	Gas	−6.3	−183	0.595
Butylene Oxide	Liquid	63	<−50	0.826
N-Butyl Mercaptan	Liquid	98.5	−115.7	0.841

Table 5.1 (cont.) Physical properties data

Chemical Name	Physical State	Boiling Point, °C	Freezing Point, °C	Specific Gravity
N-Butyl Methacrylate	Liquid	163	<0	0.898
1,4-Butynediol	Solid	238	58	1.070
Iso-Butyraldehyde	Liquid	64.1	−80	0.791
N-Butyraldehyde	Liquid	74.8	−96.4	0.803
N-Butyric Acid	Liquid	164	−5	0.958
Cacodylic Acid	Solid	>200	Not pertinent	>1.1
Cadmium Acetate	Solid	Not pertinent	Not pertinent	2.340
Cadmium Nitrate	Solid	Not pertinent	59	2.450
Cadmium Oxide	Solid	Not pertinent	Not pertinent	6.950
Cadmium Sulfate	Solid	Not pertinent	Not pertinent	4.700
Calcium Arsenate	Solid	Not pertinent	Not pertinent	3.620
Calcium Carbide	Solid	Not pertinent	Not pertinent	2.220
Calcium Chlorate	Solid	Decomposes	340	2.710
Calcium Chloride	Solid	Not pertinent	Not pertinent	2.150
Calcium Chromate	Solid	Not pertinent	Not pertinent	>1
Calcium Fluoride	Solid	Not pertinent	Not pertinent	3.180
Calcium Hydroxide	Solid	Not pertinent	Not pertinent	2.240
Calcium Hypochlorite	Solid	Not pertinent	Not pertinent	2.350
Calcium Nitrate	Solid	Decomposes	561	2.500
Calcium Oxide	Solid	Not pertinent	Not pertinent	3.300
Calcium Peroxide	Solid	Decomposes	Not pertinent	2.920
Calcium Phosphide	Solid	Decomposes	1,600	2.510
Camphene	Solid	154	50	0.870
Carbolic Oil	Liquid	181.8	<40.9	1.040
Carbon Dioxide	Liquifed compressed gas or solid	Not pertinent	−78.5	1.556
Carbon Monoxide	Compressed gas or liquified gas	−191.5	−199	0.791

Table 5.1 (cont.) Physical properties data

Chemical Name	Physical State	Boiling Point, °C	Freezing Point, °C	Specific Gravity
Carbon Tetrachloride	Liquid	76.5	−23	1.590
Caustic Potash Solution	Liquid	>130	Not pertinent	1.45–1.50
Caustic Soda Solution	Liquid	>130	Not pertinent	1.500
Chlordane	Liquid	Decomposes	Not pertinent	1.600
Chlorine	Liquified compressed gas	−34.1	−101	1.424
Chlorine Triflouride	Liquified compressed gas	11.6	−76.1	1.850
Chloroacetophenone	Solid	247	20–59	1.320
Chloroform	Liquid	61.2	−63.5	1.490
Chromic Anhydride	Solid	Not pertinent	Not pertinent	2.700
Chromyl Chloride	Liquid	116	−96.5	1.960
Citric Acid	Solid	Not pertinent	153	4.540
Cobalt Acetate	Solid	Not pertinent	140	1.710
Cobalt Chloride	Solid	Not pertinent	86	1.924
Cobalt Nitrate	Solid	Not pertinent	55	1.540
Copper Acetate	Solid	Not pertinent	115	1.900
Copper Arsenite	Solid	Decomposes	Not pertinent	>1.1
Copper Bromide	Solid	Not pertinent	498	4.770
Copper Chloride	Solid	Not pertinent	Not pertinent	2.540
Copper Cyanide	Powder	Not pertinent	Not pertinent	2.920
Copper Fluoroborate	Liquid	100	Data not available	1.540
Copper Iodide	Solid	1,290	605	5.620
Copper Naphthenate	Liquid	154–202	Not pertinent	0.93~1.05
Copper Nitrate	Solid	Not pertinent	114.5	2.320
Copper Oxalate	Solid	Not pertinent	Not pertinent	>1
Copper Sulfate	Solid	Not pertinent	Not pertinent	2.290
Creosote, Coal Tar	Liquid	>180	Not pertinent	1.05–1.09

Chemical Data

Table 5.1 (cont.) Physical properties data

Chemical Name	Physical State	Boiling Point, °C	Freezing Point, °C	Specific Gravity
Cresols	Liquid or Solid	>177	Not pertinent	1.03–1.07
Cumene	Liquid	152.4	–96.1	0.866
Cyanogen	Liquified compressed gas	–21.1	–29.9	0.954
Cyanogen Bromide	Solid	Not pertinent	49 to 51	2.015
Cyclohexane	Liquid	80.7	6.6	0.779
Cyclohexanol	Solid or Liquid	161	23.6	0.947
Cyclohexanone	Liquid	155.8	31.2	0.945
Cyclopentane	Liquid	49.3	–93.9	0.740
P-Cumene	Liquid	177	–67.9	0.857
DDD	Solid	Not pertinent	112	1.476
DDT	Solid	Not pertinent	108	1.560
Decaborane	Solid	213	99	0.940
Decahydro-naphthalene	Liquid	195	–42	0.890
Decaldehyde	Liquid	207–210	18	0.830
1-Decene	Liquid	170.6	–66.3	0.741
N-Decyl Alcohol	Liquid	230	6.9	0.840
N-Decylbenzene	Liquid	300	Not pertinent	0.855
2, 4-D Esters	Liquid	Very high	Not pertinent	1.088 ~ 1.237
Dextrose Solution	Liquid	>100	<0	1.200
Diacetone Alcohol	Liquid	169.2	–42.8	0.938
Di-N-Amyl-Phthalate	Liquid	Very high	Not pertinent	0.820
Diaznon	Solid, or liquid solution	Very high, decomposes	Not pertinent	1.117
Dibenzoyl Peroxide	Solid	Not pertinent	103	1.334
Di-N-Butylamine	Liquid	159.6	–62	0.759
Di-N-Butyl Ether	Liquid	142	–95.4	0.767
Di-N-Butyl Ketone	Liquid	188	–6	0.822

Table 5.1 (cont.) Physical properties data

Chemical Name	Physical State	Boiling Point, °C	Freezing Point, °C	Specific Gravity
Dibutylphenol	Solid or liquid	253	36	0.914
Dibutyl Phthalate	Liquid	335	–35	1.049
O-Dichlorobenzene	Liquid	180.5	–17.6	1.306
P-Dichlorobenzene	Solid	174.2	53	1.458
Di-(P-Chlorobenzoyl) Peroxide	Solid	Decomposes	Not pertinent	>1.1
Dichlorobutene	Liquid	156	–48	1.112
Dichlorodifluoro-methane	Liquified compressed gas	–29.8	–157.7	1.350
1,2-Dichloroethylene	Liquid	Cis: 60, Trans: 48	Cis: –81, Trans: –50	1.270
Dichloroethyl Ether	Liquid	178	–52	1.220
Dichloromethane	Liquid	39.8	–96.7	1.322
2,4-Dichlorophenol	Solid	216	45	1.400
2,4-Dichlorophen-oxyacetic Acid	Solid	Very high	141	1.563
Dichloropropane	Liquid	96.4	–100	1.158
Dichloropropene	Liquid	77	Not pertinent	1.200
Dicyclopentadiene	Liquid	170	5	0.978
Dieldrin	Solid	Not pertinent	176	1.750
Diethanolamine	Liquid	268.4	28	1.095
Diethylamine	Liquid	55.5	–49.8	0.708
Diethylbenzene	Liquid	180	<70	0.860
Diethyl Carbonate	Liquid	126.8	–43	0.975
Diethylene Glycol	Liquid	245	–8	1.118
Diethylene Glycol Dimethyl Ether	Liquid	162	–70	0.945
Diethyleneglycol Monobutyl Ether	Liquid	231	–68	0.954
Diethyleneglycol Monobutyl Ether Acetate	Liquid	246	–33	0.985

Table 5.1 (cont.) Physical properties data

Chemical Name	Physical State	Boiling Point, °C	Freezing Point, °C	Specific Gravity
Diethylene Glycol Monoethyl Ether	Liquid	202	−76	0.990
Diethylenetriamine	Liquid	207	−39	0.954
Di(2-Ethylhexyl) Phosphoritc Acid	Liquid	Decomposes	<−60	0.977
Diethyl Phthalate	Liquid	298.5	−3	1.120
Diethylzinc	Liquid	124	−28	1.207
1,1-Difluoroethane	Liquified compressed gas	11.3	−117	0.950
Difluorophosphoric Acid, Anhydrous	Liquid	116	−95	1.583
Diheptyl Phthalate	Liquid	Not pertinent	Not pertinent	1.000
Diisobutylcarbinol	Liquid	178	−65	0.812
Diisobutylene	Liquid	101.5	−93.5	0.715
Diisobutyl Ketone	Liquid	163	−42	0.806
Diisodecyl Phthalate	Liquid	Very high	−50	0.967
Diisopropanolamine	Liquid or Solid	248.7	42	0.990
Diisopropylamine	Liquid	83.9	−96.3	0.717
Diisopropylbenzene Hydroperoxide	Liquid	Not pertinent (decomposes)	<−9	0.956
Dimethylacetamide	Liquid	166	−20	0.943
Dimethylamine	Compressed gas	6.9	−92.2	0.671
Dimethyl Ether	Liquid under pressure	−24.7	−141.5	0.724
Dimethyl Sulfate	Liquid	188.8	−31.8	1.330
Distillates: Flashed Feed Stocks	Liquid	14–135	Not pertinent	0.71–0.75
Dodecyl-trichlorosilane	Liquid	>149	Not pertinent	1.030

Table 5.1 (cont.) Physical properties data

Chemical Name	Physical State	Boiling Point, °C	Freezing Point, °C	Specific Gravity
Ethane	Liquid or compressed gas	−88.6	−183.3	0.546
Ethyl Acetate	Liquid	77	−83	0.902
Ethyl Alcohol	Liquid	78.3	−114	0.790
Ethyl Butanol	Liquid	146	−114	0.843
Ethyl Chloroformate	Liquid	94	−81	1.135
Ethylene	Compressed gas or liquified gas	−103.7	−169.1	0.569
Ethylene Dichloride	Liquid	83.5	−35.7	1.253
Ethyl Lactate	Liquid	154	Not pertinent	1.030
Ethyl Nitrate	Liquid	17	−50	0.900
Ferric Ammonium Citrate	Solid	Not pertinent (decomposes)	Not pertinent	1.800
Ferric Nitrate	Solid	Not pertinent (decomposes)	47	1.700
Ferric Sulfate	Solid	Not pertinent (decomposes)	Not pertinent	3.100
Fluorine	Compressed gas	−188	−219	1.500
Formaldehyde Solution	Liquid	Not pertinent	Not pertinent	1.100
Gallic Acid	Solid	Not pertinent (decomposes)	Not pertinent	1.700
Gasolines: Automotive	Liquid	60–199	Not pertinent	0.732
Gasolines: Aviation	Liquid	71–171	<24.4	3.400
Glycerine	Liquid	290	17.9	Not pertinent
Heptanol	Liquid	176	−34	0.822

Chemical Data

Table 5.1 (cont.) Physical properties data

Chemical Name	Physical State	Boiling Point, °C	Freezing Point, °C	Specific Gravity
N-Hexaldehyde	Liquid	128	Not pertinent	0.830
Hydrochloric Acid	Liquid	50.5	Not pertinent	1.190
Hydrogen Chloride	Compressed liquified gas	–85	–115	1.191
Hydrogen, Liquified	Liquid	–253	–259	0.071
Isobutane	Liquid under pressure	–11.8	–255.3	0.557
Isobutyl Alcohol	Liquid	107.9	–108	0.802
Isodecaldehyde	Liquid	Data not available	Data not available	0.84 (est.)
Isohexane	Liquid	60.3	–153.7	0.653
Isopropyl Alcohol	Liquid	82.3	–88.5	0.785
Isopropyl Mercaptan	Liquid	52.5	–130.5	0.814
Kerosene	Liquid	200–260	–45.6	0.800
Lactic Acid	Syrupy liquid	Not pertinent (decomposes)	Not pertinent	1.200
Lead Arsenate	Solid	Decomposes	Not pertinent	5.790
Lean Iodide	Solid	Not p	Not pertinent	6.160
Linear Alcohols	Liquid	>252	>19	0.840
Liquified Natural Gas	Liquified gas	–161	–182.2	0.415–0.45
Liquified Petroleum Gas	Liquified compressed gas	>–40	Not pertinent	0.51–0.58
Lithium, Metallic	Solid	Not pertinent	Not pertinent	0.530
Magnesium	Solid	1100	650	1.740
Mercuric Acetate	Solid	Not pertinent (decomposes)	Not pertinent	3.270
Mercuric Cyanide	Solid	Not pertinent (decomposes)	Not pertinent	4.000
Methane	Liquified gas	–161.5	–182.5	0.422

Table 5.1 (cont.) Physical properties data

Chemical Name	Physical State	Boiling Point, °C	Freezing Point, °C	Specific Gravity
Methyl Alcohol	Liquid	64.5	−97.8	0.792
Methyl Chloride	Liquified gas	−24.2	97.7	0.997
Methyl Isobutyl Carbinol	Liquid	131.8	<−90	0.807
Alpha-Methylstyrene	Liquid	165	−23.2	0.910
Mineral Spirits	Liquid	154–202	Not pertinent	0.780
Nitrous Oxide	Liquified compressed gas	−89.5	−90.8	1.266
Nonanol	Liquid	213	−5	0.827
Octane	Liquid	125.6	−56.8	0.703
Oils: Clarified	Liquid	Data not available	Not pertinent	(est.) 0.85
Oils: Crude	Liquid	32–>400	Not pertinent	Not pertinent
Oils: Diesel	Liquid	288–338	−18 to −34	0.841
Oils, Edible: Castor	Liquid	Varies depending on composition	−12	0.960
Oils, Edible: Coconut	Liquid	Not pertinent (very high)	(approx.) 24	Not pertinent
Oils, Edible: Cottonseed	Liquid	Very high	0	0.922
Oils, Edible: Fish	Liquid	Very high	Not pertinent	0.930
Oils, Edible: Lard	Solid	Not pertinent	19–37	0.861
Oils, Fuel: 2	Liquid	282–338	−29	0.879
Oils, Fuel: 4	Liquid	101 to >588	−29 to −9	Not pertinent
Oils, Miscellaneous: Coal Tar	Liquid	106–167	Not pertinent	(est.) 0.90
Oils, Miscellaneous: Motor	Liquid	Very high	34.4	(est.) 0.84
Oils, Miscellaneous: Penetrating	Liquid	Very high	Not pertinent	0.896

Chemical Data

Table 5.1 (cont.) Physical properties data

Chemical Name	Physical State	Boiling Point, °C	Freezing Point, °C	Specific Gravity
Oils, Miscellaneous: Resin	Liquid	300–400	Not pertinent	0.960
Oils, Miscellaneous: Spray	Liquid	310–371	Not pertinent	0.820
Oils, Miscellaneous: Tanner's	Data not available	Very high	Not pertinent	(est.) 0.85
Oleic Acid	Liquid	222	14	0.89
Oleum	Liquid	Decomposes	Not pertinent	1.91–1.97
Oxalic Acid	Solid	Decomposes	101.5	1.900
Oxygen, Liquified	Liquified gas	–182.9	–218	1.140
Paraformaldeyde	Solid	Decomposes	155–172	1.460
Pentaerythritol	Solid	Not pertinent	261	1.390
Pentane	Liquid	36.1	–129.4	0.626
1-Pentene	Liquid	29.9	–165	0.641
Peracetic Acid	Liquid	Not pertinent	–30	Not pertinent
Petrolatum	Liquid	Very high	38–57	Not pertinent
Phenol	Solid or liquid	181.8	40.9	1.058
Phenyldi-chloroarsine, Liquid	Liquid	257	–15.6	1.657
Phosgene	Compressed gas	8.2	–126	1.380
Phosphoric Acid	Liquid	>130	Not pertinent	1.892
Piperazine	Solid	148	106	1.100
Polybutene	Liquid	Very high	Not pertinent	0.81–0.91
Polychlorinated Biphenyl	Liquid or Solid	Very high	Not pertinent	1.3–1.8
Polypropylene	Solid	Decomposes	Not pertinent	Not pertinent
Potassium Cyanide	Solid	Very high	634.5	1.520
Potassium Iodide	Solid	Very high	681	3.130

Table 5.1 (cont.) Physical properties data

Chemical Name	Physical State	Boiling Point, °C	Freezing Point, °C	Specific Gravity
Propane	Liquified compressed gas	–42.1	–108.7	0.590
Propionaldehyde	Liquid	48.08	–80	0.805
Propylene Oxide	Liquid	34.3	–11.9	0.830
Pyridine	Liquid	115.3	–42	0.983
Pyrogallic Acid	Solid	309		1.450
Quinoline	Liquid	237	–15	1.095
Salicylic Acid	Solid	Not pertinent	157	1.44
Selenium Dioxide	Solid	Not pertinent	Not pertinent	3.95
Selenium Trioxide	Solid	Not pertinent	Not pertinent	3.6
Silicon Tetrachloride	Liquid	57.6	–70	1.48
Silver Acetate	Solid	Not pertinent	Not pertinent	3.26
Silver Carbonate	Solid	Not pertinent	Not pertinent	6.1
Silver Fluoride	Solid	1.159	Not pertinent	5.82
Silver Iodate	Solid	Not pertinent	Not pertinent	5.53
Silver Nitrate	Solid	Decomposes	212	4.35
Silver Oxide	Solid	Decomposes	Not pertinent	7.14
Silver Sulfate	Solid	Not pertinent	Not pertinent	5.45
Sodium	Soft solid or liquid	883	97.5	0.971
Soduim Alkyl-benzene-sulfonates	Solid or liquid	Decomposes	Not pertinent	1
Sodium Alkyl Sulfates	Liquid	Decomposes	Not pertinent	Data not available
Sodium Amide	Solid	400	210	1.39
Soduim Arsenate	Solid	180	57	1.87
Sodium Arsenite	Solid	Decomposes	615	1.87
Sodium Azide	Solid	Decomposes	Not pertinent	1.85
Sodium Bisulfite	Solid	Decomposes	Not pertinent	1.48
Sodium Borate	Solid	Decomposes	Not pertinent	2.367

Chemical Data

Table 5.1 (cont.) Physical properties data

Chemical Name	Physical State	Boiling Point, °C	Freezing Point, °C	Specific Gravity
Sodium Borohydride	Solid	Decomposes	Not pertinent	1.074
Sodium Cacodylate	Solid or water solution	Not pertinent	Not pertinent	>1
Sodium Chlorate	Solid	Decomposes	248	2.49
Sodium Chromate	Solid	Decomposes	Not pertinent	2.723
Sodium Cyanide	Solid	Very high	564	1.6
Sodium Dichromate	Solid	Decomposes	357	2.35
Sodium Hydride	Solid	Very high	Not pertinent	Data not available
Sodium Hydrosulfide Solution	Liquid	100	17	1.3
Sodium Hydroxide	Solid	Very high	318	2.13
Sodium Hypochlorite	Liquid	Decomposes	Not pertinent	1.06
Sodium Methylate	Solid	Decomposes	Not pertinent	>1
Sodium Oxalate	Solid	Decomposes	Not pertinent	2.27
Sodium Phosphate	Granular or powdered solid	Decomposes	Not pertinent	1.8–2.5
Sodium Silicate	High-viscosity liquid	Decomposes	Not pertinent	1.1–1.7
Sodium Silicofluoride	Solid	Decomposes	Not pertinent	2.68
Sodium Sulfide	Solid	Very high	Not pertinent	1.856
Sodium Sulfite	Solid	Decomposes	Not pertinent	2.633
Sodium Thiocyanate	Solid	Decomposes	300	>1
Sorbitol	Liquid	Very high	110	1.49
Stearic Acid	Solid	Decomposes	70	0.86
Styrene	Liquid	145.2	–30.6	0.906
Sucrose	Solid	Decomposes	Decomposes (160–186)	1.59
Sulfolane	Liquid	285	26	1.26
Sulfer Dioxide	Liquified gas	–10	–75.5	1.45

Table 5.1 (cont.) Physical properties data

Chemical Name	Physical State	Boiling Point, °C	Freezing Point, °C	Specific Gravity
Sulfuric Acid	Liquid	340	Not pertinent	1.84
Sulfuric Acid, Spent	Liquid	100	Not pertinent	1.39
Titanium Tetrachloride	Liquid	136	–24	1.221
Toluene	Liquid	110.6	–95	0.867
Toluene 2,4-Diisocyanate	Liquid	250	20–22	1.22
P-Toluenesulfonic Acid	Solid	Decomposes	104–105	1.45
O-Toludine	Liquid	200	–24	0.998
Toxamphene	Waxy solid	Decomposes	65–90	1.6
Trichloroethylene	Liquid	87	–86.4	Not pertinent
Trichlorofluoro-methane	Liquid	23.8	–111	1.49
Trichlorosilane	Liquid	32	–127	1.344
Tridecanol	Liquid	274	Not pertinent	0.846
1-Tridecene	Liquid	233	–24	0.765
Triethylaluminum	Liquid	186.6	–46	0.836
Triethylamine	Liquid	89.5	–114.7	0.729
Triethylene Glycol	Liquid	288	–4.3	1.125
Tripropylene Glycol	Liquid	273	–45	1.022
Turpentine	Liquid	150–160	Not pertinent	0.86
Undecanol	Liquid	245	15.9	0.835
1-Undecene	Liquid	192.7	49	0.75
N-Undecyl-benzene	Liquid	316	–5	0.855
Uranyl Acetate	Solid	Decomposes	Not pertinent	2.89
Uranyl Nitrate	Solid	Decomposes	60.2	2.81
Uranyl Sulfate	Solid	Not pertinent	Not pertinent	3.28
Urea	Solid	Decomposes	133	1.34
Urea Peroxide	Solid	Decomposes	Not pertinent	0.8
Vanadium Oxytrichloride	Liquid	126	–77	1.83

Chemical Data

Table 5.1 (cont.) Physical properties data

Chemical Name	Physical State	Boiling Point, °C	Freezing Point, °C	Specific Gravity
Vanadium Pentoxide	Solid	Decomposes	Not pertinent	3.36
Vanadyl Sulfate	Solid	Decomposes	Not pertinent	2.5
Vinyl Acetate	Liquid	72.9	−92.8	0.934
Vinyl Chloride	Liquified gas	−13.8	.153.8	0.969
Vinyl Fluoride, Inhibited	Liquified compressed gas	−72	−161	0.707
Vinylidene Chloride, Inhibited	Liquid	31.6	−122	1.21
Vinyl Methyl Ether, Inhibited	Liquified compressed gas	5.5 (decomposes)	−122	0.777
Vinyltoluene	Liquid	167.7	−77	0.897
Vinyltrichlorosilane	Liquid	90.6	−95	1.26
Waxes: Carnauba	Liquid	Very high	80–86	0.998
Waxes: Paraffin	Liquid to hard solid	Very high	48–65	0.78–0.79
M-Xylene	Liquid	131.9	−47.9	0.864
Xylenol	Liquid or solid	212	(−)45 to (+)40	1.01
Zinc Acetate	Solid	Not pertinent	Not pertinent	1.74
Zinc Ammonium Chloride	Solid	340	Not pertinent	1.81
Zinc Bromide	Solid	Decomposes	Not pertinent	4.22
Zinc Chloride	Solid	Very high	283	2.91
Zinc Chromate	Solid	Not pertinent	Not pertinent	3.43
Zinc Fluoroborate	Liquid	100	Data not available	1.45
Zinc Nitrate	Solid	Decomposes	36	2.07
Zinc Sulfate	Solid	Decomposes	50–100 (decomposes)	1.96
Zirconium Acetate	Liquid	Not pertinent	Not pertinent	1.37
Zirconium Nitrate	Solid	Decomposes	Not pertinent	>1

Table 5.2 Chemical synonyms

Chemical Name	Synonyms				
	1	2	3	4	5
Acetaldehyde	Acetic Aldehyde	Ethanal	Ethyl Aldehyde		
Acetic Acid	Ethanoic Acid	Glacial Acetic Acid	Vinegar Acid		
Acetic Anhydride	Ethanoic Anhydride				
Acetone	Dimethyl Ketone	2-Propane			
Acetone Cyanohydrin	Alpha-Hydroxyiso-Butyronitrile	1-Methylla-ctonitrile			
Acetonitrile	Ethanenitrile	Ethyl Nitrate	Cynomethane	Methyl cyanide	
Acetyl Peroxide	Diacetyl Peroxide Solution				
Acridine	10-Azaanthracene	Benzo (b) Quiline	Dibenzo (b, e) Pyridine		
Acrolein	Acraldehyde	Acrylic Aldehyde	2-Propenal	Acrylaldehyde	

Chemical Data

Acrylamide	Acrylic Amide 50%	Propenamide 50%		
Acrylic Acid	Propenic Acid			
Acrylonitrile	Cyanoethylene	Fumigrain	Ventox	Vinyl Cyanide
Aldrin	endo-, exo-, 1,2,3,4, 10,10-Hexachloro 1,4,4a,5,8,8a-Hexahydro-1,4, 5,8-dimethano-naphtalene	HHDN		
Allyl Alcohol	2-Propane-1-Ol-Vinylcarbinol			
Allyl Chloroformate	Allyl Carbonate			
Allyltrlorosilane	Allylsilicone	Trichloride		
Aluminum Chloride	Anhydrous Aluminum Chloride			
Aluminum Nitrate	Aluminum Nitrate Nonahydrate	Nitric Acid	Aluminum Salt	

Table 5.2 (cont.) Chemical synonyms

Chemical Name	Synonyms 1	Synonyms 2	Synonyms 3	Synonyms 4	Synonyms 5
Ammonium Bifluoride	Acid Ammonium Fluoride	Ammonium Acid Fluoride	Ammonium Hydrogen Fluoride		
Amonium Carbonate	Hartshorn	Salt Volatile			
Ammonium Dichromate	Ammonium Bichromate				
Ammonium Fluoride	Neutral Ammonium Fluoride				
Ammonium Hydroxide	Ammonia Water	Aqueous Ammonia	Household Ammonia		
Ammonium Lactate	Ammonium Lactate Syrup	Dl-Lactic Acid	Ammonium Salt		
Ammonium Nitrate	Nitram				
Ammonium Oxalate	Ammonium Oxalate Hydrate	Diammonium Oxalate	Oxalic Acid	Diammonium Salt	

Chemical Data

Ammonium Perchlorate	Ammonium Peroxydisulfate	Ammonium Peroxydisulfuric Acid	Diammonium Salt
Ammonium Silicofluoride	Ammonium Fluosilicate		
Amyl Acetate	Amyl Acetate, Mixed Isomers	Pentyl Acetates	
Aniline	Aminobenzene	Aniline Oil	Blue Oil	Phenylamine
Anisoyl Chloride	P-Anisoyl Chloride		
Anthracene	Anthracin	Green Oil	Paranaphtalene
Antimony Pentachloride	Antimony (V) Chloride	Antimony Perchloride	
Antimony Potassium Tartrate	Potassium Antimonyl Tartrate	Tartar Emetic	Tartarized Antimony	Tartrated Antimony
Antimony Trichloride	Antimony Butter	Antimony (Iii) Chloride	Butter Of Antimony

Table 5.2 (cont.) Chemical synonyms

Chemical Name	Synonyms 1	2	3	4	5
Anitmony Trioxide	Diantimony Trioxide	Exitelite	Flowers Of Antimony	Senarmontite	Valentinite
Arsenic Acid	Arsenic Pentoxide	Orthoarsenic Acid			
Arsenic Disulfide	Realgar	Red Arsenic Glass	Red Arsenic Sulfide	Red Ointment	Ruby Arsenic
Arsenic Trichloride	Arsenic (Iii) Trichloride	Arsenic Chloride	Arsenous Chloride	Butter Of Arsenic	Caustic Arsenic Chloride
Arsenic Trioxide	Arsenous Acid Anhydride	Arsenous Oxide	Arsenic Sesquioxide	White Arsenic	
Arsenic Trisulfide	Arsenic Yellow	King's Gold	King's Yellow	Ointment	Yellow Arsenic Sulfide
Asphalt	Asphalt Cements	Asphalt Bitumen	Bitumen	Petroleum Asphalt	
Atrazine	2-Chloro-4-Ethylamino-6-Isopropylamino-S-Trizine	Aertex Herbicide			

Chemical Data

Azinphos-methyl	O-O-Dimethyl S-[(4-Oxo-1, 2, 3-Benzotiazine-3(4h)-Yl)Methyl] Phosphoro-Dithide	Gurthion Insecticide	Gusathion Insecticide
Barium Chlorate	Barium Chlorate Monohydrate		
Barium Perchlorate	Barium Perchlorate Trihydrate		
Barium Peroxide	Barium Dioxide	Barium Superoxide	Barium Binoxide
Benzaldehyde	Benzoic Aldehyde	Oil Of Bitter Almond	
Benzene	Benzol	Benzole	
Benzene Hexachloride	Bhc, 1,2,3,4,5,6-Hexachloro-Cyclohexane Lindane		
Benzene Phosphorous Dichloride	Phenyl Phosphorous Dichloride	Phenylphosphine Dichloride	Dichlorophnyl-Phosphine

Table 5.2 (cont.) Chemical synonyms

Chemical Name	Synonyms				
	1	2	3	4	5
Benzene Phosphorous Thiodichloride	Benzene-Phosphonyl Chloride	Phenylosphonothioic Dichloride	Phenylphosphine Thiochloride		
Benzoic Acid	Benzene-Carboxylic Acid	Carboxybenzene	Dracyclic Acid		
Benzonitrile	Benzoic Acid Nitrile	Cyanobenzene	Phenylcyanide		
Benzophenone	Benzoylbenzene	Diphenyl Ketone	Diphenyl Methanone	Alpha-Oxodiphenyl-methane	Alpha-Oxoditane
Benzoyl Chloride	Benzenecarbonyl Chloride				
Benzyl Alcohol	Benzenecarbinol	Alpha-Hydroxytoluene	Phenylcarbinol	Phenyl-Methanol	Phenylmethyl Alcohol
Benzylamine	Alpha-Aminotoluene	Phenylmethyl Amine			

Chemical Data

Benzyl Bromide	Alpha-Bromotoluene	Omega-Bromotoluene	Bromotoluene Alpha	
Benzyl Chloride	Alpha-Chlorotoluene	Omega-Chlorotoluene	Chlorotoluene, Alpha	
Beryllium Nitrate	Beryllium Nitrate Trihydrate			
Beryllium Oxide	Beryllia	Bromellite		
Beryllium Sulfate	Beryllium Sulfate Tetrahydrate			
Bismuth Oxychloride	Basic Bismuch Chloride	Bismuth Chloride Oxide	Bismuth Subchloride	Bismuthyl Chloride Pearl White
Bisphenol A	2,2-Bis (4-Hydroxy-Phenyl) Propane	P,P-Dihydroxy-diphenyl-dimethylmethane	4, 4-Isopropylidenediphenol	Ucan Bisphenol Hp
Boric Acid	Boracic Acid	Orthoboric Acid		
Boron Trichloride	Boron Chloride			

Table 5.2 (cont.) Chemical synonyms

Chemical Name	Synonyms				
	1	2	3	4	5
Bromobenzene	Monobromo-Benzene	Phenyl Bromide	Bromobenzol		
Butadiene, Inhibited	Biethylene	1,3-Butadiene	Bivinyl	Divinyl	Vinylethylene
Butane	N-Butane				
N-Butyl Acetate	Acetic Acid	Butyl Ester	Butyl Acetate	Butyl Ethanoate	
Sec-Butyl Acetate	Acetic Acid	Sec-Butyl Ester			
Iso-Butyl Acrylate	Acrylic Acid	Isobutyl Ester			
N-Butyl Acrylate	Acrylic Acid	Butyl Ester	Butyl Acrylate	Butyl 2-Propenoate	
N-Butyl Alcohol	Butanol	Butyl Alcohol	1-Butanol	1-Hydroxybutane	Propylcarbinol
Sec-Butyl Alcohol	2-Butanol	Butylene Hydrate	2-Hydroxy-Butane	Methylethyl-Carbinol	

Chemical Data

Tert-Butyl Alcohol	2-Methyl-2-Propanol	Trimethylcarbinol		
N-Butylamine	1-Aminobutane	Butlamine	Mono-N-Butylamine	Norvalamine
Tert-Butylamine	2-Aminoisobutane	2-Amino-2-Methylpropane	1,1-Dimethyl-Ethylamine	Tba	Trimethyl-aminomethane
Butylene	1-Butene			
Butylene Oxide	1-Butene Oxide	1,2-Butylene Oxide	Alpha-Butylene Oxide	1,2-Epoxybutane
N-Butyl Mercaptan	1-Butanethiol	Thiobutyl Alcohol		
N-Butyl Methacrylate	Methacrylic Acid	Butyl Ester	Butyl Methacrylate	Butyl 2-Methacrylate	N-Butyl Alpha-Methyl Acrylate
1,4-Butynediol	2-Butyne-1,4-Diol	1,4-Dihydroxy-2-Butyne		
Iso-Butyraldehyde	Isobutyric Aldehyde	Isobutyraldehyde	Isobutylaldehyde	2-Methylpropanal
N-Butyraldehyde	Butanal	Butyraldehyde	Butyric Aldehyde	Butyl Aldehyde

Table 5.2 (cont.) Chemical synonyms

Chemical Name	Synonyms				
	1	2	3	4	5
N-Butyric Acid	Butanic Acid	Butanoic Acid	Butyric Acid	Ethylacetic Acid	Propane-Carboxylic Acid
Cacodylic Acid	Hydroxy-Dimethylarsine Oxide	Dimethylarsenic Acid	Ansar	Silvisar 510	
Cadmium Acetate	Cadmium Acetate Dihydrate				
Cadmium Nitrate	Cadmium Nitrate Tetrahydrate				
Cadmium Oxide	Cadmium Fume				
Calcium Arsenate	Cucumber Dust	Taricalcium Arsenate	Tricalcium Orthoarsenate		
Calcium Carbide	Acetylenogen	Carbide			

Chemical Data

Calcium Chloride	Calcium Chloride, Anhydrous	Calcium Chloride Hydrates		
Calcium Chromate	Calcium Chromate (Vi)	Calcium Chromate Dihydrate	Gelbin Yellow Ultramarine	Steinbuhl Yellow
Calcium Fluoride	Fluospar	Fluorspar		
Calcium Hydroxide	Slaked Lime			
Calcium Hypochlorite	HTH	HTH Dry Chlorine	Neutral Anhydrous Calcium Hypochlorite	Sentry
Calcium Nitrate	Calcium Nitrate Tetrahydrate			
Calcium Oxide	Quicklime	Unslaked Lime		
Calcium Peroxide	Calcium Dioxide			
Calcium Phosphide	Photophor			

Table 5.2 (cont.) Chemical synonyms

Chemical Name	Synonyms				
	1	2	3	4	5
Camphene	2,2-Dimethyl-3-Methylene-norbornane	3,3-Dimethyl-2-Methylene-Norcamphane			
Carbolic Oil	Middle Oil	Liquefied Phenol			
Carbon Dioxide	Carbonic Acid Gas	Carbonic Anhydride			
Carbon Monoxide	Monoxide				
Carbon Tetrachloride	Benzinoform	Necatorina	Perchloro-methane	Tetrachloromethane	
Caustic Potash Solution	Potassium Hydroxide Solution	Lye			
Caustic Soda Solution	Sodium Hydroxide Solution	Lye			

Chemical Data

Chlordane	Chlordan	1,2,4,5,6,7,8,8-Octachloro-2,3,3a,4,7,7a-Hexahydro-4,7-Methano-indene	Texichlor				
Chlorine Triflouride	CTF						
Chloro-acetophenone	Phenacyl Chloride	Omega-Chloro-acetophenone	Alpha-Chloro-acetophenone	Phenyl Chloromethyl Ketone	Tear Gas		
Chloroform	Tricloromethane						
Chromic Anhydride	Chromic Acid	Chromic Oxide	Chromium Trioxide				
Chromyl Chloride	Chromium (Vi) Dioxyxhloride	Chromium Oxyxhloride					
Citric Acid	2-Hydroxy-1,2,3-Propane-Tricarboxylic Acid	Beta-Hydroxy-tricarballylic Acid	Beta-Hydroxy-tricarboxylic Acid				
Cobalt Acetate	Cobalt (Ii) Acetate	Cobalt Acetate Tetrahydrate	Cobaltous Acetate				

Table 5.2 (cont.) Chemical synonyms

Chemical Name	_____ Synonyms _____				
	1	2	3	4	5
Cobalt Chloride	Cobalt (Ii) Chloride	Cobaltous Chloride	Cobaltous Chloride Dihydrate	Cobaltous Chloride Hexahydrate	
Cobalt Nitrate	Cobalt (Ii) Nitrate	Cobaltous Nitrate	Cobaltous Nitrate Hexahydrate		
Copper Acetate	Acetic Acid	Cupric Salt	Crystallized Verdigris	Cupric Acetate Monohydrate	Neutral Verdigris
Copper Arsenite	Cupric Arsenite	Swedish Green	Scheele's Green	Cupric Green	Copper Orthoarsenite
Copper Bromide	Cupric Bromide	Anhydrous			
Copper Chloride	Cupric Chloride Dehydrate	Eriochalcite (Anhydrous)			
Copper Cyanide	Cupricin	Cuprous Cyanide			
Copper Fluoroborate	Copper Borofluoride Solution	Copper (Ii) Fluoroborate Solution	Cupric Fluoroborate Solution		

Chemical Data

Copper Iodide	Cuprous Iodide	Marshite			
Copper Naphthenate	Paint Drier				
Copper Nitrate	Cupric Nitrate Trihydrate	Gerhardite			
Copper Oxalate	Cupric Oxalate Hemihydrate				
Copper Sulfate	Blue Vitriol	Copper Sulfate Pentahydrate	Cupric Sulfate	Sulfate Of Copper	
Creosote, Coal Tar	Creosote Oil	Dead Oil			
Cresols	Cresylic Acids	Hydroxytoluenes	Methylphenold	Oxytoluenes	Tar Acids
Cumene	Cumol	Isopropylbenzene			
Cyanogen	Ethanedenitrate	Dicyan	Oxalic Acid Dinitrile	Oxalonitrile	Dicyanogen
Cyclohexane	Hexa-hydrobenzene	Hexamethylene	Hexanaphthene		
Cyclohexanol	Adronal	Anol	Cyclohexyl Alcohol	Hexalin	Hexa hydrophenol
Cyclohexanone	Anone	Hytrol O	Nadone	Pimelic Ketone	Sextone

Table 5.2 (cont.) Chemical synonyms

Chemical Name	Synonyms				
	1	2	3	4	5
Cyclopentane	Pentamethylene				
P-Cumene	Cymol	P-Isopropyl-Toluene	Isopropyltoluol 1-Methyl-4-Isopropylbenzene	Methyl Propyl Benzene	
DDD	1,1-Dichloro-2,-bis (P-Chlorophenyl) Ethane	Diclorodiphenyl-chloroethane	TDE		
DDT	Dichlorophenyl-Trichloroethane	P,P'-Ddt	1,1,1-Trichloro-2,2-Bis (P-Chlorophenyl) Ethane		
Decahydro-naphthalene	Bicyclo [4.4.0] Decane	Naphthalene	Perhydro-naphthalene	Dec	Decalin
Decaldehyde	Aldehyde C-10	Capraldehyde	Capric Aldehyde	Decanal	N-Decyl Aldehyde
1-Decene	Alpha-Decene				
N-Decyl Alcohol	Alcohol C-10	Capric Alcohol	1-Decanol	Dytol S-91	Lorol-22
N-Decylbenzene	Decylbenzene	1-Phenyldecane			

Chemical Data

2, 4-D Esters	Butoxyethyl	2, 4-Dichloro-phenoxyacetate	Butyl 2, 4-Dichloro-phenoxyacetate	2, 4-Dichloro-phenoxyacetic Acid	Butoxyethyl Ester
Dextrose Solution	Corn Sugar Solution	Glucose Solution	Grape Sugar Solution		
Diacetone Alcohol	Diacetone	4-Hydroxy-4-Methyl-2-Pentanone	Tyranton		
Di-N-Amyl-Phthalate	Diamyl Phthalate	Dipentyl Phthalate	Phthalic Acid, Diamyl Ester	Phthalic Acid, Dipentyl Ester	
Diaznon	O,O-Diethyl O-(-Isopropyl-6-Methyl-4-Pyrimidinyl)	O,O-Diethyl O-2-Isopropyl-4-Methyl-6-Pyrimidyl Thiophosphate	Diethyl 2-Isopropyl-4-Methyl-6-Pyrimidyl Thionophosphate	Alpha-Tox	Saralex
Dibenzoyl Peroxide	Benzoyl Peroxide	Benzoyl Superoxide	BP	BPO	Lucidol-70
Di-N-Butylamine	1-Butanamine, -butyl	Dibutylamine			
Di-N-Butyl Ether	N-Dibutyl Ether	N-Butyl Ether	Butyl Ether	1-Butoxybutane	Dibutyl Ether
Di-N-Butyl Ketone	5-Nonanone				

Table 5.2 (cont.) Chemical synonyms

Chemical Name	Synonyms				
	1	2	3	4	5
Dibutylphenol	2,6-Di-tert-butylphenol				
Dibutyl Phthalate	Butyl Phthalate	DBP	Phthalic Acid, Dibutyl Ester	Rc Plasticizer Dbp	Witcizer 300
O-Dichloro-benzene	1,2-Dichloro benzene	Downtherm E	Orthodichloro-Benzene		
P-Dichloro-benzene	Dichloricide	Paradow	Paradi	Parmoth	Paradichloro-benzene
Di- (P-Chlorobenzoyl) Peroxide	Bis- (P-Chloro benzoyl) Peroxide	p-Chlorobenzoyl Peroxide	P, P'-Dichloro-benzoyl Peroxide	Di-(4-chlorobenzoyl) Peroxide	Cadox Ps
Dichlorobutene	1,4-Dichloro-2-butene	2-Butylene Dichloride	1,4-Dichloro-2-Butylene	Cis-1, 4-Dichloro-2-Butene	Trans-1,4-Dichloro-2-Butene
Dichloro difluoro-methane	Eskimon 12	Genetron 12	F-12	Halon 122	Freon 12
1, 2-Dichloro-ethylene	Acetylene Dichloride	Sym-Dichloro-ethylene	Dioform	Cis- or Trans-1,2-Dichloro-Ethylene	

Chemical Data

Dichloroethyl Ether	Bis (2-Chloroethyl) Ether	2,2'-Dichloroethyl Ether	Dichlorodiethyl Ether	Di-(2-chloroethyl) Ether	Chlorex
Dichloro-methane	Methylene Chloride	Methylene Dichloride			
2,4-Dichloro-phenoxyacetic Acid	2,4-D				
Dichloro-propane	1,2-Dichloro-Propane	Propylene Dichloride			
Dichloro-propene	1,3-Dichloro-propene	Telone			
Dicyclo-pentadiene	DICY	3a,4,7,7a-Tetrahydro-4,7-Methanoindene			
Dieldrin	HEOD	endo, exo-1, 2, 3, 4, 10,10-Hexachloro-6,7-epoxy-1, 4, 4a, 5, 6, 7, 8, 8a-octahydro-1,4:5, 8-dimethano-naphthalene			
Diethanolamine	Bis(2-hydroxy-Ethyl)Amine	Dead Oil	2,2'-Dihydroxy-diethyl Amine	Di(2-Hydroxyethyl) Amine	2, 2'-Imino-diethanol

Table 5.2 (cont.) Chemical synonyms

Chemical Name	\	\	Synonyms	\	\
	1	2	3	4	5
Diethylamine	DEN				
Diethyl Carbonate	Carbonic Acid Diethyl Ester	Ethyl Carbonate	Eufin		
Diethylene Glycol	Bis (2-Hydroxyethyl) Ether	Diglycol	β,β-Dihydroxy-Diethyl Ether	3-Oxa-1,5-Pentanediol	2,2'-Oxybis-ethanol
Diethylene Glycol Dimethyl Ether	Bis (2-Methoxyethyl) Ether	Poly-Solv			
Diethyleneglycol Monobutyl Ether	Butoxydiethylene Glycol	Butoxydiglycol	Diglycol Monobutyl Ether	Butyl "Carbitol"	Dowanol DB
Diethyleneglycol Monobutyl Ether Acetate	2-(2-Butoxyethyoxyl) ethyl Acetate	Diglycol Monobutyl Ether Acetate	Butyl "Carbitol" Acetate	Ektasolve DB Acetate	
Diethylene Glycol Monoethyl Ether	Carbitol	Diethylene Glycol Ethyl Ether	Dowanol DE	2-(2-Ethoxyethoxy) ethanol	Ethoxy Diglycol

Diethylene-triamine	Bis(2-Aminoethyl) Amine	2,2'-Diamino-Diethylamine						
Di(2-Ethylhexyl) Phosphoric Acid	Bis-(2-Ethylhexyl) Hydrgoen Phosphate	Di-(2-ethylhexyl) Phosphate	Di-(2-Ethylhexyl) Acid Phosphate					
Diethyl Phthalate	1,2-Benzene-dicarboxylic Acid, Diethyl Ester	Ethyl Phthalate	Phthalic Acid, Diethyl Ester					
Diethylzinc	Zinc Diethyl	Ethyl Zinc	Zinc Ethyl					
1,1-Difluoro-ethane	Ethylidene Di fluoride	Ethylidene Fluoride	Refigerant 152a					
Difluoro-phosphoric Acid, Anhydrous	Dilfuoro-phosphorous Acid							
Diheptyl Phthalate	Phthalic Acid, Diheptyl Ester							
Diisobutyl-carbinol	2,6-Dimethyl-4-heptanol							
Diisobutylene	2,4,4-Trimethyl-1-pentene							

Table 5.2 (cont.) Chemical synonyms

Chemical Name	Synonyms				
	1	2	3	4	5
Diisobutyl Ketone	DIBK	Sym-Diisopropylacetone	2,6-Dimethyl-4-Heptanone	Isovalerone	
Diisodecyl Phthalate	Phthalic Acid, Bis (8-Methylonyl) Ester	Phthalic Acid, Diisodecyl Ester	Plasticizer DDP		
Diisopropanol-amine	2,2'-Dihydroxy-Dipropylamine	1,1'-Iminodi-2-Propanol			
Diisopropyl-benzene Hydroperoxide	Isopropylcumyl Hydroperoxide				
Dimethyl-acetamide	N,N-Dimethyl-Acetamide, Acetic Acid, Dimethylamide				
Dimethyl Ether	Methyl Ether	Wood Ether			
Distillates: Flashed Feed Stocks	Petroleum Distillate				
Ethane	Methylmethane				

Ethyl Acetate	Acetic Acid	Ethyl Ester	Acetic Ester	Acetic Ether	Ethyl Ethanoate
Ethyl Alcohol	Alcohol	Cologne Spirit	Denatured Alcohol	Ethanol	Formentation Alcohol
Ethyl Butanol	2-Ethyl-1-Butanol	2-Ethylbutyl Alcohol	Sec-Hexyl Alcohol	Sec-Pentyl Carbinol	Pseudohexyl Alcohol
Ethyl Chloroformate	Chloroformic Acid, Ethyl Ester	Ethyl Chlorocarbonate			
Ethylene	Ethene				
Ethylene Dichloride	Brocide	1,2-Dichloroethane	Dutch Liquid	Edc	Ethylene Chloride
Ethyl Lactate	Ethyl Alpha-Hydroxy-Propionate	Ethyl 2-Hydroxy-Propanoate	Ethyl Dl-Lactate	Lactic Acid Ethyl Ester	
Ethyl Nitrate	Nitrous Ether	Spirit Of Ether Nitrate	Sweet Spirit If Nitrate		
Ferric Ammonium Citrate	Ammonium Ferric Citrate	Ferric Ammonium Citrate (Brown)	Ferric Ammonium Citrate (Green)		

Table 5.2 (cont.) Chemical synonyms

Chemical Name	Synonyms				
	1	2	3	4	5
Ferric Nitrate	Ferric Nitrate Monohydrate	Nitric Acid	Iron (+3) Salt		
Ferric Sulfate	Iron Sesquisulfate	Iron (Iii) Sulfate	Iron Tersulfate		
Formaldehyde Solution	Formalin	Fyde	Formalith	Methanal	Formic Aldehyde
Gallic Acid	Gallic Acid Monohydrate	3, 4, 5-Trihydroxy-Benzoic Acid			
Gasolines: Automotive	Motor Spirit	Petrol			
Glycerine	Glycerol	1, 2, 3-Propanetriol	1, 2, 3-Trihydroxy-propane		
Heptanol	Enanthic Alcohol	1-Heptanol	Heptyl Alcohol	1-Hydroxyheptane	
N-Hexaldehyde	Caproaldehyde	Caproic Aldehyde	Capronaldehyde	N-Caproylaldehyde	Hexanal
Hydrochloric Acid	Muriatic Acid				

Chemical Data

Hydrogen Chloride	Hydrochloric Acid	Anhydrous		
Hydrogen, Liquified	Liquid Hydrogen	Para Hydrogen		
Isobutane	2-Methylpropane			
Isobutyl Alcohol	Isobutanol	Isopropylcarbinol	2-Methyl-1-Propanol	Fermentation Butyl Alcohol
Isodecaldehyde	Isodecaldehyde, Mixed Isomers	Trimethylheptanals		
Isohexane	2-Methylpentane			
Isopropyl Alcohol	Dimethylcarbinol	2-Propanol	Isopropanol	Sec-Propyl Alcohol	Petrohol
Isopropyl Mercaptan	2-Propanethiol	Propane-2-Thiol		
Kerosene	No. 1 Fuel Oil	Kerosine	Illuminating Oil	Range Oil	JP-1
Lactic Acid	2-Hydroxy-Propanoic Acid	Alpha-Hydroxypropionic Acid	Milk Acid	Racemic Acid
Lead Arsenate	Lead Arsenate, Acid	Plumbous Arsenate		

Table 5.2 (cont.) Chemical synonyms

Chemical Name	Synonyms				
	1	2	3	4	5
Linear Alcohols	Dodecanol	Tridecanol	Tetradecanol	Pentadecanol	
Liquified Natural Gas	LNG				
Liquified Petroleum Gas	Bottled Gas	Propane-Butane-(Propylen) Pyrofax	LPG		
Mercuric Cyanide	Cianurina	Mercury Cyanide	Mercury (Ii) Cyanide		
Methane	Marsh Gas				
Methyl Alcohol	Colonial Spirit	Wood Alcohol	Columbian Spirit	Wood Naphtha	Methanol
Methyl Chloride	Arctic	Chloromethane			
Methyl Isobutyl Carbinol	Isobutyl Methyl Carbinol	Methyl Alcohol	MAOH	4-Methyl-2-Pentanol	MIBC
Alpha-Methylstyrene	Isopropenyl-Benzene	1-Methyl-1-phenylethylene	Phenylpropylene		
Mineral Spirits	Naphtha	Petroleum Spirits			

CHEMICAL DATA

Nitrous Oxide	Dinitrogen monoxide				
Nonanol	1-Nonanol	Octylcarbinol	Nonilalcohol	Pelargonic Alcohol	
Octane	N-Octane				
Oils: Clarified					
Oils: Crude	Petroleum				
Oils: Diesel	Fuel Oil 1-D	Fuel Oil 2-D			
Oils, Edible: Castor					
Oils, Edible: Coconut	Coconut Butter	Coconut Oil	Copra Oil		
Oils, Edible: Lard	Kettle-Rendered Lard	Leaf Lard	Prime Steam Lard		
Oils, Fuel: 2	Home Heating Oil				
Oils, Fuel: 4	Residual Fuel Oil	No. 4			
Oils, Miscellaneous: Coal Tar	Light Oil				
Oils, Miscellaneous: Motor	Crankcase Oil	Lubricating Oil	Transmission Oil		

Table 5.2 (cont.) Chemical synonyms

Chemical Name	Synonyms				
	1	2	3	4	5
Oils, Miscellaneous: Penetrating	Protective Oil	Water Displacing Oil			
Oils, Miscellaneous: Resin	Codoil	Retinol	Rosin Oil	Rosinol	
Oils, Miscellaneous: Spray	Dormant Oil	Foliage Oil	Kerosene, Heavy	Plant Spray Oil	
Oils, Miscellaneous: Tanner's	Sulfated Neatsfoot Oil	Sodium Salt			
Oleic Acid	Cis-8-Heptadecylene-Carboxylic Acid	Cis-9-Ocadecenoic Acid	Cis-9-Octadecylenic Acid	Red Oil	
Oleum	Fuming Sulfuric Acid				
Oxalic Acid	Ethanedioc Acid				
Oxygen, Liquified	Liquid Oxygen	LOX			

Chemical Data

Paraformaldehyde	Formaldehyde Polymer	Polyformaldehyde	Polyfooxy-methylene	Polyoxymethylene Glycol	
Pentaerythritol	Mono PE	PE	Pentaerythrite	Pentek	Tetrahydroxy-methylmethane
1-Pentene	Alfa-N-Amylene	Propylethylene			
Peracetic Acid	Acetyl Hydroperoxide	Peroxyacetic Acid			
Petrolatum	Petrolatum Jelly	Petroleum Jelly	Vaseline	Yellow Petrolatum	
Phenol	Carbonic Acid	Hydrobenzene	Phenic Acid Phenyl Hydroxide		
Phenyldichloroarsine, Liquid	Phenylarsenic dichloride				
Phosgene	Carbonyl Chloride	Chloroformyl Chloride			
Phosphoric Acid	Orthophosphoric Acid				
Piperazine	Diethylene-Diamine	Hexahydro-1,4-diazine	Hexahydropyrazine	Lumbrical	Pyrazine Hexahydride

Table 5.2 (cont.) Chemical synonyms

Chemical Name	Synonyms				
	1	2	3	4	5
Polybutene	Butene Resins	Polybutylene Plastics	Polybutylene Resins	Polybutylene Waxes	
Polychlorinated Biphenyl	Chlorinated Biphenyl	Halogenated Waxes	PCB	Polychloro-Polyphenyls	
Polypropylene	Propene Polymer				
Potassium Cyanide	Cyanide				
Propane	Dimethylmethane				
Propionaldehyde	Melhylacetal-Dehyde	Propaldehyde	Propanal	Propionic Aldehyde	Propylaldehyde
Propylene Oxide	1,2-Epoxypropane	Methyloxirane Propene Oxide			
Pyrogallic Acid	1,2,3-Benzenetriol	Pyrogallol	1, 2, 3-Trihydroxy-Benzene		
Quinoline	1-Azanapthalene	1-Benzazine	Benzo(B)Pyridine	Chinoline	Leucol

Chemical Data

Salicylic Acid	O-Hydroxybenzoic Acid	Retarder W	
Selenium Dioxide	Selenious Anhydride	Selenium Oxide	
Selenium Trioxide	Selenic Anhydride		
Silicon Tetrachloride	Silicone Chloride		
Silver Fluoride	Argentous Fluoride	Silver Monofluoride	
Silver Nitrate	Lunar Caustic		
Silver Oxide	Argentous Oxide		
Soduim Alkylbenzene-sulfonates	Alkylbenzene-Sulfonic Acid	Sodium Salt	Sulfonated Alkelbenzene
Sodium Alkyl Sulfates	Sodium Hydrogen Alkyl Sulfate		
Sodium Amide	Sodamite		
Soduim Arsenate	Disodium Arsenate Heptahydrate	Sodium Arsenate, Diabasic	

Table 5.2 (cont.) Chemical synonyms

Chemical Name	Synonyms				
	1	2	3	4	5
Sodium Arsenite	Sodium Metaarsenite				
Sodium Azide	Hydrazoic Acid	Sodium Salt			
Sodium Bisulfite	Sodium Acid Sulfite	Sodium Metebisulfite	Sodium Pyrosulfite		
Sodium Borate	Borax Anhydrous	Sodium Biborate	Sodium Pyroborate	Sodium Tetraborate Anhydrous	
Sodium Cacodylate	Arsecodile	Phytar 160	Phytar 560	Sodium dimethyl-arsonate	
Sodium Chlorate	Chlorate of Soda				
Sodium Chromate	Neutral Sodium Chromate, Anhydrous	Sodium Chromate (Vi)			
Sodium Cyanide	Hydrocyanic Acid				

Sodium Hydrosulfide Solution	Sodium Bisulfide	Sodium Hydrogen Sulfide	Sodium Sulfhydrate			
Sodium Hydroxide	Caustic Soda					
Sodium Hypochlorite	Clorox					
Sodium Methylate	Sodium Methoxide					
Sodium Oxalate	Ethanedioic Acid	Disodium Salt				
Sodium Phosphate	Monosodium Phosphate (Msp, Sodium Phosphate, Monobasic)	Disodium Phosphate (Dsp, Sodium Phosphate Dibasic)	Trisodium Phosphate (TSP, Sodium phosphate, Tribasic)	Sodium Acid Pyrophosphate (ASPP, SAPP, Disodium Pyrophosphate (TSPP)	Sodium Metaphosphate (Insoluble Sodium Metaphosphate)	
Sodium Silicate	Water Glass	Soluble Glass				
Sodium Silicofluoride	Salufer	Sodium Fluosilicate	Sodium Hexafluorosilicate			
Sodium Thiocyanate	Rhodanate	Sodium Rhodanide	Sodium Sulfocyanate			

Table 5.2 (cont.) Chemical synonyms

Chemical Name	Synonyms				
	1	2	3	4	5
Sorbitol	D-Glucitol	Heyahydric Alcohol	1,2,3,4,5,6-Hexane-hexol	Sorbit	Sorbo
Stearic Acid	1-Heptadecance-Carboxylic Acid	Octadecanoic Acid	N-Octadecylic Acid	Stearophoic Acid	
Styrene	Phenethylene	Phenylethylene	Styrol	Styrolene	Vinylbenzene
Sucrose	Beet Sugar	Cane Sugar	Saccharose	Saccharum	Sugar
Sulfolane	Sulfolane-W	Tetrahydro-thiophene-1,1-Dioxide	Tetramethylene Sulfone		
Sulfuric acid	Battery Acid	Chamber Acid	Fertilizer Acid	Oil of Vitriol	
Toluene	Methylbenzene	Methylbenzol	Toluol		
Toluene 2,4-Diisocyante	Hylene T	Mondur Tds	Nacconate 100	2,4-Tolylene Diisocyanate	TDE
P-Toluene-sulfonic Acid	Methylbenzenene-Sulfonic Acid	Tosic Acid	P-TSA		

Chemical Data

O-Toludine	2-Amino-1-Methyl-Benzene	2-Aminotoluene	2-Methylaniline	O-Methylaniline	
Toxamphene	Octachloro-Camphene				
Trichloro-ethylene	Algylen	Clorilen	Gemalgene	Threthylene	Trihloran
Trichlorofluoro-methane	Arcton 9	Freon 11	Eskimon 11	Frigen 11	Isotran 11
Trichlorosilane	Silichlorofom	Trichloro-Monosilane			
Tridecanol	Isotridecanol	Isotridecyl Alcohol	1-Tridecanol		
1-Tridecene	Undecylethylene				
Triethyl-aluminum	ATE	Aluminum Triethyl	TEA		
Triethylamine	TEN				
Thiethylene Glycol	Di-Beta-Hydroxy-ethoxy-Ethane	2,2-Ethylenedioxy-diethanol	Ethylene gycol dihydroxydiethyl ether	TEG	Triglycol

Table 5.2 (cont.) Chemical synonyms

Chemical Name	Synonyms				
	1	2	3	4	5
Turpentine	D.D.Turpentine	Gum Turpentine	Spirits Of Turpentine	Sulphate Turpentine	Turps
Undecanol	Hendecanoic Alcohol	1-Hendecanol	N-Hendecylenic Alcohol	1-Undecanol	Undecyl Alcohol
1-Undecene	N-Nonylethylene				
N-Undecylbenzene	1-Phenylundecane				
Uranyl Acetate	Bis(Acetato) Dixouranium	Uranium Acetate	Uranium Acetate Dihydrate	Uranium Oxyacetate	
Uranyl Nitrate	Uranium Nitrate				
Uranyl Sulfate	Uranium Sulfate	Uranium Sulfate Trihydrate	Uranyl Sulfate Trihydrate		
Urea	Carbamide	Carbonyldiamide			
Urea Peroxide	Carbamide Peroxide	Carbonyldiamine	Hydrogen Peroxide Carbamide	Percarbamide	Perhydrol-Urea
Vanadium Oxytri-chloride	Trichloroxo Vanadium	Vanadyl Chloride	Vanadyl Trichloride		

Chemical Data

Vanadium Pentoxide	Vanadic Anhydride	Vanadium Pentaoxide		
Vanadyl Sulfate	Vanadium Oxysulfate	Vanadyl Sulfate Dihydrate		
Vinyl Acetate	VAM	Vinyl A Monomer	Vy Ac	
Vinyl Chloride	Chloroethene	Chloroethylene	Vinyl C Monomer	VCL	VCM
Vinyl Fluoride Inhibited	Fluoroethylene	Monofluoro Ethylene		
Vinylidene Chloride Inhibited	1,1-Dichloro-Ethylene	Unsym-Dichloroethylene		
Vinyl Methyl Ether, Inhibited	Methoxyethylene	Methyl Vinyl Ether		
Vinyltoluene	Methylstyrene			
Vinyltrichloro-silane	Trichloro-Vinylsilane	Trichloro-Vinylsilicane	Vinylsilicon	
Waxes:Paraffin	Petroleum Wax			
M-Xylene	1,3-Dimethyl-Benzene	Xylol		

Table 5.2 (cont.) Chemical synonyms

Chemical Name	Synonyms				
	1	2	3	4	5
Xylenol	Cresylic Acid	2,6-Dimethylphenol	2-Hydroxy-M-Xylene	2,6-Xylenol	Vic-M-Xylenol
Zinc Acetate	Acetic Acid	Zinc Salt	Dicarbo-methoXyzine	Zinc Acetate Dehydrate	Zinc Diacetate
Zinc Ammonium Chloride	Ammonium Pentachloro-Zincate	Ammonium Zinc Chloride			
Zinc Chromate	Buttercup Yellow	Zinc Chromate (Vi) Hydroxide	Zinc Yellow		
Zinc Fluoroborate	Zinc Fluoroborate Solution				
Zinc Nitrate	Zinc Nitrate Hexahydrate				
Zinc Sulfate	White Vitriol	Zinc Sulfate Heptahydrate	Zinc Vitriol		
Zirconium Acetate	Zirconium Acetate Solution				

6
Chemical Safety Data

Table 6.1 Recommended emergency response to acute exposures and direct contact

Chemical Name	PPE	Emergency Treatment Actions for Exposure by		
		Inhalation	Skin	Eyes
Acetaldehyde	Rubber gloves, eye goggles, and other equipment to prevent contact with body. Organic canister or air pack as required.	Remove victim to fresh air; if breathing has stopped give artificial respiration; if breathing is difficult give oxygen; call a physician at once.	Wash with soap and water.	Flush with water.
Acetic Acid	Protective clothing should be worn when skin contact can occur. Respiratory protection is necessary when exposed to vapor. Complete eye protection is recommended.	Move the victim immediately to fresh air. If breathing becomes difficult, give oxygen and get medical attention immediately.	Flush immediately with lots of clean running water.	Wash eyes immediately.
Acetuc Anhydride	Protective clothing should be worn when skin contact can occur. Respiratory protection is necessary for all exposures. Complete eye protection is recommended.	Move the victim immediately to fresh air. If breathing becomes difficult, give oxygen and get medical attention immediately.	Flush immediately with lots of clean running water.	Wash eyes for at least 15 minutes and seek medical attention immediately.

263

Table 6.1 (cont.) Recommended emergency response to acute exposures and direct contact

Chemical Name	PPE	Emergency Treatment Actions for Exposure by			
		Inhalation	Skin	Eyes
Acetone	Organic vapor canister or air-supplied respirator, synthetic rubber gloves, chemical safety goggles or face splash shield.	Vapor irritating to eyes and mucous membranes; acts as an anesthetic in very high concentrations.	Wash with clean running water.	Flush with water for at least 15 minutes and consult a physician.
Acetone Cyanohydrin	Air-supplied respirator or chemical cartridge respirator approved for use with acrylonitrile in less than 2% concentrations, rubber or plastic gloves, cover goggles or face mask, rubber boots, chemical protective suit, safety helmet.	Remove victim to fresh air. First responders/rescuers should wear suitable respiratory protection. If breathing has stopped, give artificial respiration until physician arrives.	Remove contaminated clothing and wash affected skin thoroughly with soap and water. Use copious amounts of water.	Hold eyelids apart and wash with continuous, gentle stream of water for at least 15 minutes.
Acetonitrile	Must wear self contained breathing apparatus (SCBA).	Remove victim from contaminated atmosphere. Apply artificial respiration and oxygen if respiration is impaired.		

Chemical Safety Data

Acetyl Bromide	Safety goggles, gloves, adequate ventilation, provisions for flushing eyes or skin with water.	Remove victim from contaminated atmosphere. Apply artificial respiration if breathing has stopped and oxygen if respiration is impaired.	Flush with water and treat chemical burns as needed.	Flush with water for at least 15 minutes and consult a physician.
Acetyl Chloride	Safety goggles, gloves, self contained breathing apparatus (SCBA).	Remove victim from exposure and seek immediate medical care.		Flush with copious amounts of fresh running water.
Acetyl Peroxide	Protective goggles, rubber apron, and gloves.		Flush with water and wash thoroughly with soap and water, seek medical attention.	Wash with plenty of water and seek medical attention.
Acridine	Dust respirator, chemical goggles, rubber gloves.	Remove victim to fresh air, and if breathing has stopped give artificial respiration, if breathing is difficult give oxygen.	Wash with large amounts of water for 20 minutes.	Wash with copious amounts of water for 20 minutes and seek medical attention.

Table 6.1 (cont.) Recommended emergency response to acute exposures and direct contact

Chemical Name	PPE	Emergency Treatment Actions for Exposure by		
		Inhalation	Skin	Eyes
Acrolein	Chemical safety goggles and full face shield, self-contained breathing apparatus (SCBA), positive pressure hose mark, airline mask, rubber safety shoes, chemical protective clothing.	Remove victim to fresh air, and if breathing has stopped give artificial respiration, if breathing is difficult give oxygen.	Flush at once with large amounts of water. Wash thoroughly with soap and large amounts of water.	Immediately flush with plenty of water for at least 15 minutes. If medical attention is not immediately available, continue eye irrigation for another 15 minutes. Period. Upon completion of first 15 minutes, eye irrigation period, it is permissible to instill 2 or 3 drops of an effective aqueous local eye anesthetic for relief of pain. No oils or ointments s

Chemical Safety Data

				hould be used unless instructed by a physician.
Acrylamide	Safety glasses with side shields, clean body-covering clothing, rubber gloves, boots, apron as dictated by circumstances, in absence of proper control use approved dust respirator.	If ill, immediately get patient to get fresh air, keep him quiet and warm, and get medical help.	Immediate, continuous, and thorough washing in flowing water is imperative, preferably deluge shower with abundant soap. If burns are present, get medical help.	Immediately flush with plenty of water for at least 15 minutes and get medical attention promptly.
Acrylic Acid	Chemical respirator at ambient temperatures to avoid inhalation of noxious fumes, rubber gloves if exposed to wet materials, acid goggles or face shield for splash exposure, safety shower and/or eye fountain may be required.	Get medical attention promptly for all exposures.	Flush with water for at least 15 minutes.	Flush with water for at least 15 minutes.

Table 6.1 (cont.) Recommended emergency response to acute exposures and direct contact

Chemical Name	PPE	Emergency Treatment Actions for Exposure by		
		Inhalation	Skin	Eyes
Acrylonitrile	Air-supplied mask (industrial chemical type with approved canister for acrilonitrile in low (less than 2% concentrations), rubber or plastic gloves, cover goggles or face mask, rubber boots, slicker suit, safety helmet.	Remove victim to fresh air (wear an oxygen or fresh-air-supplied mask when entering contaminated area).	Remove contaminated clothing and wash affected area thoroughly with soap and water.	Hold eyelids apart and wash with continuous gentle steam of water for at least 15 minutes. If victim is not breathing give artificial respiration until the physician arrives.
Aldrin	During prolonged exposure to mixing and loading operations, wear clean synthetic rubber gloves and mask or respirator of the type passed by the U.S. Bureau of Mines for aldrin protection.		Wash with soap and running water.	Wash immediately for at least 15 minutes and get medical attention.

Chemical Safety Data

Allyl Alcohol	Organic vapor canister or air-supplied respirator, synthetic rubber gloves, chemical safety goggles and other protective equipment as required to prevent all body contact.	Remove victim from contaminated area and administer oxygen. Get medical attention immediately.	Remove liquid with soap and water.	Flush with continuous stream of water for 15 minutes.
Allyl Chloroformate	Vapor-proof protective goggles and face shield, plastic or rubber gloves, shoes, and clothing, gas mask or self-contained breathing apparatus.	Remove from exposure, support respiratory if necessary, call physician.	Wash with large amounts of water for at least 15 minutes.	If irritated by either vapor or liquid, flush with water for at least 15 minutes.
Allyltrlorosilane	Acid-vapor-type respiratory protection, rubber gloves, chemical goggles, other equipment necessary to protect skin and eyes.	Remove from exposure, support respiratory if necessary.	Flush with water.	Flush with water for 15 minutes.

Table 6.1 (cont.) Recommended emergency response to acute exposures and direct contact

Chemical Name	PPE	Emergency Treatment Actions for Exposure by		
		Inhalation	Skin	Eyes
Aluminum Chloride	Safety clothing, including fully closed goggles, rubber or plastic coated gloves, rubber shoes and coverall of acid-resistant material. An acid canister mask should be carried in case of emergency.		Flush immediately with lots of clean running water.	Flush with water for 15 minutes.
Aluminum Nitrate	Goggles or face shield, dust respirator, rubber gloves.		Flush with water.	Flush with water for 15 minutes.
Ammonium Bifluoride	Bu: Mines approved respirator, rubber gloves, safety goggles.	Remove victim to fresh air.	Flush with water. Treat burns.	Flush with water for at least 15 minutes.
Ammonium Carbonate	Dust respirator, protection against ammonia vapors.	Leave contaminated area.	Flush with water.	Flush with copious amounts of fresh running water.

Ammonium Dichromate	Dust respirator, protective goggles, gloves.	Remove to clean air and summon medical attention.	Immediately flush with water for at least 15 minutes. If skin irritation develops, get medical attention.	Immediately flush out with water and consult a physician.
Ammonium Fluoride	Dust mask, goggles or face shield, rubber gloves.	Remove to fresh air.	Shower immediately with large quantities of water. Remove all contaminated clothing. Consult a physician.	Flush with water for at least 15 minutes. Consult a physician.
Ammonium Hydroxide	Rubber boots, gloves, apron, and coat, broad brimmed rubber or felt hat, safety goggles, and protective oil to reduce skin irritation from ammonia.	Move the victim immediately to fresh air. If breathing becomes difficult, give oxygen.	Wash with plenty of water.	Wash with plenty of water.

Table 6.1 (cont.) Recommended emergency response to acute exposures and direct contact

Chemical Name	PPE	Inhalation	Skin	Eyes
Ammonium Lactate	Dust mask, goggles or face shield, rubber gloves.	Remove to fresh air.		Flush with water.
Ammonium Nitrate	Self contained breathing apparatus.			
Ammonium Nitrate-Sulfate Mixture	Self contained breathing apparatus must be used when fighting fires. At all other times a dust mask is adequate.	Move to fresh air.		Flush with water for at least 15 minutes.
Ammonium Oxalate	Approved dust respirator, rubber or plastic coated gloves, chemical goggles.	Move to fresh air.	Flush with water.	Flush with water and seek medical attention.
Ammonium Perchlorate	Data not available.	Data not available.	Data not available.	Data not available.

Emergency Treatment Actions for Exposure by

Chemical Safety Data

Ammonium Perchlorate	U.S. Bu. Mines approved toxic dust mask, chemical goggles, rubber gloves, neoprene-coated shoes.	Move to fresh air.	Wash with water.	Wash with water for 20 minutes.
Ammonium Silicofluoride	Dust respirator, acid resistant clothing and hat, rubber gloves, goggles, and safety shoes.	Move to fresh air.	Wash with soap and water.	Wash with water for 20 minutes.
Amyl Acetate	Air-supplied mask or chemical cartridge respirator, protective gloves, goggles, safety shower, and eye bath.	Move to fresh air.	Flush with water.	Flush with water.
Aniline	Respirator for organic vapors, splash-proof goggles, rubber gloves, boots.	Move to fresh air and call physician.	Flush with water.	Flush with water.
Anisoyl Chloride	Goggles or face shield, protective clothing, plastic gloves.	Move to fresh air.	Wash well with soap and water.	Flush with water for at least 15 minutes.
Anthracene	Dust mask, goggles or face shield, rubber gloves.	Move to fresh air.		Flush with water for at least 15 minutes.

Table 6.1 (cont.) Recommended emergency response to acute exposures and direct contact

Chemical Name	PPE	Emergency Treatment Actions for Exposure by		
		Inhalation	Skin	Eyes
Antimony Pentachloride	Organic vapor-acid gas type canister mask, rubber gloves, chemical safety goggles, plus face shield where appropriate.	Move to fresh air. Rinse mouth and gargle with water.	Flush thoroughly with water. Remove all contaminated clothing and wash effected area with soap and water.	Flush eyes and eyelids thoroughly with large amounts of water, and get prompt medical attention.
Antimony Pentafluoride	Acid-gas-type respiratory protection, rubber gloves, chemical goggles, and protective clothing.	Move to fresh air. Rinse mouth and gargle with water.	Flush with copious amounts of water, wash well with soap and water.	Irrigate with copious amounts of water for at least 15 minutes.
Animony Potassium Tartrate	Dust respirator, rubber of plastic coated gloves, chemical goggles, tightly woven close fitting clothes.	Move to fresh air.	Flush with copious amounts of water, wash well with soap and water.	Irrigate with copious amounts of water for at least 15 minutes. Consult a physician.

CHEMICAL SAFETY DATA

Antimony Trichloride	Bu. Mines approved respirator, face shield, leather or rubber safety shoes, rubber apron, rubber gloves, safety goggles.	Move to fresh air. Keep patient warm but not hot and get medical attention immediately.	Flush thoroughly with water. Remove all contaminated clothing and wash effected area with soap and water.
Anitimony Trifluoride	Approved respirator, rubber gloves.		
Anitmony Trioxide	Rubber gloves, safety goggles, dust mak.		Wash well with soap and water. Flush with water for at least 15 minutes.
Arsenic Acid	Calamine lotion and zinc oxide powder on hands and other skin areas, rubber gloves, Bu.		Wash well with soap and water.
Arsenic Disulfide	Approved respirator, rubber gloves, goggles, protective clothing.	Move to fresh air.	Wash well with water. Flush with water for at least 15 minutes.
Arsenic Trichloride	Safety goggles and face shied, acid-type canister gas mask, rubber gloves, protective clothing.	Move to fresh air.	Wash well with water. Flush with water for at least 15 minutes.

Table 6.1 (cont.) Recommended emergency response to acute exposures and direct contact

Chemical Name	PPE	Emergency Treatment Actions for Exposure by		
		Inhalation	Skin	Eyes
Arsenic Trioxide	Chemical cartridge approved respirator, protective gloves, eye protection, full protective coveralls.		Flush thoroughly with water. Remove all contaminated clothing and wash effected area with soap and water.	
Arsenic Trisulfide	Self-contained breathing apparatus, goggles, rubber gloves, clean protective clothing.	Move to fresh air.	Wash well with water.	Flush with water for at least 15 minutes.
Asphalf	Protective clothing, face and eye protection when handling hot material.			
Atrazine	Dust mask, goggles, rubber gloves.		Wash well with water.	Flush with water for at least 15 minutes.

Chemical Safety Data

Azinphosmethyl	Dust mask, goggles, rubber gloves.	Move the victim to fresh air, keep warm, and call a physician.	Wash well with soap and water.	Flush with water for at least 15 minutes.
Barium Carbonate	Dust respirator.			
Barium Chlorate	Dust respirator, rubber shoes and gloves, coveralls or other suitable clothing.	Move to fresh air.	Flush with water.	Flush with water for at least 15 minutes and get medical attention.
Barium Nitrate	Goggles or face shield, dust respirator, rubber gloves and shoes, suitable clothing.	Move to fresh air.	Flush with water.	Flush with water for at least 15 minutes and get medical attention.
Barium Perchlorate	Goggles or face shield, dust respirator, rubber gloves and shoes, suitable clothing.	Move to fresh air.	Flush with water.	Flush with water for at least 15 minutes and get medical attention.
Barium Permanganate	Goggles or face shield, dust respirator, rubber gloves and shoes.	Move to fresh air.	Flush with water.	Flush with water for at least 15 minutes and get medical attention.

Table 6.1 (cont.) Recommended emergency response to acute exposures and direct contact

Chemical Name	PPE	Emergency Treatment Actions for Exposure by		
		Inhalation	Skin	Eyes
Barium Peroxide	Toxic gas respirator, liquid-proof PVC gloves, chemical safety goggles, full cover clothing.	Move to fresh air.	Flush with water.	Flush with water for at least 15 minutes and get medical attention.
Benzaldehyde	Chemical goggles and chemical protective clothing.		Move victim to air and contact doctor immediately.	Move victim to air and contact doctor immediately.
Benzene		Move the victim to fresh air and call a physician.	Wash well with soap and water.	Flush with water for at least 15 minutes.
Benzene Hexachloride	Respiratory protection, ensure handling in a well-ventilated area.			
Benzene Phosphorous Dichloride	Self-contained breathing apparatus, goggles and face shield, rubber gloves, chemical protective clothing, acid-type canister mask.	Move the victim to fresh air. If breathing has stopped, start giving mouth-to-mouth resuscitation.	Wash well with soap and water.	Flush with water for at least 15 minutes.

Chemical Safety Data

Benzene Phosphorous Thiodi-chloride	Self-contained breathing apparatus (SCBA) or acid-type canister mask.	Move to fresh air.	Wash well with soap and water.	Flush with water for at least 15 minutes.
Benzoic Acid	Dust respirator, eye protection and organic respirator for fumes when melted material is present.			Flush eyes with water.
Benzonitrile	Rubber gloves, chemical resistant splash-proof goggles, rubber boots, chemical protective clothing, chemical cartridge type respirator or other protection against vapor must be worn when working in poorly ventilated areas or where overexposure by inhalation could occur.	Move to fresh air and call physician.	Wash well with soap and water without scrubbing.	Flush with water for at least 15 minutes.
Benzophenone	Goggles or face shield, rubber gloves.	Move to fresh air.	Wash well with soap and water.	Flush with water for at least 15 minutes.

Table 6.1 (cont.) Recommended emergency response to acute exposures and direct contact

Chemical Name	PPE	Emergency Treatment Actions for Exposure by		
		Inhalation	Skin	Eyes
Benzoyl Chloride	Full protective clothing including full-face respirator for acid gases and organic vapors (yellow GMC canister), close fitting goggles, nonslip rubber gloves, plastic apron, face shield.	Remove to fresh air. Administer oxygen with patient in sitting position.	Wash well with soap and water.	Flush with water for at least 15 minutes.
Benzyl Alcohol	Rubber gloves, chemical safety goggles.	Remove victim from contaminated area and call physician immediately.	Flush with water, wash with soap and water, obtain medical attention in case of irritation or central nervous system depression.	Flush with water for at least 15 minutes and contact a physician.
Benzylamine	Self-contained breathing apparatus, goggles or face shield, rubber gloves.	Move the victim immediately to fresh air. If breathing becomes difficult, give	Flush with water for at least 15 minutes.	Flush with water for at least 15 minutes.

Chemical Safety Data

Benzyl Bromide	Self-contained breathing apparatus, goggles or face shield, rubber gloves, protective clothing.	oxygen, if breathing has stopped, give artificial respiration. Move to fresh air.	Flush with water.	Flush with water for at least 15 minutes.
Benzyl Chloride	Chemical safety goggles or full face shield, self-contained breathing apparatus (SCBA), positive pressure hose mark, industrial canister-type gas mask, or chemical cartridge respirator, rubber gloves, chemical protective clothing.	Remove victim from contaminated atmosphere. If breathing has ceased start mouth-to-mouth resuscitation. Oxygen if available administered only by an experienced person when authorized by a physician. Keep patient warm and comfortable. Call a physician immediately.	Immediately flush affected area with water. Remove contaminated clothing under shower and continue washing with water. Do not attempt to neutralize with chemical agents. Obtain medical attention if irritation persists.	Immediately flush with large quantities of running water for a minimum of 15 minutes; hold eyelids apart during irrigation to ensure flushing of the entire surface of the eye and lids with water. Do not attempt to neutralize with chemical agents.

Table 6.1 (cont.) Recommended emergency response to acute exposures and direct contact

Chemical Name	PPE	Emergency Treatment Actions for Exposure by		
		Inhalation	Skin	Eyes
				Obtain medical attention as soon as possible. Oils and ointments should not be used before consulting a physician. If physician is not available, continue to flush for an additional 15 minutes.
Beryllium Fluoride	Respiratory protection, gloves, goggles.	Move to fresh air, chest x-ray should be taken immediately to detect pneumonitis if exposure has been severe.	Flush with water and get medical attention if skin has been broken.	Flush with water for at least 15 minutes and get medical attention.

Chemical Safety Data

Beryllium Metallic	Chemical cartridge, clean work clothes daily, gloves and eye protection.	Acute disease may require hospitalization with administration of oxygen, chest x-ray should be taken immediately.		Flush with water followed by washing with soap and water. All cuts, scratches, and injuries should get medical attention.
Beryllium Nitrate	Respiratory protection, gloves, clean clothes, chemical safety goggles.	Move to fresh air, chest x-ray should be taken immediately to detect pneumonitis.	Cuts or puncture wounds in which beryllium may be imbedded must be flushed with water for at least 15 minutes.	Flush with water for at least 15 minutes and get medical attention.
Beryllium Oxide	Respiratory protection, gloves, clean clothes, chemical safety goggles.	Move to fresh air, chest x-ray should be taken immediately to detect pneumonitis if exposure has been severe.	Cuts or puncture wounds in which beryllium may be imbedded must be cleaned immediately by a physician.	Flush with water for at least 15 minutes and get medical attention.

Table 6.1 (cont.) Recommended emergency response to acute exposures and direct contact

Chemical Name	PPE	Emergency Treatment Actions for Exposure by		
		Inhalation	Skin	Eyes
Beryllium Sulfate	Respiratory protection, gloves, clean clothes, chemical safety goggles.	Move to fresh air, chest x-ray should be taken immediately to detect pneumonitis if exposure has been severe.	Cuts or puncture wounds in which beryllium may be imbedded must get attention.	Flush with water.
Bismuth Oxychloride	Goggles or face shield, protective gloves, dust mask.			Flush with water.
Bisphenol A	Approved dust mask, clean body-covering clothing sufficient to prevent excessive or repeated exposure to dust, fumes, or solutions.		Wash with soap and plenty of water.	Flush with water and seek medical attention.
Boric Acid	Chemical goggles.	Remove victim from contaminated atmosphere.	Immediately flush affected area with water, remove	Immediately flush with large quantities of running water

for a minimum of 15 minutes; hold eyelids apart during irrigation to ensure flushing of the entire surface of the eye and lids with water. Do not attempt to neutralize with chemical agents. Obtain medical attention as soon as possible.	Oils and ointments should not be used before consulting a physician. If physician is not available, continue to flush for an additional 15 minutes.
contaminated clothing in the shower, continue to wash with water, do not attempt to neutralize with chemical agents, obtain medical attention unless burn is minor.	

Table 6.1 (cont.) Recommended emergency response to acute exposures and direct contact

Chemical Name	PPE	Emergency Treatment Actions for Exposure by		
		Inhalation	Skin	Eyes
Boron Trichloride	Chemical goggles, rubber protective clothing and gloves.	Remove to fresh air, give oxygen or apply artificial respiration.	Wash off with plenty of water.	Wash with plenty of water for at least 15 minutes.
Bromine	Chemical safety goggles, face shield, self-contained air-line canister mask, rubber suit.	Induces sever irritation of respiratory passages and pulmonary edema. Probable lethal oral dose for an adult is 1ml.	Wash well with water and sodium bicarbonate solution.	Wash well with water and sodium bicarbonate solution.
Bromine Trifluoride	Self-contained breathing apparatus, complete protective clothing, safety glasses, face shield.	Remove from exposure and support respiration.	Wash with large amounts of water for at least 15 minutes and then rinse with sodium bicarbonate or lime solution.	Irrigate with copious amounts of water for at least 15 minutes.
Bromobenzene	Goggle or face shield; rubber gloves and apron.		Wipe off; wash with soap and water.	Flush with water for at least 15 minutes.

Chemical Safety Data

Butadiane, Inhibited	Chemical-type safety goggles; rescue harness and lifeline for those entering a tank or enclosed strange space; hose mask with hose inlet in a vapor-free atmosphere; self-contained breathing apparatus; rubber suit.	If breathing is irregular or stopped, start resuscitation, administer oxygen.	Remove contaminated clothing and wash affected skin area.	Irrigate with water for at least 15 minutes.
Butane	Self-contained breathing apparatus and safety goggles.	Guard against self-injury if stuporous, confused, or anesthetized. Apply artificial respiration if not breathing. Avoid administration of epinephrine or other sympathomimetic amines. Prevent aspirations of vomitus by proper positioning of the head. Give symptomatic and supportive treatment.		

Table 6.1 (cont.) Recommended emergency response to acute exposures and direct contact

Chemical Name	PPE	Emergency Treatment Actions for Exposure by		
		Inhalation	Skin	Eyes
N-Butyl Acetate	All-purpose canister mask, chemical safety goggles, rubber gloves.	Remove from exposure immediately. Call a physician. If breathing is irregular or stopped, start resuscitation, administer oxygen.		In case of contact, flush with water for at least 15 minutes.
Sec-Butyl Acetate	Organic vapor canister or air-supplied mask; chemical goggles or face splash shield.	If victim is overcome by vapors, remove from exposure immediately; call a physician; if breathing is irregular or stopped, start resuscitation and administer oxygen.		Flush with water for at least 15 minutes.
Iso-Butyl Acrylate	Self-contained breathing apparatus; rubber gloves; chemical goggles.	Move victim to fresh air; give oxygen if breathing is difficult or artificial respiration if breathing has stopped; call a doctor.	Remove chemical by flushing with plenty of clean, running water; remove contaminated clothing and	Remove chemical by flushing with plenty of clean, running water; remove contaminated clothing and

Chemical Safety Data

N-Butyl Acrylate	Self-contained breathing apparatus; rubber gloves; acid goggles.	Remove to fresh air; administer artificial respiration or oxygen if indicated; call a physician.	wash exposed skin with soap and water. Wash with plenty of water.	wash exposed skin with soap and water. Wash with plenty of water.
N-Butyl Alcohol	Organic vapor canister or air-supplied mask; chemical goggles or face splash shield.	Remove from exposure immediately; call a physician; if breathing is irregular or has stopped, start resuscitation and administer oxygen.		Flush with water for at least 15 minutes.
Sec-Butyl Alcohol	Organinc vapor canister or air-supplied mask; chemical goggles or face splash shield.	Remove from exposure immediately; call a physician; if breathing is irregular or has stopped, start resuscitation and administer oxygen.		Flush with water for at least 15 minutes.
Tert-Butyl Alcohol	Air pack or organic canister mask, rubber gloves, and goggles.	Remove victim from exposure and restore breathing.	Remove liquid from skin with water.	Flush eyes with water.

Table 6.1 (cont.) Recommended emergency response to acute exposures and direct contact

Chemical Name	PPE	Emergency Treatment Actions for Exposure by		
		Inhalation	Skin	Eyes
N-Butylamine	Air-supplied mask; rubber gloves; coverall goggles; face shield; butyl apron.	Remove victim to fresh air; call a physician; give oxygen if breathing is difficult; if not breathing give artificial respiration.	Remove contaminated clothing; flush skin with plenty of water at least 15 minutes.	Flush with water for at least 15 minutes; get medical care.
Sec-Butylamine	Safety goggles; rubber gloves and apron; respiratory protective equipment; non-sparking shoes.	Remove patient from exposure; keep him quiet; contact physician.	Remove all contaminated clothing; flood all affected areas with large quantities of water; consult a physician.	Flush thoroughly with water for 15 minutes; call physician immediately.
Tert-Butylamine	Self-contained breathing apparatus; goggles or face shield; rubber gloves.	Move to fresh air; give artificial respiration if breathing has stopped`	Flush with water; wash with soap and water.	Immediately flush with water for at least 15 minutes; get medical attention.

Butylene	Chemical goggles, gloves, self-contained breathing apparatus or organic canister.	Remove victim to fresh air and supply resuscitation.	Flush with water for at least 15 minutes.	Flush with water for at least 15 minutes.
Butylene Oxide	Clean protective clothing; rubber gloves; chemical worker's goggles; self-contained breathing apparatus.	If any ill effects occur, immediately remove person to fresh air and get medical help; if breathing stops, start artificial respiration.	Promptly flush with plenty of water; remove all contaminated clothing and wash before reuse.	Promptly flush with plenty of water for 15 minutes and get medical help.
N-Butyl Mercaptan	Plastic gloves, goggles; self-contained breathing apparatus.	Remove victim from contaminated atmosphere; give artificial respiration and oxygen if needed; observe for signs of pulmonary edema.	Wash with soap and water.	Wash with plenty of water; see a physician.
N-Butyl Methacrylate	Self-contained breathing apparatus; impervious gloves; chemical splash goggles.	Remove to fresh air; give oxygen or artificial respiration as required.	Wash with soap and water.	Flush with copious amounts of water for 15 minutes and consult physician.

Table 6.1 (cont.) Recommended emergency response to acute exposures and direct contact

Chemical Name	PPE	Emergency Treatment Actions for Exposure by		
		Inhalation	Skin	Eyes
1,4-Butynediol	Neoprene rubber gloves and safety goggles or face shield.		Wash affected skin area thoroughly with water.	Immediately wash with water for at least 15 minutes and get medical attention.
Iso-Butyral-dehyde	Appropriate clothing, including rubber gloves, rubber shoes and protective eyewear.			Immediately flush with plenty of water for at least 15 minutes.
N-Butyral-dehyde	Protective goggles, gloves, and organic canister mask.	Remove victim to fresh air; if breathing has stopped give artificial respiration; if breathing is difficult give oxygen; call a doctor at once.	Immediately flush with water for at least 15 minutes; remove contaminated clothing and wash underlying skin.	Immediately flush with water for at least 15 minutes; get medical attention.

Chemical Safety Data

N-Butyric Acid	Self-contained breathing apparatus; rubber gloves, vapor-proof plastic goggles; impervious apron and boots.	Remove victim to fresh air; give oxygen if breathing is difficult; call a physician.	Flush affected areas immediately and thoroughly with water.	Irrigate with water for 15 minutes and get medical attention.
Cacodylic Acid	Dust respirator; goggles, protective clothing.		Flush with water.	Flush with water.
Cadmium Acetate	Dust mask; goggles or face shield; rubber gloves.	Remove victim to fresh air, seek medical attention.		Flush with water for at least 15 minutes.
Cadmium Nitrate	Rubber gloves, safety goggles, dust mask.	Remove patient to fresh air; seek medical attention.	Wash with soap and water.	Flush with copious amounts of water for 15 minutes; consult physician.
Cadmium Oxide		If there has been known exposure to dense cadmium oxide fume or if cough, chest tightness, or respiratory distress occur after possible exposure, place patient at bed rest and call a physician.		Flush with water at least 15 minutes.

Table 6.1 (cont.) Recommended emergency response to acute exposures and direct contact

Chemical Name	PPE	Emergency Treatment Actions for Exposure by		
		Inhalation	Skin	Eyes
Cadmium Sulfate	Respirator, goggles, rubber gloves.	Remove victim from exposure and call physician.	Wash with soap and water.	Flush with water for at least 10 minutes; consult physician.
Calcium Arsenate	Dust mask; goggles or face shield; protective gloves.	Move to fresh air.	Flush with water, wash with soap and water.	Flush with water for at least 15 minutes.
Calcium Carbide	Chemical safety goggles and (for those exposed to unusually dusty operations) a respirator such as those approved by the U.S. Bureau of Mines for "nuisance dusts"	Remove from further exposure and call a doctor.		Flush with clean running water in an eye wash fountain for at least 15 minutes and get medical attention.
Calcium Chlorate	Goggles or face shield; dust respirator; coveralls or other protective clothing.	Remove to fresh air.	Flush with water.	Flush with water for 15 minutes.

Chemical Safety Data

Calcium Chloride	Safety glasses or face shield, dust-type respirator, rubber gloves.	Move to fresh air, if discomfort persists, get medical attention.	Flush with water.	Promptly flood with water and continue washing for at least 15 minutes; consult an ophthalmologist.
Calcium Chromate	Dust mask; goggles or face shield; protective gloves.	Remove to fresh air.	Treat local injuries like acid burns; scrub with dilute (2%) sodium hyposulfite solution.	Flush with water for at least 15 minutes.
Calcium Fluoride	For dust only.			
Calcium Hydroxide.	Dust-proof goggles and mask.		Wash off the lime and consult a physician.	Flush with a gentle stream of water for at least 10 minutes. And consult an ophthalmologist for further treatment without delay.

Table 6.1 (cont.) Recommended emergency response to acute exposures and direct contact

Chemical Name	PPE	Emergency Treatment Actions for Exposure by		
		Inhalation	Skin	Eyes
Calcium Hypochlorite	Protective goggles, dust mask.		Wash with liberal quantities of water and apply a paste of baking soda.	
Calcium Nitrate	Dust respirator and rubber gloves.		Flush with water.	Flush with water.
Calcium Oxide	Protective gloves, goggles and any type of respirator prescribed for fine dust.		Flush with water and seek medical help.	Flush with water and seek medical help.
Calcium Peroxide	Toxic dust respirator; general-purpose gloves; chemical safety goggles; full cover clothing.	Remove to fresh air.	Flush with water.	Flush with water for 15 minutes. And consult physician.
Calcium Phosphide	Dust respirator; protective gloves and clothing; goggles.	Remove to fresh air, call a physician and alert to possibility of phosphine poisoning.	Flush with water, call a physician and alert to possibility of phosphine poisoning.	Flush with water, call a physician and alert to possibility of phosphine poisoning.

Chemical Safety Data

Camphene	Gloves and face shield.	Move to fresh air; call a physician immediately.	Wash with alcohol, follow with soap and water wash.	Flush immediately with clean, cool water.
Carbolic Oil	Fresh air mask for confined areas; rubber gloves; protective clothing; full face shield.	Remove victim to fresh air, keep quiet and warm. If breathing stops, start artificial respiration.	Remove contaminated clothing; wash skin with soap and water.	Flush eyes with water for 15 minutes or until physician arrives.
Carbon Dioxide	Self-contained breathing apparatus in excessively high CO_2 concentration areas. For handling liquid or solid, wear safety goggles or face shield, insulated gloves, long-sleeved shirt, and trousers worn outside boots or over high-top shoes to shed spilled liquid.	Move victim to fresh air.	Treat burns from contact with solid in same way as frostbite.	
Carbon Monoxide	Self-contained breathing apparatus; safety glasses and safety shoes; Type D or Type N canister mask.	Remove from exposure; give oxygen if available; support respiration; call a doctor.	If burned by liquid, treat as frostbite.	

Table 6.1 (cont.) Recommended emergency response to acute exposures and direct contact

Chemical Name	PPE	Emergency Treatment Actions for Exposure by		
		Inhalation	Skin	Eyes
Carbon Tetrachloride	Organic vapor canister with full face mask; protective clothing; rubber gloves.	Immediately remove to fresh air, keep patient warm and quiet and get medical attention promptly. Start artificial respiration if breathing stops.	Flush with plenty of water. Remove contaminated clothing and wash before reuse.	Flush with plenty of water; for eyes get medical attention.
Caustic Potash Solution	Wide-brimmed hat and close-fitting safety goggles equipped with rubber side shields; long-sleeved cotton shirt or jacket with buttoned collar and buttoned sleeves; rubber or rubber-coated canvas gloves. (Shirt sleeves should be buttoned over the gloves so that any spilled material will run down the outside).		(Act quickly!) Flush with water for at least 15 minutes.	(Act quickly!) Flush with water, and then rinse with dilute vinegar (acetic acid).

Chemical Safety Data

	Rubber safety-toe shoes or boots and cotton overalls. (Trouser cuffs should be worn outside of boots). Rubber apron.			
Caustic Soda Solution	Wide-brimmed hat and safety goggles equipped with rubber side shields; tight fitting cotton clothing; rubber gloves under shirt cuffs. Rubber boots and apron.		(Fast response is important) Flush with water thoroughly and then rinse with dilute vinegar (acetic acid).	(Fast response is important) Flush with water for at least 15 minutes.
Chlordane	Use respirators for spray, fogs mists or dusts; goggles; rubber gloves.	Administer victim oxygen and give fluid therapy; do not give epinephrine, since it may induce ventricular fibrillation; enforce complete rest.	Wash off skin with large amounts of fresh running water and wash thoroughly with soap and water. Do not scrub infected area of skin.	Flush with water for at least 15 minutes.

Table 6.1 (cont.) Recommended emergency response to acute exposures and direct contact

Chemical Name	PPE	Emergency Treatment Actions for Exposure by		
		Inhalation	Skin	Eyes
Chlorine	Quick-opening safety shower and eye fountain; respiratory equipment approved for chlorine service. Wear safety goggles at all times when in vicinity of liquid chlorine.	Remove victim from source of exposure; call a doctor; support respiration; administer oxygen.		Flush with copious amounts of water for at least 15 minutes.
Chlorine Triflouride	Neoprene gloves and protective clothing made of glass fiber and Teflon, including full hood; self contained breathing apparatus with full face mask.	Remove victim to fresh air and keep him quiet; give artificial if breathing has stopped; give oxygen; enforce rest for 24 hours.	Flush with water, then with 2–3% aqueous ammonia, then again with water; apply ice-cold pack of saturated Epsom salt or 70% ethyl alcohol.	Flush with water for at least 15 minutes; get medical care, but do not interrupt flushing for at least 10 minutes.

Chloroaceto-phenone	Full-face organic canister mask; self-contained breathing apparatus; rubber gloves; protective clothing.	Remove victim from contaminated atmosphere at once; give artificial respiration and oxygen, if necessary; watch for pulmonary edema for several days.	Flush with water.	Flush with water; do not rub.
Chloroform	Chemical goggles. 50 ppm to 2%: suitable full-face gas mask. Above 2%: suitable self-contained system.	If ill effects develop, get victim to fresh air, keep him warm and quiet, and get medical attention. If breathing stops, start artificial respiration.	Wash with soap and water, remove contaminated clothing and free of chemical.	Flush with plenty of water for at least 15 minutes and get medical attention.
Chromic Anhydride	Goggles and respirator (Special chromic acid filters are available for respirators to prevent inhalation of dust or mist)		Flush contacted skin area with water; remove contaminated clothing and wash before reuse.	Wash eyes thoroughly for at least 15 minutes.
Chromyl Chloride	SCBA (full face); rubber gloves; protective clothing.	Remove from exposure; support respiration.	Flush with water for 15 minutes.	Flush with copious quantities of water for 15 minutes.

Table 6.1 (cont.) Recommended emergency response to acute exposures and direct contact

Chemical Name	PPE	Emergency Treatment Actions for Exposure by		
		Inhalation	Skin	Eyes
Citric Acid	Dust mask; goggle or face shield; protective gloves.	Move to fresh air.	Flush with water.	Flush immediately with physiological saline or water; get medical care if irritation persists.
Cobalt Acetate	Dust respirator; rubber gloves; goggles or face shield; protective clothing.	Move to fresh air; if breathing has stopped, begin artificial respiration.	Wash with soap and water.	Flush with water for at least 15 minutes.
Cobalt Chloride	Rubber gloves; side shield goggles; Bu. of Mines respirator; protective clothing.	Move victim to fresh air; if breathing has stopped, begin artificial respiration and call a doctor.	Flush with water.	Flush with water for at least 15 minutes; consult physician if irritation persists.
Cobalt Nitrate	Bu. of Mines approved respirator; rubber gloves; safety goggles; protective clothing.	Move to fresh air; if breathing has stopped, begin artificial respiration and call a doctor.	Flush with water.	Flush with water for at least 15 minutes.

Chemical Safety Data

Copper Acetate	Dust mask; goggles or face shield; protective gloves.	Move to fresh air.	Flush with water.	Flush with water for at least 15 minutes; get medical attention if injury was caused by solid.
Copper Arsenite	Dust respirator; rubber gloves; goggles or face shield.		Wash with soap and water.	Flush with water for at least 15 minutes.
Copper Bromide	Dust mask; goggles or face shield; protective gloves.	Move victim to fresh air.	Flush with water.	Flush with water for at least 15 minutes; get medical attention if injury was caused by solid.
Copper Chloride	Bu. of Mines approved respirator; rubber gloves; safety goggles.	Move to fresh air.	Flush with water.	Flush with water for at least 15 minutes; consult physician if the injury was caused by solid.
Copper Cyanide	Dust respirator; protective goggles or face mask; protective clothing.	Remove victim to fresh air.	Flush with water; wash with soap and water.	Flush with water for at least 15 minutes.

Table 6.1 (cont.) Recommended emergency response to acute exposures and direct contact

Chemical Name	PPE	Emergency Treatment Actions for Exposure by		
		Inhalation	Skin	Eyes
Copper Fluoroborate	Goggles or face shield; rubber apron and gloves.	Move to fresh air.	Flush with water.	Flush with water for at least 15 minutes; get medical attention if irritation persists.
Copper Iodide	Dust mask; goggles or face shield; protective gloves.	Move to fresh air.	Flush with water; wash with soap and water.	Flush with water for at least 15 minutes.
Copper Naphthenate	Goggles or face shield; plastic gloves (as for gasoline)	Remove victim to fresh air.	Wipe off and wash with soap and water.	Wash with copious amounts of water for at least 15 minutes.
Copper Nitrate	Dust mask; goggles or face shield; protective gloves.	Move to fresh air.	Flush with water.	Flush with water for at least 15 minutes; get medical attention if injury was caused by solid.

Chemical Safety Data

Copper Oxalate	Dust mask; goggles or face shield; protective gloves.	Remove to fresh air; if exposure has been prolonged, watch for symptoms of oxalate poisoning (nausea, shock, collapse, and convulsions)	Flush with water.	Flush with water for at least 15 minutes.
Copper Sulfate	Filtering mask to minimize inhalation of dust.		Wash affected tissues with water.	Wash affected tissues with water.
Creosote, Coal Tar	All-service canister mask; rubber gloves; chemical safety goggles and/or shield; overalls or a neoprene apron; barrier creams.	Remove victim to fresh air; if he is not breathing, give artificial respiration, preferably mouth-to-mouth; if breathing is difficult, give oxygen; call a physician.	Wipe with vegetable oil or margarine; then wash with soap and water.	Flush immediately with water for at least 15 minutes. And call a physician.

Table 6.1 (cont.) Recommended emergency response to acute exposures and direct contact

| Chemical Name | PPE | Emergency Treatment Actions for Exposure by |||
		Inhalation	Skin	Eyes
Cresols	Organic vapor canister unit; rubber gloves; chemical safety goggles; face shield; coveralls and/or rubber apron; rubber shoes or boots.	Remove to fresh air.	Flush immediately with plenty of water for at least 15 minutes; remove contaminated clothing immediately and wash before reuse; discard contaminated shoes.	Flush immediately with plenty of water for at least 15 minutes.
Cumene	As necessary to avoid skin exposure. If concentration in air is greater than 1000 ppm, use self-contained breathing apparatus.	Move patient immediately to fresh air; administer artificial respiration or oxygen if necessary; seek medical attention.	Wash exposed skin surfaces thoroughly.	Flush eyes thoroughly with water for 15 minutes.

Chemical Safety Data

Cyanogen	Self-contained breathing apparatus; rubber gloves; rubber protective clothing; rubber-soled shoes.	Move victim to fresh air and let him lie down; do not permit him to exert himself; remove contaminated clothing but keep patient covered and comfortably warm; summon a physician; break an amyl nitrite pearl in a cloth and hold lightly under the victim's nose for 15 seconds; repeat five times at about 15 second intervals; use artificial respiration if breathing has stopped.	Flush with water for at least 15 minutes.
Cyanogen Bromide	Chemical cartridge respirator, goggles, protective clothing, rubber gloves.	Remove victim to fresh air; if he is not breathing, give artificial respiration, preferably mouth-to-mouth; if symptoms of cyanide poisoning are observed, administer amyl nitrite as instructed for HCN.	

Table 6.1 (cont.) Recommended emergency response to acute exposures and direct contact

Chemical Name	PPE	Emergency Treatment Actions for Exposure by		
		Inhalation	Skin	Eyes
Cyclohexane	Hydrocarbon vapor canister, supplied-air or hose mask, hydrocarbon-insoluble rubber or plastic gloves, chemical goggles or face splash shield, hydrocarbon-insoluble rubber or plastic apron.	Remove victim to fresh air; if breathing stops, apply artificial respiration and administer oxygen.	Remove contaminated clothing and gently flush affected areas with water for 15 minutes. Call a physician.	Remove contaminated clothing and gently flush affected areas with water for 15 minutes. Call a physician.
Cyclohexanol	Goggles or face shield.	Remove victim to fresh air.		Flush eyes with water.
Cyclohexanone	Chemical goggles.			Immediately flush eyes with plenty of water; call a physician.
Cyclopentane	Hydrocarbon canister, supplied air or hose mask; rubber or plastic gloves; chemical goggles or face shield.	Remove to fresh air; if breathing stops, apply artificial respiration and administer oxygen.	Flush well with water, then wash with soap and water.	Flush with water for at least 15 minutes; call a physician.

Chemical Safety Data

P-Cumene	Self-contained or air-line breathing apparatus; solvent-resistant rubber gloves; chemical splash goggles.	Remove victim from contaminated area; administer artificial respiration if necessary; call physician.	Wipe off liquid; wash well with soap and water.	Flush with water.
DDD	Dust mask; goggles or face shield; rubber gloves.			
DDT	Data not available.			
Decaborane	Self-contained breathing apparatus or positive-pressure hose mask; rubber boots or overshoes; clothing made of material resistant to decaborane; rubber gloves; chemical types goggles or face shield.	Move patient to fresh air; keep him warm and quiet.	Immediately wash with soap and plenty of water.	Flush with water for at least 15 minutes.
Decahydronaph-thalene	Air mask or self-contained breathing apparatus if in enclosed tank; rubber gloves or protective cream; goggles or face shield.	Remove to fresh air.	Wash with water and mild soap.	Flush with water for at least 15 minutes.

Table 6.1 (cont.) Recommended emergency response to acute exposures and direct contact

Chemical Name	PPE	Emergency Treatment Actions for Exposure by		
		Inhalation	Skin	Eyes
Decaldehyde	Protective clothing and chemical goggles.		Wash with water for 15 minutes.	Wash with water for 15 minutes.
1-Decene	Organic canister or air-supplied mask; goggles or face shield.		Skin areas should be washed with soap and water. Contaminated clothing should be laundered before reuse.	Splashes in the eye should be removed by thorough flushing with water.
N-Decyl Alcohol.	Goggles or face shield.			Flush with water for 15 minutes.
N-Decylbenzene	Goggles or face shield, rubber gloves.	Move to fresh air.	Wipe off, flush with water, and wash with soap and water.	Flush with water.
2, 4-D Esters	Goggles or face shield, rubber gloves.		Flush with water, wash with soap and water.	Flush with plenty of water and see a doctor.
Dextrose Solution	None needed.	None needed.	None needed.	None needed.

Chemical Safety Data

Diacetone Alcohol	Air pack or organic canister, rubber gloves, goggles.	Remove victim to fresh air. Give artificial respiration if breathing has stopped.	Wash affected skin areas with water.	Flush eyes with water and get medical attention.
Di-N-Amyl-Phthalate	Goggles or face shield; rubber gloves.	Move to fresh air.	Wipe off; flush with water; wash with soap and water.	Flush with water.
Diaznon	Goggles or face shield; rubber gloves; protective clothing.	Remove to fresh air; keep warm; get medical attention at once.	Wash contaminated area with soap and plenty of water.	Flush with plenty of water for at least 15 minutes and get medical attention.
Dibenzoyl Peroxide	Safety goggles, face shield, rubber gloves.		Do not use oils or ointments; flush eyes with plenty of water, get medical attention; wash skin with plenty of soap and water.	Do not use oils or ointments; flush eyes with plenty of water, get medical attention; wash skin with plenty of soap and water

Table 6.1 (cont.) Recommended emergency response to acute exposures and direct contact

Chemical Name	PPE	Emergency Treatment Actions for Exposure by		
		Inhalation	Skin	Eyes
Di-N-Butylamine	Goggles or face shield; rubber gloves.	Move from exposure; if breathing has stopped, start artificial respiration.	Wash with large amounts of water for 15 minutes.	Irrigate with water for 15 minutes, get medical attention for possible eye damage.
Di-N-Butyl Ether	Goggles or face shield; rubber gloves.	Remove to fresh air.	Wipe off, wash well with soap and water.	After contact with liquid, flush with water for at least 15 minutes.
Di-N-Butyl Ketone	Rubber gloves; goggles or face shield.	Remove to fresh air; administer artificial respiration if needed.	Flush with water.	Flush with water for at least 15 minutes.
Dibutylphenol	Goggles or face shield.		Wipe off, wash well with soap and water.	Flush with water for at least 15 minutes.
Dibutyl Phthalate	Eye protection.	Remove to fresh air.	Wash affected skin areas with water.	Flush eyes with water.

Chemical Safety Data

O-Dichlorobenzene	Organic vapor-acid gas respirator; neoprene or vinyl gloves; chemical safety spectacles, face shield, rubber footwear, apron, protective clothing.	Remove victim to fresh air, keep him quiet and warm, and call a physician promptly.	Flush with plenty of water; get medical attention for eyes; remove contaminated clothing and wash before reuse.	Flush with plenty of water; get medical attention for eyes; remove contaminated clothing and wash before reuse.
P-Dichlorobenzene	Full face mask fitted with organic vapor canister for concentrations over 75 ppm; clean protective clothing; eye protection.	If any ill effects develop, remove patient to fresh air and get medical attention. If breathing stops, give artificial respiration.	Likely no problems.	Flush with plenty of water and get medical attention if ill effects develop.
Di-(P-Chlorobenzoyl) Peroxide	Goggles or face shield; rubber gloves; protective clothing.		Wash with soap and water.	Wash with water for at least 15 minutes; consult a doctor.
Dichlorobutene	Rubber gloves; chemical splash goggles; rubber boots and apron; barrier cream; organic canister mask.	Remove from exposure; provide low-pressure oxygen if required; keep under observation until edema is ruled out.	Wash immediately and thoroughly with soap and water; treat as a chemical burn.	Irrigate immediately for 15 minutes; call a physician.

Table 6.1 (cont.) Recommended emergency response to acute exposures and direct contact

Chemical Name	PPE	Emergency Treatment Actions for Exposure by		
		Inhalation	Skin	Eyes
Dichlorodi-fluoromethane	Rubber gloves; goggles or face shield.	Remove patient to non-contaminated area and apply artificial respiration if breathing has stopped; call a physician immediately; oxygen may be given.		
1,2-Dichloro-ethylene	Rubber gloves; safety goggles; air supply mask or self-contained breathing apparatus.	Remove from further exposure; if breathing is difficult, give oxygen; if victim is not breathing, give artificial respiration, preferably mouth-to-mouth; give oxygen when breathing is resumed; call a physician.	Wash well with soap and water.	Flush with water for at least 15 minutes.

CHEMICAL SAFETY DATA

Dichloroethyl Ether	Goggles or face shield; rubber gloves; protective clothing.	Remove from exposure; support respiration; call physician if necessary.	Wipe off, wash well with soap and water.	Irrigate with copious quantities of water for 15 minutes; call a physician.
Dichloromethane	Organic vapor canister mask, safety glasses, protective clothing.	Remove from exposure. Give oxygen if needed.	Remove contaminated clothing; wash skin or eyes if affected.	Remove contaminated clothing; wash skin or eyes if affected.
2,4-Dichlorophenol	Bu. of Mines approved respirator, rubber gloves, chemical goggles.	Rest.		
2,4-Dichlorophenoxyacetic Acid	Protective dust mask, rubber gloves; chemical gloves.		Wash well with soap and water.	Flush with water for at least 15 minutes.
Dichloropropane	Air supply in confined area, rubber gloves, chemical goggles, protective coveralls and rubber footwear.	Remove to fresh air.	Wash skin thoroughly with soap and water.	Flush eyes with water for 15 minutes. Call a doctor.

Table 6.1 (cont.) Recommended emergency response to acute exposures and direct contact

Chemical Name	PPE	Emergency Treatment Actions for Exposure by		
		Inhalation	Skin	Eyes
Dichloropropene	An approved full face mask equipped with a fresh black canister meeting specifications of the U.S. Bureau of Mines for organic vapors, a full face self-contained breathing apparatus, or full face air-supplied respirator.	Remove patient to fresh air, keep warm and quiet; call physician immediately; give artificial respiration if breathing has stopped.	Immediately remove contaminated clothing and shoes. Wash skin with soap and plenty of water.	For eyes, flush immediately with plenty of water for at least 15 minutes. Call a physician.
Dicyclopentadiene	Air-supplied mask in confined areas, rubber gloves, safety glasses.	Remove victim from contaminated area and call physician if unconscious; if breathing is irregular or stopped, give oxygen and start resuscitation.	Flush with plenty of water for 15 minutes.	Flush with plenty of water for 15 minutes.

CHEMICAL SAFETY DATA

Dieldrin	U.S. bureau of Mines approved respirator; clean rubber gloves; goggles or face shield.	Move to fresh air; give oxygen and artificial respiration as required.		Flush with plenty of water.
Diethanolamine	Full face mask or amine vapor mask only, if required; clean body covering clothing, chemical goggles.	No problem likely. Get medical attention if ill effects develop.	Flush with water. Wash contaminated clothing before reuse.	Flush with plenty of water for at least 15 minutes and get medical attention promptly.
Diethylamine	Chemical safety goggles, rubber gloves, and apron.		In case of contact, flush skin or eyes with plenty of water for at least 15 minutes.	In case of contact, flush skin or eyes with plenty of water for at least 15 minutes; for eyes, get medical attention.
Diethylbenzene	Self-contained breathing apparatus, safety goggles.	Remove to fresh air and start artificial respiration.	Flush with water for at least 15 minutes. Wash skin with soap and water.	Flush with water for at least 15 minutes. Wash skin with soap and water.

Table 6.1 (cont.) Recommended emergency response to acute exposures and direct contact

Chemical Name	PPE	Emergency Treatment Actions for Exposure by		
		Inhalation	Skin	Eyes
Diethyl Carbonate	Protective clothing; rubber gloves and goggles; organic vapor canister or air mask.	Remove from exposure; administer artificial respiration and oxygen if needed.		Flush with water for at least 15 minutes.
Diethylene Glycol	Full face mask with canister for short exposures to high vapor levels; rubber gloves; goggles.	No problem likely. If any ill defects do develop, get medical attention.	Flush with water. If any ill defects occur, get medical attention.	Flush with water. If any ill defects occur, get medical attention.
Diethylene Glycol Dimethyl Ether	Vinyl (not rubber) gloves; safety goggles.			
Diethyleneglycol Monobutyl Ether	Safety goggles or face shield.	Remove to fresh air; if ill effects are observed, call a doctor.	Wash well with soap and water.	Immediately flush with plenty of water for at least 15 minutes.

Diethyleneglycol Monobutyl Ether Acetate	Face shield or safety glasses; protective gloves; air mask for prolonged exposure to vapor.	Move victim to fresh air; if breathing has stopped, begin artificial respiration.	Wash skin with large amounts of water for 15 minutes; call a physician if needed.	Flush with water for at least 15 minutes.
Diethylene Glycol Monoethyl Ether	Goggles.		Flush with water.	Flush with water.
Diethylenetriamine	Amine respiratory cartridge mask; rubber gloves; splash-proof goggles.	Remove victim to fresh air.	Flush with plenty of water for at least 15 minutes and get medical attention.	
Di(2Ethylhexyl) Phosphoritc Acid	Goggles or face shield; rubber gloves; protective clothing.		Immediately flush with plenty of water for at least 15 minutes; see a physician.	Immediately flush with plenty of water for at least 15 minutes; see a physician.
Diethyl Phthalate	Rubber gloves; goggles or face shield.	Remove to fresh air.	Flush with water, wash well with soap and water.	Flush with water.

Table 6.1 (cont.) Recommended emergency response to acute exposures and direct contact

Chemical Name	PPE	Emergency Treatment Actions for Exposure by		
		Inhalation	Skin	Eyes
Diethylzinc	Cartridge-type or fresh air mask for fumes or smoke; PVC fire-retardant or asbestos gloves; full face shield, safety glasses or goggles; fire-retardant coveralls as standard wear; for special cases, use asbestos coat or rain suit.	Move victim to fresh air and call doctor immediately; give mouth-to-mouth resuscitation if needed; keep victim warm and comfortable; oxygen should be given only by experienced person, and only on doctor's instructions.	Flush affected area with large amounts of water; do not use chemical neutralizers; get medical attention if irritation persists.	Flush with large amounts of running water for at least 15 minutes, holding eyelids apart to insure thorough washing; get medical attention as soon as possible; do not use chemical neutralizers, and avoid oils or ointments unless prescribed by doctor.
1,1 Difluoro-ethane	Individual breathing devices with air supply; neoprene gloves; protective clothing; eye protection.	Remove to fresh air; use artificial respiration if necessary.	Soak in lukewarm water (for frostbite)	Get medical attention promptly if liquid has entered eyes.

Chemical Safety Data

Difluorophosphoric Acid, Anhydrous	Air line mask or self-contained breathing apparatus; full protective clothing.	Remove victim from exposure and support respiration.		Wash with copious volumes of water for at least 15 minutes.
Diheptyl Phthalate	Goggles or face shield; rubber gloves.	Move to fresh air.	Wipe off; flush with water; wash with soap and water.	Flush with water.
Diisobutyl-carbinol	Air-supplied mask for prolonged exposure; plastic gloves; goggles.		Flush with water.	Flush with water.
Diisobutylene	Protective goggles.	Remove from exposure; support respiration.		
Diisobutyl Ketone	Air-supplied mask in confined areas; plastic gloves; face shield and safety glasses.	Move to fresh air; give oxygen if breathing is difficult; call a physician.	Wipe off; flush with plenty of water; wash with soap and water.	Flush with plenty of water.
Diisodecyl Phthalate	Goggles or face shield; rubber gloves.		Wipe off; wash with soap and water.	Flush with water; call physician.

Table 6.1 (cont.) Recommended emergency response to acute exposures and direct contact

Chemical Name	PPE	Emergency Treatment Actions for Exposure by		
		Inhalation	Skin	Eyes
Diisoprpo-anolamine	Full face mask or amine vapor mask only if required; clean, body covering clothing, rubber gloves, apron boots and face shield.	If ill effects occur, remove person to fresh air and get medical help.	Immediately flush with plenty of water for at least 15 minutes. For eyes, get medical help promptly. Remove and wash contaminated clothing before reuse.	Immediately flush with plenty of water for at least 15 minutes. For eyes, get medical help promptly. Remove and wash contaminated clothing before reuse.
Diisopro-pylamine	Air-supplied mask; plastic gloves; goggles; rubber apron.	Move victim to fresh air and keep him quiet and comfortably warm; give oxygen if breathing is difficult; call a physician.	Flush with water; remove contaminated clothing and wash skin; if there is any redness or evidence of burning.	Immediately flush with plenty of water for at least 15 minutes, then get medical care.

CHEMICAL SAFETY DATA

Diisopropyl-benzene Hydroperoxide	Solvent-resistant gloves; chemical-resistant apron; chemical goggles or face shield; self-contained breathing apparatus.	Move to fresh air; call a doctor.	Wash several times with soap and water.	Flush with water for 15 minutes; holding eyelids open; call physician.
Dimethyl-acetamide	Goggles or face shield; rubber gloves.		Flush with plenty of water for 15 minutes.	Flush with plenty of water for 15 minutes; get medical attention.
Dimethylamine	Chemical goggles and full face shield; molded rubber acid gloves; self-contained breathing apparatus.	Remove victim to fresh air and call a physician; if breathing has stopped, administer artificial respiration and oxygen; keep victim warm and quiet; do not give stimulants.	Remove contaminated clothing immediately; flush affected area with large amounts of water and then wash with soap and water.	Flush continuously and thoroughly with water for at least 15 minutes.
Dimethyl Ether	Mask for organic vapors; plastic or rubber gloves; safety glasses.	Remove from exposure and support respiration; call a physician.	Treat frostbite by use of warm water or by wrapping the affected part in blanket.	Wash with water for at least 15 minutes; consult an eye specialist.

Table 6.1 (cont.) Recommended emergency response to acute exposures and direct contact

Chemical Name	PPE	Emergency Treatment Actions for Exposure by		
		Inhalation	Skin	Eyes
Dimethyl Sulfate	Chemical goggles; self-contained breathing apparatus; safety hat; rubber suit; rubber shoes; rubber gloves; safety shower and eye wash fountain.	Get victim to fresh air immediately; administer 100% oxygen, even if no injury is apparent, and continue for 30 minutes each hour for 6 hours; give artificial respiration if breathing is weak or fails, but do not interrupt oxygen therapy; if victim's coughing prevents use of a mask, use oxygen tent under atmospheric pressure.	Wash thoroughly.	Flush with running water for at least 15 minutes.
Distillates: Flashed Feed Stocks	Data not available.	Maintain respiration, administer oxygen.	Wipe off and wash with soap and water.	Wash with copious amounts of water.

Chemical Safety Data

Dodecyltri-chlorosilane	Acid-vapor-type respiratory protection, rubber gloves, chemical worker's goggles, other equipment as necessary to protect skin and eyes.	Remove from exposure; support respiration; call a physician if needed.	Flush with water; obtain medical attention if skin is burned.	Flush with water for 15 minutes; obtain medical attention immediately.
Ethane	Self-contained breathing apparatus for high vapor concentrations.	Remove from exposure; support respiration.		
Ethyl Acetate	Organic vapor canister or sir mask; goggles or face shield.	If victim is overcome, move him to fresh air immediately and call a physician; if breathing is irregular or stopped, start resuscitation and administer oxygen.		Flush with water for at least 15 minutes.
Ethyl Alcohol	All-purpose canister; safety goggles. Avoid contact with liquid and irritation of vapors.	If breathing is affected, remove victim to the fresh air; call physician; administer oxygen. Speed is of primary importance.	Flush with water.	Flush with water.

Table 6.1 (cont.) Recommended emergency response to acute exposures and direct contact

Chemical Name	PPE	Emergency Treatment Actions for Exposure by		
		Inhalation	Skin	Eyes
Ethyl Butanol	Fresh-air mask; plastic gloves; coverall goggles; safety shower and eye bath.	Remove to fresh air.	Remove and wash contaminated clothing. Wash affected skin areas with water.	Flush eyes with water for 15 minutes and get medical care.
Ethyl Chloroformate	Air-line mask, self-contained breathing apparatus, or organic and canister mask; full protective clothing.	Remove patient to fresh air; if breathing stops give artificial respiration. Call doctor, keep victim quiet and administer oxygen if needed.	Wash liberally with water for at least 15 minutes, then apply dilute solution of sodium bicarbonate or commercially prepared neutralizer.	Flush with water for at least 15 minutes; see a doctor.
Ethylene	Organic vapor canister or air-supplied mask.	Remove victim to fresh air, give artificial respiration and oxygen is breathing has stopped, and call a physician.		

CHEMICAL SAFETY DATA

Ethylene Dichloride	Clean, body-covering clothing and safety glasses with side shields. Respiratory protection: up to 50 ppm, none; 50 ppm to 2%, 1/2 hr or less, full face mask and canister; greater than 2%, self-contained breathing apparatus.	If victim is overcome, move him to fresh air, keep him quiet and warm, and get medical attention.	Remove clothing and wash skin thoroughly with soap and water; wash contaminated clothing before reuse.	Flush immediately with copious amounts of flowing water for 15 minutes.
Ethyl Lactate	Goggles or face shield; rubber gloves.	Remove victim to fresh air.	Flush well with water.	Flush well with water.
Ethyl Nitrate	Self-contained breathing apparatus; goggles or face shield; rubber gloves.	Remove victim from exposure; if breathing has stopped, give artificial respiration; call physician.		Flush with water, wash with soap and water.
Ferric Ammonium Citrate	Approved respirator for nuisance dust, chemical goggles or face shield.		Flush with water.	Flush with water.
Ferric Nitrate	Dust mask; goggles or face shield; protective gloves.	Move to fresh air.	Flush with water.	Flush with water; get medical attention if irritation persists.

Table 6.1 (cont.) Recommended emergency response to acute exposures and direct contact

Chemical Name	PPE	Emergency Treatment Actions for Exposure by		
		Inhalation	Skin	Eyes
Ferric Sulfate	Dust mask; goggles or face shield; protective gloves.	Move to fresh air.	Flush with water.	Flush with water; get medical attention if irritation persists.
Fluorine	Tight-fitting chemical goggles, special clothing, not easily ignited by fluorine gas.	Administer artificial respiration and oxygen is required.	Flush all affected parts with water for at least 15 minutes. Do not use ointments.	Flush all affected parts with water for at least 15 minutes. Do not use ointments.
Formaldehyde Solution	Self-contained breathing apparatus, chemical goggles, protective clothing, synthetic rubber or plastic gloves.	Remove victim to fresh air; give oxygen is breathing is difficult; call physician.	Flush immediately with plenty of water for at least 15 minutes; remove contaminated clothing; call a physician for eyes.	Flush immediately with plenty of water for at least 15 minutes; remove contaminated clothing; call a physician for eyes.

Gallic Acid	Bu. Mines approved respirator, rubber gloves, and safety goggles.		Wash skin with soap and water.	Flush with water for at least 10 minutes; consult physician if irritation persists.
Gasolines: Automotive	Protective goggles, gloves, and organic canister mask.	Maintain respiration, administer oxygen; enforce bed rest if liquid is in lungs.	Wipe off and wash with soap and water.	Wash with copious amount of water.
Gasolines: Aviation	Protective goggles, gloves.	Maintain respiration, give oxygen if needed.	Wipe off, wash well with soap and water.	Wash with copious amounts of water.
Glycerine	Rubber gloves, goggles.	No hazard.	No hazard.	No hazard.
Heptanol	Chemical goggles or face shield.		Flush all affected parts with plenty of water.	Flush all affected parts with plenty of water.
N-Hexaldehyde	Goggles or face shield; rubber glove.		Wipe off; wash with soap and water.	Flush with water for at least 15 minutes.

Table 6.1 (cont.) Recommended emergency response to acute exposures and direct contact

Chemical Name	PPE	Emergency Treatment Actions for Exposure by		
		Inhalation	Skin	Eyes
Hydrochloric Acid	Self-contained breathing equipment, air-line mask, or industrial canister-type gas mask; rubber coated gloves, apron, coat, overalls, shoes.	Remove person to fresh air; keep him warm and quiet and get medical attention immediately; start artificial respiration if breathing stops.	Immediately flush skin while removing contaminated clothing; get medical attention promptly; use soap and water and wash area for at least 15 minutes.	Immediately flush with plenty of water for at least 15 minutes and get medical attention; continue flushing for another 15 minutes if physician does not arrive promptly.
Hydrogen Chloride	Full face mask and acid gas canister; self-contained breathing apparatus; chemical goggles; rubber apron and gloves; acid proof clothing; safety shower.	Immediately remove patient to fresh air, keep him warm and quiet, and call a physician immediately; if a qualified person is available to give oxygen, such treatment may be helpful.	Immediately flush with plenty of water for at least 15 minutes; for eyes get medical attention promptly; air contaminated clothing and wash before reuse.	Immediately flush with plenty of water for at least 15 minutes; for eyes get medical attention promptly; air contaminated clothing and wash before reuse.

Hydrogen, Liquefied	Safety goggles or face shield; insulated gloves and long sleeves; cuff less trousers worn outside boots or over high-top shoes to shed spilled liquid; self-contained breathing apparatus containing air (never use oxygen)	If victim is unconscious (due to oxygen deficiency) move him to fresh air and apply resuscitation method; call physician.	Treat for frostbite; soak in lukewarm water; get medical attention if burn is severe.	Treat for frostbite.
Isobutane	Self-contained breathing apparatus; safety goggles.	Protect victim against self-injure if he is stupor us, confused, or anesthetized; apply artificial reparation if breathing has stopped; avoid administration of epinephrine or other symptoms amines; prevent aspiration of vomits by proper positioning of head; give symptomatic and supportive treatment.		

Table 6.1 (cont.) Recommended emergency response to acute exposures and direct contact

Chemical Name	PPE	Emergency Treatment Actions for Exposure by		
		Inhalation	Skin	Eyes
Isobutyl Alcohol.	Air pack or organic canister mask; chemical goggles.	If victim is overcome by vapors, remove him from exposure immediately; call a physician; if breathing is irregular or has stopped, start resuscitation; administer oxygen.		Flush with water for at least 15 minutes.
Isodecaldehyde	Protective clothing; chemical goggles.		Wash skin and eyes with plenty of water for at least 15 minutes.	Wash skin and eyes with plenty of water for at least 15 minutes.
Isohexane	Eye protection (as for gasoline)	Maintain respiration; give oxygen if needed.	Wipe off, wash with soap and water.	Wash with copious amounts of water.
Isopropyl Alcohol	Organic vapor canister or air-supplied mask; chemical goggles or face shield.	If victim is overcome by vapors, remove from exposure immediately; call a physician; if		Flush eyes with water for at least 15 minutes.

		breathing is irregular or stopped, start resuscitation and administer oxygen.		
Isopropyl Mercaptan	Self-contained breathing apparatus; goggles or face shield; rubber gloves.	Move victim to fresh air; start artificial respiratory and give oxygen if required; observe for signs of pulmonary edema; get medical attention.	Flush with water.	Flush with water.
Kerosene	Protective gloves; goggles or face shield.		Wipe off and wash with soap and water.	Wash with plenty of water.
Lactic Acid	Rubber gloves; goggles; self-contained breathing apparatus where high concentrations of mist are present.	Move to fresh air.	Flush with water.	Flush with water for 15 minutes.

Table 6.1 (cont.) Recommended emergency response to acute exposures and direct contact

Chemical Name	PPE	Emergency Treatment Actions for Exposure by		
		Inhalation	Skin	Eyes
Lead Arsenate	Dust respirator; protective clothing to prevent accidental inhalation or ingestion of dust.	A specific medical treatment is used for exposure to this chemical; call a physician immediately!		
Lean Iodide	Dust mask and protective gloves.	Remove immediately all cases of lead intoxication from further exposure until the blood level is reduced to a safe value; immediately place the individual under medical care.	Remove immediately all cases of lead intoxication from further exposure until the blood level is reduced to a safe value; immediately place the individual under medical care. Flush skin with water.	Remove immediately all cases of lead intoxication from further exposure until the blood level is reduced to a safe value; immediately place the individual under medical care. Flush eyes with water.

Chemical Safety Data

Linear Alcohols	Eye protection.		Wash eyes with water for at least 15 minutes.
Liquefied Natural Gas	Self-contained breathing apparatus; protective clothing if exposed to liquid.	Remove victim to open air. If the victim is overcome by gas, apply artificial resuscitation.	
Liquefied Petroleum Gas	Self-contained breathing apparatus for high concentrations of gas.	Remove victim to open air. If the victim is overcome by gas, apply artificial respiration. Guard against self-injury if confused.	
Lithium, Metallic	Rubber or plastic gloves; face shield; respirator; fire-retardant clothing.		Flush with water and treat with boric acid.	Flush with water and treat with boric acid.
Magnesium	Eye protection.		Treat as any puncture.	Flush with water to remove dust.
Mercuric Acetate	Rubber gloves, dust mask, goggles.	Have physician treat for mercury poisoning.	Have physician treat for mercury poisoning. Flush with water.	Have physician treat for mercury poisoning. Flush with water.

Table 6.1 (cont.) Recommended emergency response to acute exposures and direct contact

Chemical Name	PPE	Emergency Treatment Actions for Exposure by		
		Inhalation	Skin	Eyes
Mercuric Cyanide	Dust mask, goggles or face shield, rubber gloves.	If victim has stopped breathing, start artificial respiration immediately; using amyl nitrate pearls, administer amyl nitrate by inhalation for 15–30 seconds of every minute while sodium nitrate solution is being prepared; discontinue amyl nitrate and immediately inject intravenously 10 ml of a 3% solution of sodium nitrate (nonsterile if necessary) over a period of 2–4 minutes; without removing needle, infuse intravenously 50 ml of a 25% aqueous solution of sodium thiosulphate;	Wash with water for 15 minutes.	Wash with water for 15 minutes.

Chemical Safety Data

		injection should take about 10 minutes. (Concentrations of 5–50% may be used, but keep total dose approx. 12 gm). Oxygen therapy may be helpful in combination with the above.		
Methane	Self-contained breathing apparatus for high concentrations; protective clothing if exposed to liquid.	Remove to fresh air. Support respiration.		
Methyl Alcohol	Approved canister mask for high vapor concentrations; safety goggles; rubber gloves.	Remove victim from exposure and apply artificial respiration is breathing has ceased.	Flush with water for 15 minutes.	
Methyl Chloride	Approved canister mask; leather or vinyl gloves; goggles or face shield.	Remove to fresh air. Call a doctor and have patient hospitalized for observation of slowly developing symptoms.	Flush with water for 15 minutes.	

Table 6.1 (cont). Recommended emergency response to acute exposures and direct contact

Chemical Name	PPE	Emergency Treatment Actions for Exposure by		
		Inhalation	Skin	Eyes
Methyl Isobutyl Carbinol.	Air pack or organic canister mask, rubber gloves, goggles or face shield.	Remove to fresh air; give artificial respiration if needed; call a doctor.	Flush with water.	Flush with water for at least 15 minutes; consult a doctor.
Alpha-Methylstyrene	Neoprene gloves; splash proof goggles or face shield.	Remove victim to fresh air; if he is not breathing, give artificial respiration; contact a physician; keep victim quiet and warm.	Wash area with soap and water.	Flush with water for at least 15 minutes; get medical attention.
Mineral Spirits	Plastic gloves; goggles or face shield (as for gasoline)	Remove victim to fresh air.	Wipe off and wash with soap and water.	Wash with copious amounts of water.
Nitrous Oxide	Self-contained breathing apparatus for high vapor concentrations.	Remove to fresh air.	Treat frostbite burn; soak in lukewarm water.	Get medical attention for frostbite burn.
Nonanol	Goggles or face shield; rubber gloves.		Flush eyes and skin with water for at least 15 minutes.	Flush eyes and skin with water for at least 15 minutes.

CHEMICAL SAFETY DATA

Octane	Self-contained breathing apparatus for high vapor concentrations; goggles or face shield; rubber gloves.	Remove victim from exposure; apply artificial respiration if breathing has stopped; call a physician if needed.	Flush with water; wash with soap and water.	Irrigate with copious quantities of water for 15 minutes.
Oils: Clarified	Goggles or face shield.		Wipe off; wash with soap and water.	Wash with water for at least 15 minutes.
Oils: Crude	Goggles or face shield; rubber gloves and boots.		Wipe off; wash with soap and water.	Flush with water for at least 15 minutes.
Oils: Diesel	Goggles or face shield.		Wipe off; wash with soap and water.	Wash with copious amounts of water for at least 15 minutes.
Oils, Edible: Castor	Goggles or face shield.		Wipe off, wash with soap and water.	Flush with water for at least 15 minutes.
Oils, Edible: Coconut	Goggles or face shield; rubber gloves.			Flush with water for at least 15 minutes.

Table 6.1 (cont.) Recommended emergency response to acute exposures and direct contact

Chemical Name	PPE	Emergency Treatment Actions for Exposure by		
		Inhalation	Skin	Eyes
Oils, Edible: Cottonseed.	Goggles or face shield.			Wash with water for at least 15 minutes.
Oils, Edible: Fish	Goggles or face shield.			Wash with water for at least 15 minutes.
Oils, Edible: Lard	Goggles of face shield; rubber gloves.		Wipe off, get medical attention for burn.	Wash with water for at least 15 minutes.
Oils, Fuel: 2	Protective gloves; goggles or face shield.		Remove solvent by wiping and wash with soap and water.	Wash with copious quantity of water.
Oils, Fuel: 4	Protective gloves; goggles or face shield.		Wipe off and wash with soap and water.	Wash with copious quantity of water.
Oils, Miscellaneous: Coal Tar	Protective gloves; goggles or face shield.		Wipe off and wash with soap and water.	Flush with water for at least 15 minutes.

CHEMICAL SAFETY DATA

Oils, Miscellaneous: Motor	Protective gloves; goggles or face shield.		Wipe off and wash with soap and water.	Wash with copious quantity of water.
Oils, Miscellaneous: Penetrating	Protective gloves; goggles or face shield.		Wipe off and wash with soap and water.	Wash with copious quantity of water.
Oils, Miscellaneous: Resin	Data not available.	Data not available.	Data not available.	Data not available.
Oils, Miscellaneous: Spray	Protective gloves; goggles or face shield.		Wipe off and wash with soap and water.	Wash with copious quantity of water.
Oils, Miscellaneous: Tanner's	Data not available.	Data not available.	Data not available.	Data not available.
Oleic Acid	Impervious gloves; goggles or face shield; impervious apron.		Wash thoroughly with soap and water.	If eye irritation occurs, flush with water and get medical attention.
Oleum	Respirator approved by U.S. Bureau of Mines for acid mists, rubber gloves, splash proof goggles, eyewash fountain and safety shower, rubber footwear, face shield.		Flush with plenty of water.	Flush with water for at least 15 minutes.

Table 6.1 (cont.) Recommended emergency response to acute exposures and direct contact

Chemical Name	PPE	Emergency Treatment Actions for Exposure by		
		Inhalation	Skin	Eyes
Oxalic Acid.	Respirator for dust or mist protection, rubber gloves, chemical safety glasses, rubber safety shoes, apron, or impervious clothing for splash protection.	Rinse mouth and/or gargle repeatedly with cold water.	Flush thoroughly with water.	Flush thoroughly with water.
Oxygen, Liquefied	Safety goggles or faces shield, insulated gloves, long sleeves, trousers worn outside boots or over high-top shoes to shed spilled liquid.	In all but the most severe cases (pneumonia) recovery is rapid after reduction of oxygen pressure. Supportive treatment should include immediate sedation, anticonvulsive therapy if needed and rest.	Treat frostbite. Soak in lukewarm water.	Treat frostbite burns.
Para-formaldeyde	Goggles or face shield, protective clothing.		Rinse with copious amounts of water.	Rinse with copious amounts of water.

Chemical Safety Data

Pentaerythritol	Dust mask, goggles.			
Pentane	Goggles or face shield.	Remove from exposure. Support respiration if needed.		
1-Pentene	Goggles or face shield.	Remove from exposure.	Wash with soap and water.	Flush with water.
Peracetic Acid	Self-contained breathing apparatus, impervious gloves, full protective clothing.	Remove victim to fresh air and apply artificial respiration and/or oxygen if not breathing. Call a doctor.		
Petrolatum	Goggles or face shield.			Wash with water.
Phenol	Fresh air mask for confined areas; rubber gloves.	If victim shows any ill effects, move him to fresh air, keep him quiet and warm, and call a doctor immediately. If breathing stops, give artificial respiration.	Immediately remove all clothing while in the shower and wash affected area with water and soap for at least 15 minutes.	Flush with water for at least 15 minutes; continue for another 15 minutes if a doctor has not taken over.

Table 6.1 (cont.) Recommended emergency response to acute exposures and direct contact

Chemical Name	PPE	Inhalation	Skin	Eyes
		Emergency Treatment Actions for Exposure by		
Phenyldi-chloroarsine, Liquid	Full protective clothing, gas mask or self contained breathing apparatus.	Remove victim from exposure, give artificial respiration if breathing has ceased.	Flush with water. Wash well with soap and water. Compound can be absorbed through skin and cause toxic system effects.	Wash with copious amounts of water for at least 15 minutes.
Phosgene	Approved by U.S. Bureau of Mines respirator, protective clothing.	Remove victim from contaminated area. Enforce rest and call a physician.		
Phosphoric Acid	Goggles or face shield, rubber gloves, protective clothing.		Flush with water for at least 15 minutes.	Flush with water for at least 15 minutes.
Piperazine	Monogoggles or face shield, rubber gloves, dust mask.	Move to fresh air.	Wash with soap and water.	Flush with plenty of water for at least 15 minutes and get medical attention

CHEMICAL SAFETY DATA

Polybutene	Goggles or face shield.	Remove victim from exposure.	
Polychlorinated Biphenyl	Gloves and protective clothing.		Wash with soap and water.
Polypropylene	Filter respirator.		
Potassium Cyanide	Wear dry cotton gloves and U.S. Mines approved dust respirator when handling solid potassium cyanide. Wear rubber gloves and approved chemical safety goggles when handling solutions.		
Potassium Iodide	Goggles or face shield.		
Propane	Self-contained breathing apparatus for high concentrations of gas.		
Propion-aldehyde	Air-supplied mask for high vapor concentrations, plastic gloves, goggles.	Remove victim to fresh air and give oxygen if breathing is difficult. Call a physician.	Flush with water.	Flush with water for at least 15 minutes and call a physician.

Table 6.1 (cont.) Recommended emergency response to acute exposures and direct contact

Chemical Name	PPE	Emergency Treatment Actions for Exposure by		
		Inhalation	Skin	Eyes
Propylene Oxide	Air-supplied mask, rubber or plastic gloves, vapor proof goggles.	Remove person to fresh air immediately. Keep quiet and warm and call a physician. If breathing has stopped give artificial respiration.	Flush with water for at least 15 minutes. Remove all clothing and watch, etc, to prevent confiding product to skin.	Flush with water for at least 15 minutes.
Pyridine	Air-supplied mask or organic canister, vapor-proof goggles, rubber gloves, and protective clothing.	Remove victim from contaminated area and give artificial respiration and oxygen if necessary. Treat symptomatically.	Wash thoroughly with water.	Irrigate with water for at least 15 minutes.
Pyrogallic Acid	Rubber gloves, safety goggles, dust mask.	Remove victim to fresh air.	Wash immediately and thoroughly with soap and water; consult a physician if exposure has been severe.	Flush with water for at least 15 minutes and consult a physician.

Chemical Safety Data

Quinoline	U.S. Bu. Mines approved respirator, rubber gloves, safety goggles with side shields or chemical goggles, coveralls or approved rubber apron.	Remove victim to fresh air. If he is not breathing give artificial respiration, if breathing is difficult give oxygen.	Flush with water.	Flush with water for at least 15 minutes.
Salicylic Acid	Gloves, goggles, respirator for dust, clean body-covering clothing.	Move to fresh air.	Wash with soap and water.	Flush with water for at least 15 minutes.
Selenium Dioxide	Dust mask, rubber gloves, protective clothing.	Move to fresh air.	Flush with water.	Flush immediately and thoroughly with water.
Selenium Trioxide	Dust mask, rubber gloves, goggles or protective shield.	Remove victim to fresh air. If breathing is difficult give oxygen.		Flush with water.

Table 6.1 (cont.) Recommended emergency response to acute exposures and direct contact

Chemical Name	PPE	Emergency Treatment Actions for Exposure by		
		Inhalation	Skin	Eyes
Silicon Tetrachloride	Add-canister-type gas mask or self-contained breathing apparatus, goggles or face shield, rubber gloves, other protective clothing to prevent skin contact.	Remove victim to fresh air. If he is not breathing give artificial respiration, if breathing is difficult give oxygen.	Immediately flush affected area with water, severe or extensive burns may be cause by silicon tetrachloride producing shock symptoms (rapid pulse, sweating, and collapse). Keep patient comfortably warm.	Immediately flush with running water for at least 15 minutes. Continue irrigation for additional 15 minutes if physician is not available.
Silver Acetate	Dust mask, goggles or face shield, protective gloves.	Move to fresh air.	Flood with water.	Flush with water for at least 15 minutes.
Silver Carbonate	Dust mask, goggles or face shield, protective gloves.	Move to fresh air.	Flush with water. Wash with soap and water.	Flush with water for at least 15 minutes.

Chemical Safety Data

Silver Fluoride	Dust mask, goggles or face shield, protective gloves.	Move to fresh air.	Flush with water.	Flush with water for at least 15 minutes.
Silver Iodate	Dust mask, goggles or face shield, protective gloves.	Move to fresh air.	Flush with water. Wash with soap and water.	Flush with water for at least 15 minutes.
Silver Nitrate	Goggles or face shield, protective gloves.		Wash promptly.	
Silver Oxide	Dust mask, goggles or face shield, protective gloves.		Flush with water. Wash with soap and water.	Flush with water.
Silver Sulfate	Dust mask, goggles or face shield, protective gloves.	Move to fresh air.	Flush with water. Wash with soap and water.	Flush with water for at least 15 minutes.
Sodium	Maximum protective clothing goggles and face shield.		Brush off any metal, and then flood with water for at least 15 minutes.	
Sodium Alkylbenzene-sulfonates	Goggles or face shield, protective gloves.		Flush with copious amounts of water.	Flush with copious amounts of water.

Table 6.1 (cont.) Recommended emergency response to acute exposures and direct contact

Chemical Name	PPE	Emergency Treatment Actions for Exposure by		
		Inhalation	Skin	Eyes
Sodium Alkyl Sulfates	Goggles or face shield, protective gloves.		Wash off with water.	
Sodium Amide	Goggles or face shield, dust respirator, rubber gloves and shoes.		Flush with copious amounts of water.	Flush with copious amounts of water.
Sodium Arsenate	Dust mask, goggles or face shield, protective gloves.	Remove victim from exposure, support respiration.	Flush with water.	Flush with water for at least 15 minutes.
Sodium Arsenite	Dust mask, goggles or face shield, protective gloves.		Wash with large amounts of water.	Flush with water for at least 15 minutes.
Sodium Azide	Dust mask, protective clothing and goggles.	Move to fresh air.	Flush with water, wash with soap and water.	Flush with water for at least 15 minutes.
Sodium Bisulfite	Dust mask, goggles or face shield.	Get medical attention at once.	Wash with plenty of water.	Flush with water for at least 15 minutes.

Chemical Safety Data

Sodium Borate	Dust mask, goggles or face shield.	Move to fresh air and call physician immediately. Give mouth-to-mouth resuscitation if breathing has ceased, give oxygen if authorized by a physician. Keep victim warm.	Flush with water. Remove contaminated clothing in the shower. Do not use chemical neutralizers and get medical attention unless burn is minor.	Immediately flush with running water for at least 15 minutes. Continue irrigation for additional 15 minutes if physician is not available.
Sodium Borohydride	Goggles, rubber gloves, protective clothing.		Flood with water.	Flood with water.
Sodium Cacodylate	Goggles or face shield, dust mask, rubber gloves.	Remove victim from exposure, call a physician.	Flush with water and wash well with soap and water.	Flush with water.
Sodium Chlorate	Clean clothing (must be washed well after exposure), rubber gloves and shoes, where dusty goggles and dust respirator.			Wash thoroughly with water.

Table 6.1 (cont). Recommended emergency response to acute exposures and direct contact

Chemical Name	PPE	Emergency Treatment Actions for Exposure by		
		Inhalation	Skin	Eyes
Sodium Chromate	U.S. Bu. Mines approved respirator, rubber gloves, safety goggles with side shields or chemical goggles, coveralls or approved rubber apron, sleeves.	Remove victim from exposure, call a physician.	Immediately flush with plenty of water for at least 15 minutes, persistent dermatitis should be referred to a physician. Wash contaminated skin or clothing until chromate color disappears.	Immediately flush with running water for at least 15 minutes. Call a physician.
Sodium Cyanide	Protective gloves when handling solid sodium cyanide, rubber gloves when handling cyanide solution (wash hands and rubber gloves thoroughly with			

Chemical Safety Data

	running water after handling cyanides), U.S. Bureau of Mines approved dust respirator, approved safety goggles.		Treat like acid burns. Flush eyes for at least 15 minutes.
Sodium Dichromate	Approved dust mask, protective gloves, goggles or face shield.		Treat like acid burns.
Sodium Hydride	Face shield, rubber gloves.		Brush off all particles at once and flood the affected area with water.
Sodium Hydrosulfide Solution	Rubber protective equipment such as apron, boots, splash-proof goggles, and gloves. Canister-type respirator or self-contained breathing apparatus.	Move to fresh air and call physician immediately. Give mouth-to-mouth resuscitation if breathing has ceased.	Immediately flush affected areas with water. Obtain medical attention as if irritation persists.
			Immediately flush with large quantities of water for at least 15 minutes and obtain medical attention as soon as possible. While awaiting instructions from

Table 6.1 (cont.) Recommended emergency response to acute exposures and direct contact

Chemical Name	PPE	Emergency Treatment Actions for Exposure by		
		Inhalation	Skin	Eyes
				physician, patient may be kept in a dark room and ice compresses applied to the eyes and forehead.
Sodium Hydroxide	Chemical safety goggles, lace shield, tiller or dust-type respirator, rubber boots, rubber gloves.	Remove from exposure, support respiration, call physician.	Wash immediately with large quantities of water under emergency safety shower while removing clothing. Continue washing until medical help arrives. Call a physician.	Irrigate immediately with copious amounts of water for at least 15 minutes and call a physician.
Sodium Hypochlorite	Rubber gloves, goggles.		Wash off contacted area.	Flush with plenty of water for at least 15 minutes and consult a physician.

Chemical Safety Data

Sodium Methylate	Self-contained breathing apparatus, rubber gloves and apron, goggles or face shield.	Remove victim from contamination and keep him quiet and warm. Rest is essential. Hot tea or coffee may be given as a stimulant if the patient is conscious, if breathing has ceased give artificial respiration, if available, oxygen should be administered by experienced personnel.	Wash well with water, then with dilute with vinegar.	Wash well with water then with 3% boric acid solution and additional water.
Sodium Oxalate	Dust mask, goggles or face shield, rubber gloves.	Move to fresh air. If exposure to dust is severe get medical attention.	Flush with water.	Flush with water.

Table 6.1 (cont.) Recommended emergency response to acute exposures and direct contact

Chemical Name	PPE	Emergency Treatment Actions for Exposure by		
		Inhalation	Skin	Eyes
Sodium Phosphate	U.S. Bu. Mines toxic dust mask, protective gloves, chemical-type goggles, full-cover clothing.	Give large amounts of water or warm salty water to induce vomiting. Repeat until comitus is clear. Milk, eggs, or olive oil may then be given to soothe stomach.	Flush with water and avoid chemical neutralizers.	Immediately flush with large amounts of water for at least 15 minutes, holding eyelids to ensure flushing of entire surface. Avoid chemical neutralizers.
Sodium Silicate	Goggles or face shield.			
Sodium Silicofluoride	Dust respirator, goggles or face shield, protective gloves.	Move to fresh air. If exposure to dust is severe get medical attention.	Flush with water and wash well with soap and water.	Flush with water for at least 15 minutes.
Sodium Sulfide	Goggles or face shield.		Wash with water for at least 15 minutes.	Wash with water for at least 15 minutes.
Sodium Sulfite	Dust mask, goggles or face shield.			

Chemical Safety Data

Sodium Thiocyanate	Rubber or plastic gloves, goggles, rubber or plastic apron.	Move to fresh air. If exposure is severe consult physician. Hemodialysis is recommended as the treatment of choice.	Flush with water for at least 15 minutes.	Flush with water for at least 15 minutes.
Sorbitol	Goggles or face shield, protective clothing for hot liquid.			
Stearic Acid	For prolonged expose to vapors use air-supplied mask of chemical cartridge respirator, impervious gloves, goggles, impervious apron.		Wash thoroughly with soap and water.	Flush with water. If irritation persists consult a physician.
Styrene	Air-supplied mask or approved canister, rubber or plastic gloves, boots, goggles or face shield.	Remove to fresh air; keep warm and quiet, use artificial respiration if needed.	Flush with plenty of water.	Flush with plenty of water. Get medical attention.
Sucrose	Dust mask and goggles or face shield.			Flush with water.
Sulfolane	Goggles or face shield, rubber gloves.		Flush with water.	Flush with water.

Table 6.1 (cont.) Recommended emergency response to acute exposures and direct contact

Chemical Name	PPE	Emergency Treatment Actions for Exposure by		
		Inhalation	Skin	Eyes
Sulfur Dioxide	Air-supplied mask or approved canister, rubber or plastic gloves, boots, goggles or face shield, rubber clothing where contact with liquid is possible.	Remove from exposure, support respiration, call physician.	Flush with water.	Flush with water for at least 15 minutes and call a physician.
Sulfuric Acid	Safety shower, eyewash fountain, safety goggles, face shield, approved respirator, rubber safety shoes, rubber apron.	Observe victim for delayed pulmonary reaction.	Wash with large amounts of water.	Wash with large amounts of water. Do not use oils or ointments.
Sulfuric Acid, Spent	Chemical safety goggles and face shield, rubber gloves, boots, and apron.		Flush affected parts with large amounts of water for at least 15 minutes.	Flush affected parts with large amounts of water for at least 15 minutes.

Chemical Safety Data

Titanium Tetrachloride	Goggles and face shield, air-supplied mask or approved canister, rubber gloves, protective clothing.	Remove victim to fresh air and if symptoms persist call a doctor.	Flush with water. Obtain medical attention if irritation persists.	Immediately flush with copious amounts of water for at least 15 minutes and call a doctor.
Toluene	Air-supplied mask, goggles and face shield, plastic gloves.	Remove victim to fresh air and give artificial respiration and oxygen if needed. Call a doctor.	Wipe off and wash with soap and water.	Flush with water for at least 15 minutes.
Toluene 2, 4-Diisocyante	Organic vapor canister goggles and face shield, rubber gloves, boots, and apron.	Remove victim to fresh air and give artificial respiration and oxygen if needed. Call a doctor.		Flush with water, Wipe off with rubbing alcohol, wash with soap and water.
P-Toluene-sulfonic Acid	Chemical goggles and face shield, rubber gloves.		Wash thoroughly with large amounts of water for at least 15 minutes.	Wash thoroughly with copious amounts of water for at least 15 minutes.

Table 6.1 (cont.) Recommended emergency response to acute exposures and direct contact

Chemical Name	PPE	Emergency Treatment Actions for Exposure by		
		Inhalation	Skin	Eyes
O-Toluidine	Chemical safety goggles, face shield, Bu.	Move to fresh air.	Remove all contaminated clothing; wash affected areas immediately and thoroughly with plenty of warm water and soap.	Flush with copious amounts of water.
Toxamphene	Chemical-type respirator, rubber gloves, chemical goggles and face shield.			
Trichloro-ethylene	Organic vapor-acid gas canister, self-contained breathing apparatus for emergencies, neoprene or vinyl gloves, chemical safety goggles, face-shield, neoprene safety shoes.	Move victim to fresh air and if necessary apply artificial respiration and/or administer oxygen.	Wash thoroughly with soap and water.	Flush thoroughly with water.

Chemical Safety Data

Trichlorofluoromethane	Air-line respirator, rubber gloves, monogoggles.	Move victim to fresh air and if necessary apply artificial respiration and/or administer oxygen. Call physician immediately.	If frostbite has occurred, flush areas with warm water.	Flush with water for at least 15 minutes and get medical attention.
Trichlorosilane	Remove victim from exposure.	Move victim to fresh air and if necessary apply artificial respiration and/or administer oxygen. Call physician immediately.	Flush with water for at least 15 minutes and get medical attention.	Flush with water for at least 15 minutes and get medical attention.
Tridecanol	Synthetic rubber gloves, chemical goggles.		Wash exposed area with soap and water.	Promptly flush with clean water for at least 15 minutes and see a physician.
1-Tridecene	Goggles or face shield.			Flush with water for at least 15 minutes.

Table 6.1 (cont.) Recommended emergency response to acute exposures and direct contact

Chemical Name	PPE	Emergency Treatment Actions for Exposure by		
		Inhalation	Skin	Eyes
Triethyl-aluminum	Full protection clothing preferably of aluminized glass cloth, goggles, face shield, gloves, in case of fire all purpose canister or self-contained breathing apparatus.	Only fumes from fire need to be considered, metal fume fever is not critical lasting less than 36 hours.	Wash with water. Treat burns if fire occurred and gets medical attention.	Flush with copious amounts of water with lids held open and treat burns if fire occurred and get medical attention.
Triethylamine	Air-supplied mask, goggles or face shield, rubber gloves.	Move victim to fresh air and if necessary apply artificial respiration and/or administer oxygen. Call physician immediately.	Flush with water for at least 30 minutes.	Flush with water for at least 30 minutes.
Thiethylene Glycol	Goggles, plastic gloves.			
Tripropylene Glycol	Plastic gloves, safety glasses or face shield.		Flush with water and get medical attention if ill effects develop.	Flush with water and get medical attention if ill effects develop.

Chemical Safety Data

Turpentine	Goggles or face shield, rubber gloves.	Move victim to fresh air and if necessary apply artificial respiration and/or administer oxygen. Call physician immediately.	Wipe off and wash with soap and water.	Flush with water for at least 15 minutes.
Undecanol	Goggles and face shield.			Flush with water for at least 15 minutes.
1-Undecene	Goggles or face shield, rubber gloves.	Remove victim to fresh air.	Wipe off and wash with soap and water.	Flush with water for at least 15 minutes.
N-Undecyl-benzene	Goggles or face shield, rubber gloves.	Remove victim to fresh air.	Remove spills on skin and clothing by washing with soap and water.	Flush with water.
Uranyl Acetate	Approved dust respirator, goggles or face shield, protective clothing.	Remove victim to fresh air.	Flush with water.	Flush with water for at least 15 minutes.
Uranyl Nitrate	Dust mask, gloves, goggles.	Remove victim to fresh air.	Wash thoroughly with soap and water.	Flush with water for at least 15 minutes.

Table 6.1 (cont.) Recommended emergency response to acute exposures and direct contact

Chemical Name	PPE	Emergency Treatment Actions for Exposure by		
		Inhalation	Skin	Eyes
Uranyl Sulfate	Approved dust respirator, goggles or face shield, protective clothing.		Flush with water.	Flush with water for at least 15 minutes.
Urea	Goggles or face shield, dust mask.			
Urea Peroxide	Rubber gloves, goggles.	Remove victim to fresh air.		Wash thoroughly for at least 15 minutes. Call a physician.
Vanadium Oxytrichloride	Acid vapor mask, rubber gloves, face shield, acid-resistant clothing.	Move victim to fresh air and if necessary apply artificial respiration.	Wipe exposed areas free of the chemical with a dry cloth then flush thoroughly with water.	Flush with water for at least 15 minutes.
Vanadium Pentoxide	Bu. Mines approved respirator, rubber gloves, and goggles for prolonged exposure.	Move victim to fresh air. If dust exposure has been severe, contact a physician.	Flush with water. Wash with soap and water.	Flush with water for at least 15 minutes.

Chemical Safety Data

Vanadyl Sulfate	Dust mask, goggles or face shield, protective gloves.	Move to fresh air if exposure to dust has been severe.	Flush with water.	Flush with water for at least 15 minutes.
Vinyl Acetate	Approved canister or air-supplied mask, goggles or face shield, rubber or plastic gloves.	Move victim to fresh air and if necessary apply artificial respiration.		Flush with water for at least 15 minutes.
Vinyl Chloride	Rubber gloves and shoes, gas-tight goggles, organic vapor canister or self-contained breathing apparatus.	Remove victim to fresh air and keep him warm and quiet. Call a doctor, and give artificial respiration if breathing stops.	Flush with water for at least 15 minutes.	Flush with water for at least 15 minutes and get medical attention.
Vinyl Fluoride Inhibited	Protective goggles, safety glasses, self-contained breathing apparatus.	Remove victim to fresh air.	If frostbite has occurred, flush areas with warm water and treat burn.	

Table 6.1 (cont.) Recommended emergency response to acute exposures and direct contact

Chemical Name	PPE	Emergency Treatment Actions for Exposure by		
		Inhalation	Skin	Eyes
Vinylidene Chloride Inhibited	Approved canister or air-supplied mask, goggles or face shield, rubber gloves and boots.	If any illness develops, remove person to fresh air and keep warm and quiet, and get medical attention. If breathing stops, start giving artificial respiration.	Flush with water for at least 15 minutes. Remove contaminated clothing and wash before re-use.	Flush with water for at least 15 minutes and get medical attention.
Vinyl Methyl Ether, Inhibited	Organic-vapor mask, plastic or rubber gloves, safety glasses.	Remove victim to fresh air. Administer oxygen if difficulty breathing and contact physician.	Wash with copious amounts of water. Treat frostbite by use of warm water or blankets.	Wash with copious amounts of water and consult a specialist.
Vinyltoluene	Air-supplied mask, goggles or face shield, plastic gloves.	Remove victim to fresh air. Give artificial respiration and administer oxygen if difficulty breathing and contact physician.	Wipe off, wash with soap and water.	Flush with water for at least 15 minutes.

CHEMICAL SAFETY DATA

Vinyltrichloro-silane	Acid-vapor-type respiratory protection, rubber gloves, chemical worker's goggles, other protective equipment necessary to protect skin and eyes.	Remove victim from exposure and give artificial respiration if required.	Flush with water.	Flush with water for at least 15 minutes.
Waxes: Carnauba	Goggles or face shield, protective gloves and clothing for hot liquid.		Remove solidified wax from skin and wash with soap and water. If burned call a doctor.	Wash with soap and water.
Waxes: Paraffin	Goggles or face shield, protective gloves and clothing for hot liquid.		Remove solidified wax from skin and wash with soap and water. If burned call a doctor.	Wash with soap and water.
M-Xylene	Approved canister or air-supplied mask, goggles and face shield, plastic gloves and boots.	Remove victim to fresh air; administer artificial respiration and oxygen if required. Call a doctor.	Wipe off and wash with soap and water.	Flush with water for at least 15 minutes.

Table 6.1 (cont.) Recommended emergency response to acute exposures and direct contact

Chemical Name	PPE	Emergency Treatment Actions for Exposure by		
		Inhalation	Skin	Eyes
Xylenol	Organic canister mask, goggles and face shield, rubber gloves, other protective clothing to prevent contact with skin.	Remove patient immediately to fresh air. Irritation of nose or throat may be somewhat relieved by spraying or gargling with water until all odor is gone. 100% oxygen inhalation is indicated for cyanosis or respiratory distress. Keep patient warm but not hot.	Wash affected areas with large quantities of water or soapy water until all odor is gone then wash with alcohol or 20% glycerin solution and more water. Keep patient warm but not hot. Cover chemical burns continuously with compresses wet with saturated solution of sodium thiosulphate. Apply no salves or ointments for 24 hours after injury.	Flood with running water for 15 minutes. If physician is not immediately available continue irrigation for another 15 minutes. Do not use oil or oily ointments unless instructed by a physician.

Chemical Safety Data

Zinc Acetate	MSA respirator, rubber gloves, chemical goggles.	Move to fresh air. If exposure is severe get medical attention.	Wash with soap and water	Flush with water for at least 15 minutes.
Zinc Ammonium Chloride	Dust mask, goggles and face shield, protective gloves.	Remove dust.		
Zinc Bromide	Chemical goggles and face shield, and dust mask.	Move to air.	Wash immediately with large amounts of water.	Wash immediately with large amounts of water.
Zinc Chloride	Goggles and face shield.			Wash with water for at least 15 minutes.
Zinc Chromate	Suitable respirator, rubber gloves, chemical goggles or face shield.	Move to fresh air. If exposure is severe get medical attention.	Wash with soap and water	Flush with water.
Zinc Fluoroborate	Rubber gloves, safety glasses or face shield.		Flush with plenty of water.	Flush with plenty of water.
Zinc Nitrate	Dust mask, goggles or face shield, protective gloves.	Move to fresh air.	Wash with soap and water	Flush with water and consult a physician.

Table 6.1 (cont.) Recommended emergency response to acute exposures and direct contact

Chemical Name	PPE	Emergency Treatment Actions for Exposure by		
		Inhalation	Skin	Eyes
Zinc Sulfate	Dust mask, goggles or face shield, protective gloves.	Move to fresh air.	Flush with water.	Flush with water.
Zirconium Acetate	Rubber gloves, chemical goggles or face shield.		Flush with water.	Flush with water for at least 15 minutes.
Zirconium Nitrate	Dust mask, goggles or face shield, protective gloves.	Move to fresh air.	Flush with water.	Flush with water.

7

Recommended Safe Levels of Exposure

Table 7.1 Recommended safe levels of exposure

Chemical Name	Health Risk Information			
	TLV, ppm	STEL, ppm	LD_{50}, g/kg	Odor Threshold, ppm
Acetaldehyde	100	50	0.5–5.0	0.21
Acetic Acid	10	40	0.5–5.0	1
Acetic Anhydride	5	No data found	0.5–5.0	0.14
Acetone	1000	1000	5–15	100
Acetone Cyanohydrin	No data	No data found	<50 mg/kg (rats)	No data found
Acetonitrile	40	40	500 mg/kg (guinea pig)	40
Acetyl Bromide	No data	No data found	3,310 mg/kg	5.0×10^{-4}
Acetyl Chloride	No data	No data found	1470 mg/kg	–1
Acetyl Peroxide	No data	No data found	No data	No data found
Acridine	No data	No data found	2000 mg/kg	No data found
Acrolein	0.1	0.5–5	<50 mg/kg	0.21
Acrylamide	0.3 mg/m³	Data not available	170 mg/kg	Not pertinent
Acrylic Acid	Data not available	Data not available	0.5–5	Data not available

Table 7.1 (cont.) Recommended safe levels of exposure

Chemical Name	Health Risk Information			
	TLV, ppm	STEL, ppm	LD_{50}, g/kg	Odor Threshold, ppm
Acrylonitrile	20	40	50–5000 mg/kg	21.4
Aldrin	0.25 mg/m³	1 mg/m³	50–500 mg/kg	Data not available
Allyl Alcohol	2	5	50–500 mg/kg	0.78
Allyl Chloroformate	Data not available	Data not available	50–500 mg/kg	Data not available
Allyltrlorosilane	Data not available	Data not available	50–500 mg/kg	Data not available
Aluminum Chloride	5	5	10	1–5
Aluminum Nitrate	Data not available	Data not available	264 mg/kg	Odorless
Ammonium Bifluoride	2.5 mg/m³	Data not available	50 mg/kg	Data not available
Amonium Carbonate	Data not available	Data not available	Data not available	<1.5
Ammonium Dichromate	0.1 mg/m³	Not pertinent	Not pertinent	Not pertinent
Ammonium Fluoride	2.5 mg/m³	Data not available	Data not available	Data not available
Ammonium Hydroxide	1	100	350 mg/kg	50
Ammonium Lactate	Data not available	Data not available	Data not available	Data not available
Ammonium Nitrate	Not pertinent	Not pertinent	Data not available	Not pertinent
Ammonium Nitrate-Sulfate Mixture	Data not available	Data not available	58 mg/kg	Data not available

Recommended Safe Levels of Exposure

Table 7.1 (cont.) Recommended safe levels of exposure

Chemical Name	Health Risk Information			
	TLV, ppm	STEL, ppm	LD_{50}, g/kg	Odor Threshold, ppm
Ammonium Oxalate	Data not available	Data not available	Data not available	Data not available
Ammonium Perchlorate	Not pertinent	Not pertinent	3500 mg/kg	Not pertinent
Ammonium Perchlorate	Data not available	Data not available	820 mg/kg	Data not available
Ammonium Silicofluoride	2.5 mg/m³	Data not available	100 mg/kg	Data not available
Amyl Acetate	100	200	6.5	0.067
Aniline	5	50	5–500	0.5
Anisoyl Chloride	Data not available	Data not available	Data not available	Data not available
Anthracene	Data not available	Data not available	Data not available	Data not available
Antimony Pentachloride	0.5 mg/m³	Data not available	1,115 mg/kg	Data not available
Antimony Pentafluoride	0.5 mg/m³	Data not available	Data not available	Data not available
Antimony Potassium Tartrate	0.5 mg/m³	Data not available	1.5 mg/kg	Odorless
Antimony Trichloride	0.5 mg/m³	Data not available	675 mg/kg	Data not available
Anitimony Trifluoride	0.5 mg/m³	Not pertinent	50–500	Not pertinent
Anitmony Trioxide	0.5 mg/m³	Data not available	20,000 mg/kg	Data not available
Arsenic Acid	0.5 mg/m³	Data not available	48 mg/kg	Odorless
Arsenic Disulfide	0.5 mg/m³	Data not available	<50 mg/kg	Odorless

Table 7.1 (cont.) Recommended safe levels of exposure

Chemical Name	Health Risk Information			
	TLV, ppm	STEL, ppm	LD_{50}, g/kg	Odor Threshold, ppm
Arsenic Trichloride	0.5 mg/m³	Data not available	138 mg/kg	Data not available
Arsenic Trioxide	0.5 mg/m³	Data not available	45 mg/kg	Odorless
Arsenic Trisulfide	0.5 mg/m³	Data not available	<50 mg/kg	Odorless
Asphalf	5 mg/m³	Data not available	5–15	Data not available
Atrazine	Data not available	Data not available	3080 mg/kg	Data not available
Azinphosmethyl	0.2 mg/m³	Data not available	11 ~ 18.5 mg/kg	Data not available
Barium Carbonate	Not pertinent	Not pertinent	0.5–5	
Barium Chlorate	0.5 mg/m³	Data not available	Data not available	Not pertinent
Barium Nitrate	0.5 mg/m³	Data not available	355 mg/kg	Not pertinent
Barium Perchlorate	0.5 mg/m³	Data not available	Data not available	
Barium Permanganate	0.5 mg/m³	Data not available	Data not available	Not pertinent
Barium Peroxide	0.5 mg/m³	Data not available	Data not available	Not pertinent
Benzaldehyde	Data not available	Data not available	0.5–5	0.042
Benzene	25	75	50 ~ 500 mg/kg	4 ~ 7
Benzene Hexachloride	0.5 mg/m³	1 mg/m³	0.5–5	Data not available

Recommended Safe Levels of Exposure

Table 7.1 (cont.) Recommended safe levels of exposure

Chemical Name	Health Risk Information			
	TLV, ppm	STEL, ppm	LD_{50}, g/kg	Odor Threshold, ppm
Benzene Phosphorous Dichloride	Data not available	Data not available	Data not available	Data not available
Benzene Phosphorous Thiodichloride	Data not available	Data not available	Data not available	Data not available
Benzoic Acid	Not pertinent	Not pertinent	0.5–5	Not pertinent
Benzonitrile	Data not available	Data not available	800 mg/kg	Data not available
Benzophenone	Data not available	Data not available	≥10,000 mg/kg	Data not available
Benzoyl Chloride	Data not available	Data not available	Data not available	Data not available
Benzyl Alcohol	Data not available	Data not available	1230 mg/kg	5.5
Benzylamine	Data not available	Data not available	Data not available	Data not available
Benzyl Bromide	Data not available	Data not available	Data not available	Data not available
Benzyl Chloride	1	Data not available	1231 mg/kg	0.047
Beryllium Fluoride	0.002 mg/m³	0.025 mg/m³	100 mg/kg	Data not available
Beryllium Metallic	0.002 mg/m³	0.025 mg/m³	50 ~ 500 mg/kg	Data not available
Beryllium Nitrate	0.02 mg/m³	0.025 mg/m³	Data not available	Data not available
Beryllium Oxide	0.002 mg/m³	0.025 mg/m³	Data not available	Data not available
Beryllium Sulfate	0.002 mg/m³	0.025 mg/m³	82 mg/kg	Data not available

Table 7.1 (cont.) Recommended safe levels of exposure

Chemical Name	Health Risk Information			
	TLV, ppm	STEL, ppm	LD_{50}, g/kg	Odor Threshold, ppm
Bismuth Oxychloride	Data not available	Data not available	>21.6	Data not available
Bisphenol A	Not pertinent	Not pertinent	0.5–5	Not pertinent
Boric Acid	10 mg/m³	Data not available	2.660 mg/kg	Odorless
Boron Trichloride	Data not available	Data not available	0.5–5	Decomposes in moist air releasing hydrochloric acid and decomposition products
Bromine	0.1	0.4	Not pertinent	3.5
Bromine Trifluoride	0.1	50	Data not available	Data not available
Bromobenzene	Data not available	Data not available	Data not available	Data not available
Butadiane, Inhibited	1000	Data not available	Data not available	4 mg/m³
Butane	500	Data not available	Not pertinent	6.16
N-Butyl Acetate	150–200	300 for 30 minutes	0.5–5	10
Sec-Butyl Acetate	200	Data not available	Data not available	Data not available
Iso-Butyl Acrylate	Data not available	Data not available	Data not available	Data not available
N-Butyl Acrylate	Data not available	LD_{50} 100	0.5–5	Data not available
N-Butyl Alcohol	100	150	0.5–5	2.5

Recommended Safe Levels of Exposure

Table 7.1 (cont.) Recommended safe levels of exposure

Chemical Name	Health Risk Information			
	TLV, ppm	STEL, ppm	LD_{50}, g/kg	Odor Threshold, ppm
Sec-Butyl Alcohol	150	200	5–15	Data not available
Tert-Butyl Alcohol	100	150	0.5–5	Data not available
N-Butylamine	5	5	500 mg/kg	Data not available
Sec-Butylamine	Data not available	Data not available	380 mg/kg	Data not available
Tert-Butylamine	Data not available	Data not available	180 mg/kg	Data not available
Butylene	Data not available	Data not available	Not pertinent	Data not available
Butylene Oxide	Data not available	Data not available	1,410 mg/kg	Data not available
N-Butyl Mercaptan	0.5	Data not available	1,500 mg/kg	0.001
N-Butyl Methacrylate	Data not available	Data not available	>15	Data not available
1,4-Butynediol	Not pertinent	Not pertinent	50–500 mg/kg	Not pertinent
Iso-Butyraldehyde	Data not available	Data not available	0.5–5	0.047
N-Butyraldehyde				
N-Butyric Acid	Data not available	Data not available	2,940 mg/kg	0.001
Cacodylic Acid	Data not available	Data not available	700 mg/kg	Not pertinent
Cadmium Acetate	0.2 mg/m^3	Data not available	250 mg/kg	Not pertinent
Cadmium Nitrate	0.2 mg/m^3	Data not available	100 mg/kg	Data not available

Table 7.1 (cont.) Recommended safe levels of exposure

Chemical Name	Health Risk Information			
	TLV, ppm	STEL, ppm	LD_{50}, g/kg	Odor Threshold, ppm
Cadmium Oxide	0.1 mg/m³	0.1 mg/m³	72 mg/kg	Data not available
Cadmium Sulfate	2 mg/m³	Data not available		Data not available
Calcium Arsenate	1 mg/m³	Data not available	20 mg/kg	Data not available
Calcium Carbide	Not pertinent	Not pertinent	Data not available	Not pertinent
Calcium Chlorate	Data not available	Data not available	4,500 mg/kg	Not pertinent
Calcium Chloride		Data not available	1,000 mg/kg	Data not available
Calcium Chromate	0.1 mg/m³	Data not available	50–50 mg/kg	Data not available
Calcium Fluoride	Not pertinent	Not pertinent		Not pertinent
Calcium Hydroxide	Not pertinent	Not pertinent	5–15	Not pertinent
Calcium Hypochlorite	Not pertinent	Not pertinent	>15	Not pertinent
Calcium Nitrate	Data not available	Data not available	Data not available	Not pertinent
Calcium Oxide	5 mg/m³	10 mg/m³	Data not available	Not pertinent
Calcium Peroxide	Data not available	Data not available	Data not available	Not pertinent
Calcium Phosphide	Data not available	Data not available	Data not available	1–100 mg/m³
Camphene	Data not available	Data not available	Data not available	Data not available
Carbolic Oil	5	Data not available	0.5–5	0.05

Recommended Safe Levels of Exposure

Table 7.1 (cont.) Recommended safe levels of exposure

Chemical Name	Health Risk Information			
	TLV, ppm	STEL, ppm	LD_{50}, g/kg	Odor Threshold, ppm
Carbon Dioxide	5000	30,000	Not pertinent (gas with low boiling point)	Not pertinent
Carbon Monoxide	50	400	Not pertinent	Not pertinent
Carbon Tetrachloride	10	25	0.5–5	>10
Caustic Potash Solution	Not pertinent		365 mg/kg	Not pertinent
Caustic Soda Solution	Not pertinent	Not pertinent	500 mg/kg	Not pertinent
Chlordane	0.5 mg/m³	2 mg/m³	238 mg/kg	No data
Chlorine	1	3	Not pertinent	3.5
Chlorine Triflouride	0.1	0.1	<50 mg/kg	Data not available
Chloroacetophenone	0.05	Data not available	52 mg/kg	0.016
Chloroform	10	50	0.5–5	205–307
Chromic Anhydride	Not pertinent	Not pertinent	50–500 mg/kg	Not pertinent
Chromyl Chloride	Data not available	Data not available	<50 mg/kg	No data
Citric Acid	Data not available	Data not available	11.7	Data not available
Cobalt Acetate	0.1 mg/m³	Data not available	50–500 mg/kg	Data not available
Cobalt Chloride	0.1 mg/m³	Data not available	50–500 mg/kg	Data not available

Table 7.1 (cont.) Recommended safe levels of exposure

Chemical Name	Health Risk Information			
	TLV, ppm	STEL, ppm	LD_{50}, g/kg	Odor Threshold, ppm
Cobalt Nitrate	Data not available	Data not available	~400 mg/kg	Not pertinent
Copper Acetate	Data not available	Data not available	0.5–5	Data not available
Copper Arsenite	0.5 mg/m³	Data not available	50–500 mg/kg	Not pertinent
Copper Bromide	Data not available	Data not available	50–500 mg/kg	Data not available
Copper Chloride	Data not available	Data not available	50–500 mg/kg	Data not available
Copper Cyanide	5 mg/m³	Data not available	<50 mg/kg	Data not available
Copper Fluoroborate	Data not available	Data not available	50–500 mg/kg	Data not available
Copper Iodide	Data not available	Data not available	50–500 mg/kg	Data not available
Copper Naphthenate	500	Data not available	4–6	Data not available
Copper Nitrate	Data not available	Data not available	0.5–5	Data not available
Copper Oxalate	Data not available	Data not available	Data not available	Data not available
Copper Sulfate	Not pertinent	Not pertinent	50–500 mg/kg	Not pertinent
Creosote, Coal Tar	Data not available	Data not available	0.5–5	Data not available
Cresols	5	Data not available	0.5–5	5
Cumene	50	Data not available	50–500 mg/kg	1.2
Cyanogen	10	5 mg/m³	Data not available	Data not available

Table 7.1 (cont.) Recommended safe levels of exposure

Chemical Name	Health Risk Information			
	TLV, ppm	STEL, ppm	LD$_{50}$, g/kg	Odor Threshold, ppm
Cyanogen Bromide	0.5	Data not available	Data not available	Data not available
Cyclohexane	300	300	0.5–5	Data not available
Cyclohexanol	50	Data not available	0.5–5	Data not available
Cyclohexanone	50	Data not available	50–5	0.12
Cyclopentane	Data not available	300	0.5–5	Data not available
P-Cumene	Data not available	Data not available	4,750 mg/kg	Data not available
DDD	Data not available	Data not available	1.2 (mouse), 3.4 (rat)	Data not available
DDT	Not pertinent	Not pertinent	50–500 mg/kg	Not pertinent
Decaborane	0.05	Data not available	40 mg/kg	0.05
Decahydro-naphthalene	25	Data not available	4,170 mg/kg	Data not available
Decaldehyde	No data	No data	>33.3	0.168
1-Decene	Data not available	Data not available	Data not available	Data not available
N-Decyl Alcohol	Not pertinent	Not pertinent	5–15	Data not available
N-Decylbenzene	Data not available	Data not available	Data not available	Data not available
2, 4-D Esters	Data not available	Data not available	Data not available	Data not available
Dextrose Solution	Not pertinent	Not pertinent	Not pertinent	Not pertinent

Table 7.1 (cont.) Recommended safe levels of exposure

Chemical Name	Health Risk Information			
	TLV, ppm	STEL, ppm	LD_{50}, g/kg	Odor Threshold, ppm
Diacetone Alcohol	50	150	0.5–5	Data not available
Di-N-Amyl-Phthalate	Data not available	Data not available	Data not available	Data not available
Diaznon	Data not available	Not pertinent	76 mg/kg	Data not available
Dibenzoyl Peroxide	5 mg/m³	Not pertinent	0.5–5	Not pertinent
Di-N-Butylamine	Data not available	Data not available	360 mg/kg	Data not available
Di-N-Butyl Ether	Data not available	Data not available	7400 mg/kg	Data not available
Di-N-Butyl Ketone	Data not available	Data not available	Data not available	Data not available
Dibutylphenol	Data not available	Not pertinent	(2,6-Di-sec-butyl phenol)	Data not available
Dibutyl Phthalate	5 mg/m³	Not pertinent	5–15	Data not available
O-Dichlorobenzene	50	50	0.5–5	4.0, 50
P-Dichlorobenzene	75	50	0.5–5	15–30
Di-(P-Chlorobenzoyl) Peroxide	Data not available	Not pertinent	Data not available	Not pertinent
Dichlorobutene	Data not available	Data not available	(1,4-dichloro-2-butene) 89 mg/kg	Data not available
Dichlorodifluoro-methane	1000	5000	Not pertinent	Data not available

Recommended Safe Levels of Exposure

Table 7.1 (cont.) Recommended safe levels of exposure

Chemical Name	Health Risk Information			
	TLV, ppm	STEL, ppm	LD_{50}, g/kg	Odor Threshold, ppm
1,2-Dichloroethylene	200	Data not available	770 mg/kg	Data not available
Dichloroethyl Ether	5	35	75 mg/kg	Data not available
Dichloromethane	500	100	0.5–5	205–307
2,4-Dichlorophenol	Not pertinent	Data not available	0.5–5	Data not available
2,4-Dichlorophenoxyacetic Acid	Data not available	Data not available	375 mg/kg (rat); 80 mg/kg (human)	Not pertinent
Dichloropropane	75	Data not available	0.5–5	Data not available
Dichloropropene	Data not available	Data not available	50–500 mg/kg	Data not available
Dicyclopentadiene	75–100	Data not available	0.82	<0.003
Dieldrin	0.25 mg/m^3	1 mg/m^3	46 mg/kg (rat); 65 mg/kg (dog)	0.041
Diethanolamine	Not pertinent	Not pertinent	0.5–5	Data not available
Diethylamine	25	100	0.5–5	0.14
Diethylbenzene	Data not available	Data not available	1.2	Data not available
Diethyl Carbonate	Data not available	Data not available	Data not available	Data not available
Diethylene Glycol	100	Not pertinent	>15	Not pertinent
Diethylene Glycol Dimethyl Ether	Not pertinent	Data not available	Data not available	Data not available

Table 7.1 (cont.) Recommended safe levels of exposure

Chemical Name	Health Risk Information			
	TLV, ppm	STEL, ppm	LD$_{50}$, g/kg	Odor Threshold, ppm
Diethyleneglycol Monobutyl Ether	Data not available	Data not available	2	Data not available
Diethyleneglycol Monobutyl Ether Acetate	Data not available	Because of high boiling point (246°C), hazards from inhalation are minimal	2.34	Data not available
Diethylene Glycol Monoethyl Ether	Not pertinent	Data not available	0.5–5	Data not available
Diethylenetriamine	1	Data not available	0.5–5	10
Di(2Ethylhexyl) Phosphoritc Acid	Data not available	Data not available	0.5–5	No data
Diethyl Phthalate	Data not available	Data not available	1,000 mg/kg	Data not available
Diethylzinc	Not pertinent	Not pertinent	Not pertinent	Not pertinent
1,1 Difluoroethane	Data not available	Data not available	Not pertinent (boils at –24.7°C)	Data not available
Difluorophosphoric Acid, Anhydrous	Data not available	Data not available	Data not available	Data not available
Diheptyl Phthalate	Data not available	Data not available	Data not available	Data not available
Diisobutylcarbinol	Not pertinent	Not pertinent	0.5–5	Data not available

Recommended Safe Levels of Exposure

Table 7.1 (cont.) Recommended safe levels of exposure

Chemical Name	Health Risk Information			
	TLV, ppm	STEL, ppm	LD_{50}, g/kg	Odor Threshold, ppm
Diisobutylene	Data not available	Data not available	Data not available	Data not available
Diisobutyl Ketone	25	50	1.4 (mouse), 5.75 (rat)	Data not available
Diisodecyl Phthalate	Data not available	Data not available	Data not available	Data not available
Diisopropanolamine	Not pertinent	Not pertinent	0.5–5	Data not available
Diisopropylamine	5	5,000	0.7	Data not available
Diisopropylbenzene Hydroperoxide	Data not available	Data not available	Data not available	Data not available
Dimethylacetamide	10	Data not available	5.63	46.8
Dimethylamine	10	20	Not pertinent	0.047
Dimethyl Ether	Data not available	Data not available	Not pertinent	Data not available
Dimethyl Sulfate	1	Data not available	50–500 mg/kg	Data not available
Distillates: Flashed Feed Stocks	No single TLV available	500	0.5–5	0.25
Dodecyltrichlorosilane	Data not available	Data not available	50–500 mg/kg	Data not available
Ethane	Not pertinent	Not pertinent	Not pertinent	899
Ethyl Acetate	400	1000	0.5–5	1
Ethyl Alcohol	1000	5000	5–15	10

Table 7.1 (cont.) Recommended safe levels of exposure

Chemical Name	Health Risk Information			
	TLV, ppm	STEL, ppm	LD_{50}, g/kg	Odor Threshold, ppm
Ethyl Butanol	Data not available	Data not available	0.5–5	Data not available
Ethyl Chloroformate	Data not available	Data not available	<50 mg/kg	Data not available
Ethylene	Simple asphyxiate	Not pertinent	Not pertinent	Data not available
Ethylene Dichloride	50	200	0.5–5	100
Ethyl Lactate	Data not available	Data not available	2,580 mg/kg	Data not available
Ethyl Nitrate	Data not available	Data not available	Data not available	Data not available
Ferric Ammonium Citrate	1 mg/m^3 (as iron)	Data not available	Data not available	Data not available
Ferric Nitrate	1 mg/m^3 (as iron)	Data not available	0.5–5	Data not available
Ferric Sulfate	1 mg/m^3 (as iron)	Data not available	Data not available	Data not available
Fluorine	1	0.5	Not pertinent	0.035
Formaldehyde Solution	2	5ppm for 5 min.; 3ppm for 60 min.	0.5–5	0.8
Gallic Acid	Data not available	Data not available	0.5–5	Data not available
Gasolines: Automotive	No single TLV applies	500	0.5–5	0.25
Gasolines: Aviation	No single TLV applies	500	0.5–5	0.25

Table 7.1 (cont.) Recommended safe levels of exposure

Chemical Name	Health Risk Information			
	TLV, ppm	STEL, ppm	LD_{50}, g/kg	Odor Threshold, ppm
Glycerine	Not pertinent	Data not available	>15	Not pertinent
Heptanol	Data not available	Data not available	1.87	0.49
N-Hexaldehyde	Data not available	Data not available	4,890 mg/kg	Data not available
Hydrochloric Acid	5	5	Data not available	1–5
Hydrogen Chloride	5	5	Data not available	1–5
Hydrogen, Liquified	Gas is non-poisonous but can act as a simple asphyxiate	Not pertinent	Not pertinent (gas with low boiling point)	Not pertinent
Isobutane	Data not available	Data not available	Not pertinent	Data not available
Isobutyl Alcohol	100	200	0.5–5	Data not available
Isodecaldehyde	Data not available	Data not available	Data not available	Data not available
Isohexane	Data not available	Data not available	Data not available	Data not available
Isopropyl Alcohol	400	400	5–15	90 mg/m³
Isopropyl Mercaptan	Data not available	Data not available	1,790 mg/kg	0.25
Kerosene	200	2500 mg/m³	5–15	Data not available
Lactic Acid	Data not available	Data not available	1810 mg/kg	4.0×10^{-7}

Table 7.1 (cont.) Recommended safe levels of exposure

Chemical Name	Health Risk Information			
	TLV, ppm	STEL, ppm	LD_{50}, g/kg	Odor Threshold, ppm
Lead Arsenate	(Dust) 0.15 mg/m^3	Not pertinent	<50 mg/kg	Not pertinent
Lead Iodide	0.2 mg/m^3	Data not available	0.5–5	Data not available
Linear Alcohols	Not pertinent	Not pertinent	5–15	Data not available
Liquified Natural Gas	Data not available	Data not available	Not pertinent	Data not available
Liquified Petroleum Gas	1000	Data not available	Not pertinent	5000–20,000
Lithium, Metallic	Data not available	Data not available	Data not available	Data not available
Magnesium	Data not available	Not pertinent	LDL (lowest lethal dose) 230 mg/kg	Not pertinent
Mercuric Acetate	0.05 mg/m^3 (as mercury)	No data	76 mg/kg	Not pertinent
Mercuric Cyanide	0.05 mg/m^3 (as mercury)	Data not available	25 mg/kg	Odorless
Methane	Not pertinent (methane is an asphyxiant, and limiting factor is available oxygen)	Data not available	Data not available	200

Table 7.1 (cont.) Recommended safe levels of exposure

Chemical Name	Health Risk Information			
	TLV, ppm	STEL, ppm	LD_{50}, g/kg	Odor Threshold, ppm
Methyl Alcohol	200	260 mg/m³	5–15	100
Methyl Chloride	100	100	Not pertinent	Data not available
Methyl Isobutyl Carbinol	25	Data not available	0.5 to 5	Data not available
Alpha-Methylstyrene	100	100	0.5–5	<10
Mineral Spirits	200	4000–7000	0.5–5	Data not available
Nitrous Oxide	Data not available	56	0.5 > 15	Data not available
Nonanol	Data not available	Data not available	0.5–5	Data not available
Octane	440	500	Data not available	4
Oils: Clarified	No single TLV applicable	Data not available	50–15	Data not available
Oils: Crude	Data not available	Data not available	Data not available	Data not available
Oils: Diesel	No single TLV applicable	Data not available	5–15	Data not available
Oils, Edible: Castor	None	Not pertinent	5–15	None
Oils, Edible: Coconut	Data not available	Data not available	Data not available	Not pertinent
Oils, Edible: Cottonseed	None	Not pertinent	None	Not pertinent
Oils, Edible: Fish	Not pertinent	Not pertinent	None	Data not available

Table 7.1 (cont.) Recommended safe levels of exposure

Chemical Name	Health Risk Information			
	TLV, ppm	STEL, ppm	LD_{50}, g/kg	Odor Threshold, ppm
Oils, Edible: Lard	Not pertinent	Not pertinent	Not pertinent	Not pertinent
Oils, Fuel: 2	No single value applicable	Data not available	5–15	Data not available
Oils, Fuel: 4	Not pertinent	Not pertinent	5–15	Data not available
Oils, Miscellaneous: Coal Tar	Data not available	Data not available	Data not available	Data not available
Oils, Miscellaneous: Motor	Data not available	Data not available	5–15	Data not available
Oils, Miscellaneous: Penetrating	Data not available	Data not available	5–15	Data not available
Oils, Miscellaneous: Resin	Data not available	Data not available	Data not available	Data not available
Oils, Miscellaneous: Spray	200	Data not available	0.5–5	Data not available
Oils, Miscellaneous: Tanner's	Data not available	Data not available	Data not available	Data not available
Oleic Acid	Data not available	Data not available	>15	Data not available
Oleum	1 mg/m³	5 mg/m³		1 mg/m³
Oxalic Acid	Not pertinent	Not pertinent	50–500 mg/kg	Not pertinent
Oxygen, Liquified	Not pertinent	Not pertinent	Not pertinent	Not pertinent

Recommended Safe Levels of Exposure

Table 7.1 (cont.) Recommended safe levels of exposure

Chemical Name	Health Risk Information			
	TLV, ppm	STEL, ppm	LD$_{50}$, g/kg	Odor Threshold, ppm
Paraformaldeyde	5	Data not available	50–500 mg/kg	Data not available
Pentaerythritol	Not pertinent	Not pertinent	>15	Data not available
Pentane	500	Data not available	Data not available	10
1-Pentene	Data not available	Data not available	Data not available	Data not available
Peracetic Acid	Data not available	Data not available	10 mg/kg	Data not available
Petrolatum	Not pertinent	Not pertinent	5–15	Not pertinent
Phenol	5	Data not available	0.5–5	0.05
Phenyldichloroarsine, Liquid	Data not available	Data not available	Data not available	Data not available
Phosgene	0.1	1	Data not available	0.5
Phosphoric Acid	1 mg/m³	Not pertinent	50–500 mg/kg	Not pertinent
Piperazine	Data not available	Data not available	0.5–5	Data not available
Polybutene	Data not available	Data not available	>15	Data not available
Polychlorinated Biphenyl	0.5–1 mg/m³	Data not available	3980 mg/kg	Data not available
Polypropylene	Data not available	Data not available	Data not available	Data not available
Potassium Cyanide	Not pertinent	Data not available	<50 mg/kg	Not pertinent
Potassium Iodide	Not pertinent	Not pertinent	364 mg/kg	Not pertinent

Table 7.1 (cont.) Recommended safe levels of exposure

Chemical Name	Health Risk Information			
	TLV, ppm	STEL, ppm	LD_{50}, g/kg	Odor Threshold, ppm
Propane	1000	Data not available	Data not available	5–20
Propionaldehyde	Data not available	Data not available	50 mg/kg	1
Propylene Oxide	100	Data not available	0.5–5	200
Pyridine	5	Data not available	0.5–5	0.021
Pyrogallic Acid	Data not available	Data not available	719 mg/kg	Odorless
Quinoline	Data not available	Data not available	460 mg/kg	71
Salicylic Acid	Data not available	Data not available	0.5–5	Data not available
Selenium Dioxide	0.2 mg/m³	0.3 mg/m³	Data not available	0.0002 mg/m³
Selenium Trioxide	0.2 mg/m³	0.3 mg/m³	Data not available	Data not available
Silicon Tetrachloride	Data not available	Data not available	<50 mg/kg	Data not available
Silver Acetate	0.01 mg/m³	Data not available	Data not available	Data not available
Silver Carbonate	0.01 mg/m³	Data not available	Data not available	Data not available
Silver Fluoride	0.01 mg/m³	Data not available	Data not available	Data not available
Silver Iodate	0.01 mg/m³	Data not available	Data not available	Data not available
Silver Nitrate	0.01 mg/m³	Data not available	0.5–5	Not pertinent
Silver Oxide	0.01 mg/m³	Data not available	0.5–5	Odorless

Recommended Safe Levels of Exposure

Table 7.1 (cont.) Recommended safe levels of exposure

Chemical Name	Health Risk Information			
	TLV, ppm	STEL, ppm	LD_{50}, g/kg	Odor Threshold, ppm
Silver Sulfate	0.01 mg/m³	Data not available	Data not available	Data not available
Sodium	Not pertinent	Not pertinent	Not pertinent	Not pertinent
Sodium Alkylbenzenesulfonates	Not pertinent	Not pertinent	0.5–5	Not pertinent
Sodium Alkyl Sulfates	Not pertinent	Not pertinent	0.5–5	Not pertinent
Sodium Amide	Not pertinent	Not pertinent	Not pertinent	Not pertinent
Soduim Arsenate	0.5 mg/m³	Data not available	<50 mg/kg	Data not available
Sodium Arsenite	0.5 mg/m³	Data not available	42 mg/kg	Not pertinent
Sodium Azide	Data not available	Data not available	27 mg/kg	Not pertinent
Sodium Bisulfite	Not pertinent	Not pertinent	0.5–5	Not pertinent
Sodium Borate	Data not available	Data not available	0.5–5	Data not available
Sodium Borohydride	Not pertinent	Data not available	Data not available	Not pertinent
Sodium Cacodylate	Data not available	Data not available	2600 mg/kg	Data not available
Sodium Chlorate	Not pertinent	Not pertinent	50–500 mg/kg	Not pertinent
Sodium Chromate	Data not available	Data not available	50–500 mg/kg	Data not available
Sodium Cyanide	Not pertinent	Not pertinent	Grade 4, <50 mg/kg	Not pertinent

Table 7.1 (cont.) Recommended safe levels of exposure

Chemical Name	Health Risk Information			
	TLV, ppm	STEL, ppm	LD_{50}, g/kg	Odor Threshold, ppm
Sodium Dichromate	Not pertinent	Data not available	50–500 mg/kg	Odorless
Sodium Hydride	Not pertinent	Not pertinent	None	Not pertinent
Sodium Hydrosulfide Solution	Data not available	Data not available	0.5–5	0.0047
Sodium Hydroxide	Not pertinent	Not pertinent	500 mg/kg	Not pertinent
Sodium Hypochlorite	Data not available	Data not available	Data not available	Not pertinent
Sodium Methylate	Data not available	Data not available	Data not available	Not pertinent
Sodium Oxalate	Data not available	Data not available	50–500 mg/kg	Data not available
Sodium Phosphate	Data not available	Data not available	Data not available	Data not available
Sodium Silicate	Not pertinent	Not pertinent	0.5–5	Not pertinent
Sodium Silicofluoride	2.5 mg/m^3	Data not available	50–500 mg/kg	Data not available
Sodium Sulfide	Not pertinent	Not pertinent	50–500 mg/kg	Data not available
Sodium Sulfite	Not pertinent	Not pertinent	0.5–5	Not pertinent
Sodium Thiocyanate	Data not available	Data not available	0.5–5	Data not available
Sorbitol	Not pertinent	Not pertinent	Data not available	Not pertinent
Stearic Acid	Data not available	Data not available	>15	20

Recommended Safe Levels of Exposure

Table 7.1 (cont.) Recommended safe levels of exposure

Chemical Name	Health Risk Information			
	TLV, ppm	STEL, ppm	LD_{50}, g/kg	Odor Threshold, ppm
Styrene	100	100	0.5–5	0.148
Sucrose	Data not available	Data not available	28,500 mg/kg	Not pertinent
Sulfolane	Not pertinent	Not pertinent	0.5–5	Not pertinent
Sulfer Dioxide	5	20	Data not available	3
Sulfuric Acid	1 mg/m³	10 mg/m³	No effects except those secondary to tissue damage	>1 mg/m³
Sulfuric Acid, Spent	Not pertinent	Not pertinent	Not pertinent	Not pertinent
Titanium Tetrachloride	5	Data not available	Data not available	Data not available
Toluene	100	600	0.5–5	0.17
Toluene 2, 4- Diisocyante	0.02	0.02	0.5–5	0.4–2.14
P-Toluenesulfonic Acid	Data not available	Not pertinent	400 mg/kg	Not pertinent
O-Toludine	5	Data not available	900 mg/kg	Data not available
Toxamphene	Not pertinent	Not pertinent	50 mg/kg	Not pertinent
Trichloroethylene	100	200	50–500 mg/kg	50
Trichlorofluoro- methane	1000	Data not available	Data not available	Data not available

Table 7.1 (cont.) Recommended safe levels of exposure

Chemical Name	Health Risk Information			
	TLV, ppm	STEL, ppm	LD_{50}, g/kg	Odor Threshold, ppm
Trichlorosilane	Data not available	Data not available	1000 mg/kg	Data not available
Tridecanol	Data not available	Data not available	Data not available	Data not available
1-Tridecene	Not pertinent	Data not available	Data not available	Not pertinent
Triethylaluminum	Not pertinent	Not pertinent	Data not available	Data not available
Triethylamine	25	100	50–500 mg/kg	Data not available
Thiethylene Glycol	Not pertinent	Not pertinent	5–15	Not pertinent
Tripropylene Glycol	Data not available	Data not available	3000 mg/kg	Odorless
Turpentine	100	200	0.5–5	Data not available
Undecanol	Not pertinent	Not pertinent	0.5–5	Not pertinent
1-Undecene	Data not available	Data not available	Data not available	Data not available
N-Undecylbenzene	Data not available	Data not available	Data not available	Data not available
Uranyl Acetate	0.2 mg/m³	Data not available	Data not available	Data not available
Uranyl Nitrate	0.05 mg/m³	Data not available	50–500 mg/kg	Not pertinent
Uranyl Sulfate	0.2 mg/m³	Data not available	5–15	Data not available
Urea	Not pertinent	Not pertinent		Not pertinent

Recommended Safe Levels of Exposure

Table 7.1 (cont.) Recommended safe levels of exposure

Chemical Name	Health Risk Information			
	TLV, ppm	STEL, ppm	LD$_{50}$, g/kg	Odor Threshold, ppm
Urea Peroxide	Data not available	Data not available	Data not available	Data not available
Vanadium Oxytrichloride	Data not available	5	140 mg/kg	10
Vanadium Pentoxide	0.5 mg/m^3	Data not available	23 mg/kg	Data not available
Vanadyl Sulfate	Data not available	Data not available	50–500 mg/kg	Data not available
Vinyl Acetate	10	Data not available	5	0.12
Vinyl Chloride	200	500		260
Vinyl Fluoride, Inhibited	Data not available	Data not available	Data not available	Data not available
Vinylidene Chloride, Inhibited	25	Data not available	84 mg/kg	Data not available
Vinyl Methyl Ether, Inhibited	Data not available	Data not available	0.5–5	Data not available
Vinyltoluene	100	400	0.5–6	50
Vinyltrichlorosilane	Data not available	Data not available	1.280 mg/kg	Data not available
Waxes: Carnauba	Not pertinent	Not pertinent	Data not available	Not pertinent
Waxes: Paraffin	Not pertinent	Not pertinent	5–15	Not pertinent
M-Xylene	100	300	50–500	0.05
Xylenol	45	Data not available	1070 mg/kg	
Zinc Acetate	Data not available	Data not available	0.5–5	Not pertinent

Table 7.1 (cont.) Recommended safe levels of exposure

Chemical Name	Health Risk Information			
	TLV, ppm	STEL, ppm	LD_{50}, g/kg	Odor Threshold, ppm
Zinc Ammonium Chloride	0.5 mg/m³	Data not available	Data not available	Not pertinent
Zinc Bromide	Data not available	Data not available	0.5–15	Data not available
Zinc Chloride	Not pertinent	Not pertinent	50–500 mg/kg	Not pertinent
Zinc Chromate	0.1 mg/m³	Data not available	0.5–5	Data not available
Zinc Fluoroborate	Data not available	Data not available	0.5–5	Data not available
Zinc Nitrate	Data not available	Data not available	2.5 mg/kg	Odorless
Zinc Sulfate	Data not available	Data not available	0.5–5	Data not available
Zirconium Acetate	5 mg/m³	Data not available	0.5–5	Data not available
Zirconium Nitrate	5 mg/m³	Data not available	25	Data not available

TLV – Threshold Limit Value; STEL – Short-terms Exposure Limit; LD_{50} – Lethal Dose (50%).

8
Fire and Chemical Reactivity Data

Table 8.1 Fire and chemical reactivity data

Chemical Name	Flash Point, F CC	Flash Point, F OC	Flammable Limits in Air, % LEL	Flammable Limits in Air, % UEL	Extinguishing Agents	Extinguishing Agents NOT To Be Used	Behavior in Fire	Ignition Temp, F	Water Reactive	Avoid Contact
Acetaldehyde	−36	−59	4	60	Dry chemical, alcohol foam, carbon dioxide	Water may be ineffective	Vapors are heavier than air and may travel to a considerable distance for a source of ignition and flash back	365	No	Heat, dust, strong oxidizing agents, strong acids and bases
Acetic Acid	104	112	5.4	16	Water, alcohol foam, dry chemical or carbon dioxide	None	Not pertinent	800	No	
Acetuc Anhydride	120	136	2.7	10	Water spray, dry chemical, alcohol foam or carbon dioxide	Water and foam react, but heat generated is not enough to create a hazard. Dry chemical forced below the surface can cause foaming and boiling	Not pertinent	600	Reacts slowly with water, but considerable heat is liverated when contacted with spray water	

Table 8.1 (cont.) Fire and chemical reactivity data

Chemical Name	Flash Point, °F CC	Flash Point, °F OC	Flammable Limits in Air, % LEL	Flammable Limits in Air, % UEL	Extinguishing Agents	Extinguishing Agents NOT To Be Used	Behavior in Fire	Ignition Temp, F	Water Reactive	Avoid Contact
Acetone	0	4	2.6	12.8	Alcohol foam, dry chemical, carbon dioxide	Water is straight hose streams will scatter fire	Not pertinent	869	No	
Acetone Cyanohydrin	165		2.2	12	Water spray, dry chemical, alcohol foam or carbon dioxide	Not pertinent	Not pertinent	1270	No	
Acetonitrile		42	4.4	16	Alcohol foam, dry chemical, carbon dioxide	Water may be ineffective	Vapors are heavier than air and may travel to a considerable distance for a source of ignition and flash back	975	No	
Acetyl Bromide	Data not available	Data not available	Data not available	Data not available	Carbon dioxide	Water	Do not apply water to adjacent fires, reacts with water to produce toxic and irritating gases	Data not available	Reacts violently, forming corrosive and toxic fumes of hydrogen bromide	

Fire Hazards and Fire Fighting / *Chemical Reactivity*

Fire and Chemical Reactivity Data 401

Acetyl Chloride	40		Data not available	Data not available	Carbon dioxide, dry chemical	Water, foam	Vapors are heavier than air and may travel to a considerable distance for a source of ignition and flash back	734	Reacts vigorously with water, involving hydrogen chloride fumes (hydrochloric acid)
Acetyl Peroxide		113	Data not available	Data not available	Water, dry chemical, carbon dioxide	Not pertinent	May explode. Burns with accelerating intensity	Data not available	No
Acridine	Not pertinent	Not pertinent	Data not available	Data not available	Water, foam, mono-ammonium phosphate	Carbon dioxide and other dry chemicals may not be effective	Sublimes before melting	Data not available	No
Acrolein	-13	<0	2.8	31	Foam, dry chemical, carbon dioxide	Water may be ineffective	Vapors are heavier than air and may travel to a considerable distance for a source of ignition and flash back	453	No

Table 8.1 (cont.) Fire and chemical reactivity data

Chemical Name	Flash Point, °F CC	Flash Point, °F OC	Flammable Limits in Air, % LEL	Flammable Limits in Air, % UEL	Extinguishing Agents	Extinguishing Agents NOT To Be Used	Behavior in Fire	Ignition Temp, F	Water Reactive	Avoid Contact
Acrylamide	Not flammable	Not flammable	Not flammable	Not flammable	Not pertinent	Not pertinent	Sealed containers may burst as a result of polymerization	Not pertinent	No	
Acrylic Acid		118	2.4		Water spray, dry chemical, alcohol foam or carbon dioxide	Not pertinent	May polymerize and explode	374	No	
Acrylonitrile	30	31	3.05	17	Dry chemical, alcohol foam, carbon dioxide	Water or foam may cause frothing	Vapors are heavier than air and may travel to a considerable distance for a source of ignition and flash back. May polymerize and explode	898	No	

Fire and Chemical Reactivity Data

Aldrin	Not flammable	Not flammable	Not pertinent	Not pertinent	Water spray, dry chemical, alcohol foam or carbon dioxide	Not pertinent	Not pertinent	Not pertinent	No
Allyl Alcohol	72	90	2.5	18	Dry chemical, alcohol foam, carbon dioxide	Water may be ineffective	Vapors are heavier than air and may travel to a considerable distance for a source of ignition and flash back	829	No
Allyl Chloroformate	88	92	Data not available	Data not available	Dry chemical, alcohol foam, carbon dioxide	Water may be ineffective	Vapors are heavier than air and may travel to a considerable distance for a source of ignition and flash back	Data not available	Reacts slowly generating hydrogen chloride
Allylchlorosilane	95	100	Data not available	Data not available	Dry chemical, carbon dioxide	Water	Difficult to extinguish. Re-ignition may occur	Data not available	Reacts vigorously, generating hydrogen chloride and phosgene may form

Table 8.1 (cont.) Fire and chemical reactivity data

Chemical Name	Flash Point, °F CC	Flash Point, °F OC	Flammable Limits in Air, % LEL	Flammable Limits in Air, % UEL	Extinguishing Agents	Extinguishing Agents NOT To Be Used	Behavior in Fire	Ignition Temp, F	Water Reactive	Avoid Contact
					Fire Hazards and Fire Fighting				Chemical Reactivity	
Aluminum Chloride	Not flammable	Not flammable	Not flammable	Not flammable	Not pertinent	Water	Reacts violently with water used in extinguishing adjacent fires	Not pertinent	Reacts violently with water, liberating chloride gas and heat	
Aluminum Nitrate	Not flammable	Not flammable	Not flammable	Not flammable	Not pertinent	Not pertinent	Dissolves and forms a weak solution of nitric acid	Not pertinent	Not pertinent	
Ammonium Bifluoride	Not flammable	Not flammable	Not flammable	Not flammable	Not pertinent	Water	Not pertinent	Not pertinent	Dissolves and forms a weak solution of hydrofluoric acid	
Ammonium Carbonate	Not flammable	Not flammable	Not flammable	Not flammable	Water	Not pertinent	Decomposes but reaction is not explosive	Not pertinent	No	

Fire and Chemical Reactivity Data 405

Ammonium Dichromate	Not pertinent	Not pertinent	Not pertinent	Not pertinent	Water	Not pertinent	Decomposes at about 180°C with spectacular swelling and evolution of heat and nitrogen leaving chromic oxide residue	437	No
Ammonium Fluoride	Not flammable	Not flammable	Not flammable	Not flammable	Not pertinent	Not pertinent	May sublime when hot and condense on cool surfaces	Data not available	Dissolves and forms a weak solution of hydrofluoric acid
Ammonium Hydroxide	Not flammable	Not flammable	Not flammable	Not flammable	Not pertinent	Not pertinent	Not pertinent	Data not available	Mild liberation of heat
Ammonium Lactate	Not pertinent	Not pertinent	Not pertinent	Not pertinent	Water, foam	Not pertinent	Not pertinent	Not pertinent	No
Ammonium Nitrate	Not flammable	Not flammable	Not flammable	Not flammable	Use flooding amount of water in early stages of fire. When large quantities are involved in massive fires, control efforts should be confined to protecting from explosion	Not pertinent	May explode in fires	Not pertinent	No

Table 8.1 (cont.) Fire and chemical reactivity data

Chemical Name	Flash Point, °F CC	Flash Point, °F OC	Flammable Limits in Air, % LEL	Flammable Limits in Air, % UEL	Extinguishing Agents	Extinguishing Agents NOT To Be Used	Behavior in Fire	Ignition Temp, F	Water Reactive	Avoid Contact
Ammonium Nitrate-Sulfate Mixture	Not pertinent	Not pertinent	Not pertinent	Not pertinent	Water	Steam, inert gases, foam, dry chemical	Will increase the intensity of fire when in contact with combustible material	Not pertinent	No	
Ammonium Oxalate	Not pertinent	Not pertinent	Not pertinent	Not pertinent	Water, foam	Not pertinent	Not pertinent	Not pertinent	No	
Ammonium Perchlorate	Not flammable	Not flammable	Not flammable	Not flammable	Water	Not pertinent	May explode when involved in a fire or exposed to shock or friction	Not pertinent	No	
Ammonium Perchlorate	Not pertinent	Not pertinent	Not pertinent	Not pertinent	Water	Data not available	Decomposes with loss of oxygen that increases intensity of fire	Not pertinent	No	
Ammonium Silicofluoride	Not flammable	Not flammable	Not flammable	Not flammable	Not pertinent	Not pertinent	Not pertinent	Not pertinent	No	

Fire and Chemical Reactivity Data

Amyl Acetate	(iso-) 69 and (n-) 91	1.1	7.5	Alcohol foam, dry chemical, carbon dioxide	Water in a straight stream will scatter and spread the fire	Not pertinent		No
Aniline	158	168	1.3	Water, foam, dry chemicals, or carbon dioxide	Not pertinent	Not pertinent	1418	No
Anisoyl Chloride	Data not available	Data not available	Not pertinent	Carbon dioxide, dry chemical	Water, foam	Data not available	Data not available	Reacts slowly to generate hydrogen chloride (hyrdochloric acid)
Anthracene	Not pertinent	Not pertinent	Not pertinent	Water, foam, dry chemicals, or carbon dioxide	None	Data not available	1004	No
Antimony Pentachloride	Not flammable	Not flammable	Not flammable	Not pertinent	Water or foam on adjacent fires	Irritating fumes of hydrogen chloride given off when water or foam is used to extinguish adjacent fire	Not pertinent	Reacts to form hydrogen chloride gas (hydrochloric acid)

Table 8.1 (cont.) Fire and chemical reactivity data

Chemical Name	Flash Point, °F CC	Flash Point, °F OC	Flammable Limits in Air, % LEL	Flammable Limits in Air, % UEL	Extinguishing Agents	Extinguishing Agents NOT To Be Used	Behavior in Fire	Ignition Temp, F	Water Reactive	Avoid Contact
Antimony Penta-fluoride	Not flammable	Not flammable	Not flammable	Not flammable	Not pertinent	Water or foam on adjacent fires.	Irritating fumes of hydrogen chloride given off when water or foam is used to extinguish adjacent fire	Not pertinent	Reacts to form hydrogen fluoride gas (hydrofluoric acid)	
Antimony Potassium Tartrate	Not flammable	Not flammable	Not flammable	Not flammable	Not pertinent	Not pertinent	Not pertinent	Not pertinent	No	
Antimony Trichloride	Not flammable	Not flammable	Not flammable	Not flammable	Not pertinent	Do not apply water on adjacent fires.	No data	Not pertinent	Reacts vigorously to form a strong solution of hydrochloric acid	
Antimony Trifluoride	Not flammable	Not flammable	Not flammable	Not flammable	Not pertinent	Not pertinent	Not flammable	Not pertinent	No	
Antimony Trioxide	Not flammable	Not flammable	Not flammable	Not flammable	Not pertinent	Not pertinent	Not pertinent	Not pertinent	No	
Arsenic Acid	Not flammable	Not flammable	Not flammable	Not flammable	Not pertinent	Not pertinent	Not pertinent	Not pertinent	No	

Fire and Chemical Reactivity Data

Arsenic Disulfide	Not pertinent	Not flammable	Not pertinent	Not pertinent	Water	Not pertinent	May ignite at very high temperatures	Not pertinent	No
Arsenic Trichloride	Not flammable	Not flammable	Not flammable	Not flammable	Not pertinent	Avoid water on adjacent fires	Becomes gaseous and cases irritation. Forms hydrogen chloride (hydrochloric acid) by reaction with water used to fight adjacent fires	Not pertinent	Forms hydrogen chloride (hydrochloric acid)
Arsenic Trioxide	Not flammable	Not flammable	Not pertinent	Not flammable	Not pertinent	Not pertinent	Can volatize forming toxic fumes of arsenic trioxide	Not pertinent	No
Arsenic Trisulfide	Not pertinent	Not pertinent	Not pertinent	Not pertinent	Water	Not pertinent	May ignite at very high temperatures	Not pertinent	No
Asphalt	Not pertinent	300–350	Not pertinent	Not pertinent	Water spray, dry chemical, alcohol foam or carbon dioxide	Water or foam may cause foaming	Not pertinent	400–700	No
Atrazine	Not flammable	Not flammable	Not flammable	Not flammable	Not pertinent	Not pertinent	Not pertinent	Not pertinent	No

410 A Guide to Safe Material and Chemical Handling

Table 8.1 (cont.) Fire and chemical reactivity data

Chemical Name	Flash Point, °F CC	Flash Point, °F OC	Flammable Limits in Air, % LEL	Flammable Limits in Air, % UEL	Extinguishing Agents	Extinguishing Agents NOT To Be Used	Behavior in Fire	Ignition Temp, °F	Water Reactive	Avoid Contact
Azinphos-methyl	Not flammable	Not flammable	Not flammable	Not flammable	Not pertinent	Not pertinent	Data not available	Not pertinent	No	
Barium Carbonate	Not flammable	Not flammable	Not flammable	Not flammable	Not pertinent	Not pertinent	Not pertinent	Not flammable	No	
Barium Chlorate	Not flammable	Not flammable	Not flammable	Not flammable	Not pertinent	Data not available	May cause an explosion when involved in a fire	Not pertinent	No	
Barium Nitrate	Not flammable	Not flammable	Not flammable	Not flammable	Not pertinent	Not pertinent	Mixes with combustible materials are readily ignited and burn fiercely. Containers may explode	Not pertinent	No	
Barium Perchlorate	Not flammable	Not flammable	Not flammable	Not flammable	Not pertinent	Not pertinent	Increases the intensity of the fire	Not pertinent	No	
Barium Permanganate	Not flammable	Not flammable	Not flammable	Not flammable	Not flammable	Not flammable	Can increase the intensity of the fire	Not pertinent	No	

FIRE AND CHEMICAL REACTIVITY DATA

Barium Peroxide	Not flammable	Not flammable	Not flammable	Not flammable	Flood with water, dry powder (e.g. graphite or powdered limestone)	Not pertinent	Can increase the intensity of the fire	Not pertinent	Decomposes slowly but reaction is not hazardous
Benzaldehyde	148	163	Data not available	0.042	Water spray, dry chemical, alcohol foam or carbon dioxide	Not pertinent	Not pertinent	378	No
Benzene	12		1.3	7.9	Dry chemical, foam, and carbon dioxide	Water may be ineffective	Vapors are heavier than air and can travel considerable distance to source of ignition and flash back	1097	No
Benzene Hexachloride	Not flammable	Not flammable	Not flammable	Not flammable	Not pertinent	Not pertinent	Not pertinent	Not flammable	No

Table 8.1 (cont.) Fire and chemical reactivity data

Chemical Name	Flash Point, °F CC	Flash Point, °F OC	Flammable Limits in Air, % LEL	Flammable Limits in Air, % UEL	Extinguishing Agents	Extinguishing Agents NOT To Be Used	Behavior in Fire	Ignition Temp, F	Water Reactive	Avoid Contact
Benzene Phosphorous Dichloride		215	Not pertinent	Not pertinent	Large amounts of water	Not pertinent	Containers may rupture. The hot liquid is spontaneously flammable because of the presence of dissolved phosphorus	319	Reacts vigorously to form hydrogen chloride (hydrochloric acid)	
Benzene Phosphorous Thiodichloride		252	Not pertinent	Not pertinent	Water	Not pertinent	Containers may rupture	338	Forms hydrogen chloride (fumes). Reacts slowly unless the water is hot	
Benzoic Acid	250		Not pertinent	Not pertinent	Dry chemical, carbon dioxide, water fog, chemical foam	None	Vapor from molten benzoic acid may form explosive mixture with air	1063	No	

Fire Hazards and Fire Fighting / Chemical Reactivity

Fire and Chemical Reactivity Data

Benzonitrile	167		Data not available	Data not available	Foam, dry chemical, carbon dioxide	Water may be ineffective	Data not available	Data not available	No
Benzophenone	This is a combustible product		Data not available	Data not available	Foam, dry chemical, carbon dioxide	Water may be ineffective	Data not available	Data not available	No
Benzoyl Chloride		162	1.2	4.9	Foam, dry chemical, carbon dioxide, and water fog	Water spray	At fire temperatures compound may react violently with water or stream	185	Slow reaction with water to produce hydrochloric acid fumes. The reaction is more rapid with steam
Benzyl Alcohol	213	220	Data not available	Data not available	Alcohol foam, dry chemical, carbon dioxide	Water or foam may cause foaming	Data not available	817	No
Benzylamine		168	Data not available	Data not available	Alcohol foam, dry chemical, carbon dioxide	Water may be ineffective	Data not available	Data not available	No
Benzyl Bromide	174		Not pertinent	Not pertinent	Water, dry chemical, carbon dioxide, foam	Not pertinent	Forms vapors that are powerful tear gas	Data not available	Reacts slowly generating hydrogen bromide (hydrobromic acid)

Table 8.1 (cont.) Fire and chemical reactivity data

Chemical Name	Flash Point, °F CC	Flash Point, °F OC	Flammable Limits in Air, % LEL	Flammable Limits in Air, % UEL	Extinguishing Agents	Extinguishing Agents NOT To Be Used	Behavior in Fire	Ignition Temp, F	Water Reactive	Avoid Contact
Benzyl Chloride	140	165	1.1		Water, dry chemical, carbon dioxide, foam	Not pertinent	Forms vapors that are powerful tear gas	1161	Undergoes slow hydrolysis liberating hydrogen chloride (hydrochloric acid)	
Beryllium Fluoride	Not flammable	Not flammable	Not flammable	Not flammable	Not pertinent	Not pertinent	Not pertinent	Not flammable	No	
Beryllium Metallic	Not pertinent	Not pertinent	Not pertinent	Not pertinent	Graphite, sand, or any other inert dry powder	Water	Powder may form explosive mixture in the air	Not pertinent	No	
Beryllium Nitrate	Not combustible	Not combustible	Not combustible	Not combustible	Water	Not pertinent	May increase the intensity of fire when in contact with combustible materials	Not pertinent	Reacts to work weak solution of nitric acid, however, reaction is usually not considered hazardous	
Beryllium Oxide	Not flammable	Not flammable	Not flammable	Not flammable	Not pertinent	Not pertinent	Not pertinent	Not flammable	No	

Fire and Chemical Reactivity Data

Beryllium Sulfate	Not flammable	Not flammable	Not flammable	Not flammable	Not pertinent	Not pertinent	No
Bismuth Oxychloride	Not flammable	Not flammable	Not flammable	Not pertinent	Not pertinent	Not flammable	No
Bisphenol A			Not pertinent	Foam, dry chemical, carbon dioxide	Data not available	Not pertinent	No
Boric Acid	Not flammable	Not flammable	Not flammable	Not pertinent	Not pertinent	Not flammable	No
Boron Trichloride	Not flammable	Not flammable	Not flammable	Not pertinent	Not pertinent	Not pertinent	Toxic fumes of hydrogen chloride are generated upon contact with water used to fight adjacent fires	Reacts vigorously, liberating head and forming hydrogen chloride fumes (hydrochloric acid) and boric acid
Bromine	Not flammable	Not flammable	Not flammable	Use water spray to cool exposed containers and to wash spill away from a safe distance	Not pertinent	Not pertinent	Not flammable	No

Table 8.1 (cont.) Fire and chemical reactivity data

Chemical Name	Flash Point, °F CC	Flash Point, °F OC	Flammable Limits in Air, % LEL	Flammable Limits in Air, % UEL	Extinguishing Agents	Extinguishing Agents NOT To Be Used	Behavior in Fire	Ignition Temp, °F	Water Reactive	Avoid Contact
Bromine Trifluoride	Not flammable	Not flammable	Not flammable	Not flammable	Dry chemical, carbon dioxide	Water, foam	Forms highly toxic and irritating	Not pertinent	Reacts vigorously, liberating heat and forming hydrogen fluoride gas (hydrochloric acid)	
Bromobenzene	124		Not pertinent	Not pertinent	Water, dry chemical, foam, cardon dioxide	Not pertinent	Not pertinent	1049	No	
Butadiene, Inhibited	−105		2	11.5	Stop flow of gas	Not pertinent	Vapor is heavier than air and can travel distances to ignition source and flash back. Containers may explode in a fire due to polymerization	788	No	
Butane	−100		1.8	8.4	Stop flow of gas	Not pertinent	Not pertinent	807	No	

Fire Hazards and Fire Fighting / *Chemical Reactivity*

Fire and Chemical Reactivity Data

N-Butyl Acetate	75	99	1.7	7.6	Foam, dry chemical, carbon dioxide	Water in straight hose stream will scatter and spread fire and should be avoided	Not pertinent	760	No
Sec-Butyl Acetate	62	88	1.7	9.8	Foam, carbon dioxide or dry chemical	Water may be ineffective	Not pertinent	No data	No
Iso-Butyl Acrylate		94	1.9	8	Dry chemical, foam or carbon dioxide	Not pertinent	Not pertinent	644	No
N-Butyl Acrylate		118	1.4	9.4	Dry chemical, foam or carbon dioxide	Not pertinent	Not pertinent	534	No
N-Butyl Alcohol	84	97	1.4	11.2	Carbon dioxide, dry chemicals	Not pertinent	Not pertinent	650	No
Sec-Butyl Alcohol	75		1.7	9	Carbon Dioxide, dry chemical	Not pertinent	Not pertinent	763	No
Tert-Butyl Alcohol	52	61	2.35	8	Dry chemical, carbon dioxide	Not pertinent	Not pertinent	896	No

Table 8.1 (cont.) Fire and chemical reactivity data

Chemical Name	Flash Point, °F CC	Flash Point, °F OC	Flammable Limits in Air, % LEL	Flammable Limits in Air, % UEL	Extinguishing Agents	Extinguishing Agents NOT To Be Used	Behavior in Fire	Ignition Temp, F	Water Reactive	Avoid Contact
N-Butylamine	10	30	1.7	9.8	Alcohol foam, dry chemical, carbon dioxide	Water may be ineffective	Vapor is heavier than air and can travel distances to ignition source and flash back. Containers may explode	594	No	
Sec-Butylamine	16				Alcohol foam, dry chemical, carbon dioxide	Water may be ineffective	Vapor is heavier than air and can travel distances to ignition source and flash back. Containers may explode	712	No	
Tert-Butylamine	16		1.7	8.9	Dry chemical, alcohol foam, carbon dioxide	Water may be ineffective	No data	716	No	

Butylene	Not pertinent	Not pertinent	1.6	10	Stop flow of gas	Not pertinent	Containers may explode in fire. Vapor is heavier than air and can travel distances to ignition source and flash back	725	No
Butylene Oxide		20	1.5	18.3	Dry chemical, alcohol foam, carbon dioxide	Water may be ineffective	Containers may explode in fire	959	No
N-Butyl Mercaptan		53			Dry chemical, alcohol foam, carbon dioxide	Water	Vapor is heavier than air and can travel considerable distance to ignition source and flash back	No data	No
N-Butyl Methacrylate		150	2	8	Dry chemical, foam, and carbon dioxide	Water may be ineffective	Containers may explode	562	No
1,4-Butynediol		263	Not pertinent	Not pertinent	Water, alcohol foam, dry chemical or carbon dioxide	Not pertinent	Not pertinent	No data	No

Table 8.1 (cont.) Fire and chemical reactivity data

Chemical Name	Flash Point, °F CC	Flash Point, °F OC	Flammable Limits in Air, % LEL	Flammable Limits in Air, % UEL	Extinguishing Agents	Extinguishing Agents NOT To Be Used	Behavior in Fire	Ignition Temp, F	Water Reactive	Avoid Contact
Iso-Butyr-aldehyde	−40	13	2	10	Foam, dry chemical or carbon dioxide	Data not available	Vapors are heavier than air and may travel considerable distances to source of ignition and flash back. Fires are difficult to control because of re-ignition	385	No	
N-Butyr-aldehyde										
N-Butyric Acid	160	166	2.19	13.4	Dry chemical, alcohol foam, carbon dioxide	Water may be ineffective	No data	842	No	
Cacodylic Acid	Not flammable	Not flammable	Not flammable	Not flammable	Not pertinent	Not pertinent	May form toxic oxides of arsenic when heated	Not pertinent	No	
Cadmium Acetate	Not flammable	Not flammable	Not flammable	Not flammable	Not pertinent	Not pertinent	Not pertinent	Not pertinent	No	

Fire and Chemical Reactivity Data

Cadmium Nitrate	Not flammable but may cause fire with other materials	Not flammable	Not flammable	Not flammable	Not pertinent	Not pertinent	Can increase the intensity of the fires when in contact with combustible materials	Not pertinent	No
Cadmium Oxide	Not flammable	Not flammable	Not flammable	Not flammable	Not pertinent	Not pertinent	No data	Not pertinent	No
Cadmium Sulfate	Not flammable	Not flammable	Not flammable	Not flammable	Not pertinent	Not pertinent	No data	Not pertinent	No
Calcium Arsenate	Not flammable	Not flammable	Not flammable	Not flammable	Not pertinent	Not pertinent	No data	Not pertinent	No
Calcium Carbide	Not flammable	Not flammable	Not flammable	Not flammable	Dry powder; preferably allow fire to burn out	Water, vaporizing liquid or foam, carbon dioxide	When contacted with water, generates highly flammable acetylene gas	Not flammable	Reacts vigorously with water to form highly flammable acetylene gas which can spontaniously ignite
Calcium Chlorate	Not flammable but may cause fire with other materials	Not flammable but may cause fire with other materials	Not pertinent	Not pertinent	Flood with water, dry powder (e.g. graphite or powdered limestone)	Not pertinent	May cause an explosion. Irritating gases may also for on exposure to heat	Not pertinent	No

Table 8.1 (cont.) Fire and chemical reactivity data

Chemical Name	Flash Point, °F CC	Flash Point, °F OC	Flammable Limits in Air, % LEL	Flammable Limits in Air, % UEL	Extinguishing Agents	Extinguishing Agents NOT To Be Used	Behavior in Fire	Ignition Temp, °F	Water Reactive	Avoid Contact
Calcium Chloride	Not flammable	Not flammable	Not flammable	Not flammable	Not pertinent	Not pertinent	Not pertinent	Not pertinent	Anhydrous grade dissolves with evolution or some heat	
Calcium Chromate	Not flammable	Not flammable	Not flammable	Not flammable	Not pertinent	Not pertinent	The hydrated salt loses water when hot and changes color, however there is no increase in hazard	Not pertinent	No	
Calcium Fluoride	Not flammable	Not flammable	Not flammable	Not flammable	Not pertinent	Not pertinent	Not pertinent	Not pertinent	No	
Calcium Hydroxide	Not flammable	Not flammable	Not flammable	Not flammable	Not pertinent	Not pertinent	Not pertinent	Not pertinent	No	
Calcium Hypochlorite	Not flammable	Not flammable	Not flammable	Not flammable	Not pertinent	Not pertinent	Poisonous gases released upon exposure to heat	Not flammable	No	

Fire and Chemical Reactivity Data

Calcium Nitrate	Not flammable, however may cause fires when in contact with flammables	Not flammable, however may cause fires when in contact with flammables	Not flammable	Not flammable	Flood with water	Not pertinent	Can greatly intensify the burning of all combustible materials	Not pertinent	No
Calcium Oxide	Not flammable	Not flammable	Not flammable	Not flammable	Not pertinent	Do not use water on adjacent fires	Not pertinent	Not flammable	Heat causes ignition of combustible materials. The materials swell during the reaction
Calcium Peroxide	Not flammable but may cause fires upon contact with combustible materials	Not flammable but may cause fires upon contact with combustible materials	Not pertinent	Not pertinent	Flood with water or use dry powder such as graphite or powdered limestone	Not pertinent	Can increase the intensity and severity of fires; containers may explode	Not pertinent	Reacts slowly with water at room temperature to form limewater and oxygen gas

Table 8.1 (cont.) Fire and chemical reactivity data

Chemical Name	Fire Hazards and Fire Fighting									Chemical Reactivity	
	Flash Point, °F		Flammable Limits in Air, %		Extinguishing Agents	Extinguishing Agents NOT To Be Used	Behavior in Fire	Ignition Temp, °F	Water Reactive	Avoid Contact
	CC	OC	LEL	UEL						
Calcium Phosphide	Not flammable but can spontaneously ignite if in contact with water	Not flammable but can spontaneously ignite if in contact with water	Not flammable	Not flammable	Extinguish adjacent fires with dry chemical or carbon dioxide	Water, foam	Can cause spontaneous ignition if wetted	Not pertinent	Reacts vigorously with water, generating phosphine, which is a poisonous and spontaneously flammable gas	
Camphene	92		No data	No data	Foam, dry chemical, carbon dioxide	Water	Not data	No data	No	
Carbolic Oil	117		No data	No data	Foam, carbon dioxide, dry chemical	Not pertinent	The solid often evaporates without first melting	466	No	
Carbon Dioxide	Not flammable	Not flammable	Not flammable	Not flammable	Not pertinent	Not pertinent	Containers may explode when exposed to heat	Not pertinent	No	

Carbon Monoxide	Not pertinent	Not pertinent	12	75	Allow fire to burn out; shut of the flow of gas and cool adjacent exposures with water. Extinguish (only if wearing a SCBA) with dry chemicals or carbon dioxide	Not pertinent	Flame has very little color. Containers may explode in fires	1,128	No
Carbon Tetrachloride	Not flammable	Not flammable	Not flammable	Not flammable	Not pertinent	Not pertinent	Decomposes to chloride and phosgene	Not flammable	No
Caustic Potash Solution	Not flammable	Not flammable	Not flammable	Not flammable	Not pertinent	Not pertinent	Not pertinent	Not pertinent	No
Caustic Soda Solution	Not flammable	Not flammable	Not flammable	Not flammable	Not pertinent	Not pertinent	Not pertinent	Not pertinent	No
Chlordane	132 (In solid form the product is not flammable)	225 (In solid form the product is not flammable)	0.7	5	Dry chemical, foam, carbon dioxide	Water may be ineffective on solution fires	Not pertinent	419	No

Table 8.1 (cont.) Fire and chemical reactivity data

| Chemical Name | Fire Hazards and Fire Fighting ||||||||| Chemical Reactivity ||
| | Flash Point, °F || Flammable Limits in Air, % || Extinguishing Agents | Extinguishing Agents NOT To Be Used | Behavior in Fire | Ignition Temp, F | Water Reactive | Avoid Contact |
	CC	OC	LEL	UEL						
Chlorine	Not flammable	Not flammable	Not flammable	Not flammable	Not pertinent	Not pertinent	Most combustible materials will burn in the presence of chlorine even though chlorine itself is not flammable	Not flammable	Forms a corrosive solution	
Chlorine Triflouride	Not flammable, but can cause fire when mixed or in contact with some materials	Not flammable, but can cause fire when mixed or in contact with some materials	Not pertinent	Not pertinent	Dry chemical	Do not use water on adjacent fires unless well protected against hydrogen flouride gas	Can greatly increase intensity of fires	Not pertinent	Reacts explosively with water, producing hydrogen flouride (hydroflouric acid) and shlorine	

Fire and Chemical Reactivity Data

Chloroaceto-phenone	This is a combustible solid, but in solutions it has a flash point of 244 CC	Not flammable	Not pertinent	Not pertinent	Water	Not pertinent	Unburned material may become volatile and casue severe skin and eye irritation	No data	Reacts slowly, producing hydrogen chloride. The reaction is not hazardous
Chloroform	Not flammable	Not flammable	Not flammable	Not flammable	Not pertinent	Not pertinent	Decomposes resulting in toxic vapors	Not flammable	No
Chromic Anhydride	Not flammable	Not flammable	Not flammable	Not flammable	Water	Not pertinent	Containers may explode. Water should be applied to cool container surfaces exposed to adjacent fires	This product may ignite organic materials on contact	No
Chromyl Chloride	Not flammable	Not flammable	Not flammable	Not flammable	Dry chemical or carbon dioxide	Do not apply water on adjacent fires unless SCBA is used to protect against toxic vapors	Vapors are extremely irritating to the eyes and mucus membranes. This product may increase the intensity of fires	Not pertinent	Reacts violently with water forming hydrogen chloride (hydrochloric acid), chlorine gases, and chromic acid

Table 8.1 (cont.) Fire and chemical reactivity data

Chemical Name	Flash Point, °F CC	Flash Point, °F OC	Flammable Limits in Air, % LEL	Flammable Limits in Air, % UEL	Extinguishing Agents	Extinguishing Agents NOT To Be Used	Behavior in Fire	Ignition Temp, °F	Water Reactive	Avoid Contact
Citric Acid	Not pertinent	Not pertinent	0.28 kg/m³ as dust	2.29 kg/m³ as dust	Water, foam, dry chemicals, or carbon dioxide	Not pertinent	This product melts and decomposes as a hazardous reaction	1,850 as a powder	No	
Cobalt Acetate	Not flammable	Not flammable	Not flammable	Not flammable	Not pertinent	Not pertinent	No data	Not pertinent	No	
Cobalt Chloride	Not flammable	Not flammable	Not flammable	Not flammable	Not pertinent	Not pertinent	Not pertinent	Not pertinent	Toxic cobalt oxide fumes can form in fire situations	
Cobalt Nitrate	Not flammable	Not flammable	Not flammable	Not flammable	Not pertinent	Not pertinent	Can increase fire intensity	Not pertinent	No	
Copper Acetate	Not flammable	Not flammable	Not flammable	Not flammable	Not pertinent	Not pertinent	No data	Not pertinent	No	
Copper Arsenite	Not flammable	Not flammable	Not flammable	Not flammable	Not pertinent	Not pertinent	Not pertinent	Not pertinent	No	
Copper Bromide	Not flammable	Not flammable	Not flammable	Not flammable	Not pertinent	Not pertinent	Not pertinent	Not pertinent	No	
Copper Chloride	Not flammable	Not flammable	Not flammable	Not flammable	Not pertinent	Not pertinent	Not pertinent	Not pertinent	No	
Copper Cyanide	Not flammable	Not flammable	Not flammable	Not flammable	Not pertinent	Not pertinent	Not pertinent	Not pertinent	No	

Fire and Chemical Reactivity Data

Copper Fluoroborate	Not flammable	Not flammable	Not flammable	Not flammable	Not pertinent	Not pertinent	Not pertinent	No
Copper Iodide	Not flammable	Not flammable	Not flammable	Not flammable	Not pertinent	Not pertinent	Not pertinent	No
Copper Naphthenate	100		0.8	5	Dry chemical, foam, carbon dioxide	Water may be ineffective	Not pertinent	540	No
Copper Nitrate	Not Flammable	Not flammable	Not flammable	Not flammable	Not pertinent	Not pertinent	Can increase intensity of fire if in contact with combustible material	Not pertinent	No
Copper Oxalate	Not flammable	Not flammable	Not flammable	Not flammable	Not pertinent	Not pertinent	Not pertinent	Not pertinent	No
Copper Sulfate	Not flammable	Not flammable	Not flammable	Not flammable	Not pertinent	Not pertinent	Not pertinent	Not flammable	No
Creosote, Coal Tar	>160		Not pertinent	Not pertinent	Dry chemical, carbon dioxide, or foam	Water may be ineffective	Heavy, irritating black smoke is formed	No	
Cresols	178	175–185	1.4 (ortho); 1.1 (meta or para)		Water, dry chemical, carbon dioxide and foam	Not pertinent	Sealed closed containers can build up pressure if exposed to heat (fire)	1110 (o-cresol); 1038 (m- or p-cresol)	No
Cumene	111		0.9	6.5	Foam, carbon dioxide, or dry chemical	Not pertinent	Not pertinent	797	No

Table 8.1 (cont.) Fire and chemical reactivity data

Chemical Name	Flash Point, °F CC	Flash Point, °F OC	Flammable Limits in Air, % LEL	Flammable Limits in Air, % UEL	Extinguishing Agents	Extinguishing Agents NOT To Be Used	Behavior in Fire	Ignition Temp, F	Water Reactive	Avoid Contact
Cyanogen	Flammable gas	Flammable gas	6.6	43	Let fire burn, shut off gas flow, cool exposed areas with water	Not pertinent	Vapor is heavier than air and may travel considerable distance to a source of ignition and flash back	Data not available	No reaction, but water provides heat to vaporize liquid cyanogen	
Cyanogen Bromide	Not flammable	Not flammable	Not flammable	Not flammable	Not pertinent	Not pertinent	Not pertinent	Not flammable	No	
Cyclohexane	−4		1.33	8.35	Foam, carbon dioxide, dry chemical	Not pertinent	Not pertinent	518	No	
Cyclohexanol	154	160	Data not available	Data not available	Water, foam, carbon dioxide, or dry chemical	Not pertinent	Not pertinent	572	No	
Cyclohexanone	111	129	1.1		Water, dry chemical, or carbon dioxide	Not pertinent	Not pertinent	788	No	
Cyclopentane	<20		1.1	8.7	Dry chemical, foam, carbon dioxide	Water may be ineffective	Containers may explode	716	No	

Fire and Chemical Reactivity Data

P-Cumene	117	140	0.7	5.6	Foam, dry chemical, carbon dioxide	Water may be ineffective	Not pertinent	817	No
DDD	Not pertinent	Not pertinent	Not pertinent	Not pertinent	Water, foam, dry chemical, carbon dioxide	Data not available	Data not available	Data not available	No
DDT	162–171		Not pertinent	Not pertinent	Water, foam, dry chemical, carbon dioxide	Not pertinent	Melts and burns	Data not available	No
Decaborane	(Flammable solid)	(Flammable solid)	Not pertinent	Not pertinent	Water, foam, dry chemical, carbon dioxide	Halogenated extinguishing agents	May explode when hot	300	Reacts slowly to form flammable hydrogen gas, which can accumulate in closed area
Decahydro-naphthalene		134	0.7	5.4	Foam, dry chemical, carbon dioxide	Water	Not pertinent	482	No
Decaldehyde		185	Data not available	Data not available	Foam, dry chemical, carbon dioxide	Not pertinent	Data not available	No data	No
1-Decene		128	Not pertinent	Not pertinent	Foam, dry chemical or carbon dioxide	Not pertinent	Not pertinent	455	No
N-Decyl Alcohol		180	Data not available	Data not available	Dry chemical	Not pertinent	Not pertinent	Data not available	No

Table 8.1 (cont.) Fire and chemical reactivity data

Chemical Name	Flash Point, °F CC	Flash Point, °F OC	Flammable Limits in Air, % LEL	Flammable Limits in Air, % UEL	Extinguishing Agents	Extinguishing Agents NOT To Be Used	Behavior in Fire	Ignition Temp, F	Water Reactive	Avoid Contact
N-Decyl-benzene	225		Data not available	Data not available	Dry chemical, carbon dioxide	Water or foam may cause frothing	Data not available	Data not available	No	
2,4-D Esters		>175	Data not available	Data not available	Foam, dry chemical, carbon dioxide	Water may be ineffective	Data not available	Data not available	No	
Dextrose Solution	Not flammable	Not flammable	Not flammable	Not flammable	Not pertinent	Not pertinent	Not pertinent	Not flammable	No	
Diacetone Alcohol	125	142	1.8	6.9	Dry chemical, alcohol foam, carbon dioxide	Not pertinent	Not pertinent	1118	No	
Di-N-Amyl-Phthalate	245		Data not available	Data not available	Foam, dry chemical, carbon dioxide	Water or foam may cause frothing	Data not available	Data not available	No	
Diaznon	82–105 (solutions only, pure liquid difficult to burn)		Not pertinent	Not pertinent	Foam, dry chemical, or carbon dioxide	Water may be ineffective	Not pertinent	Not pertinent	No	

Fire and Chemical Reactivity Data

Dibenzoyl Peroxide	Highly flammable solid; explosion-sensitive to shock, heat and friction		Not pertinent	Not pertinent	Difficult to extinguish once ignited. Use water spray to cool surrounding area	Do not use hand extinguisher	May explode	Data not available	No	Avoid contamination with combustible materials, various inorganic and organic acids, alkalies, alcohols, amines, easily oxidizable materials such as ethers, or materials used as accelerators in polymerization reactions
Di-N-Butylamine		125	1.1		"Alcohol" foam, dry chemical, carbon dioxide	Water may be ineffective	Data not available	Data not available	No	

Table 8.1 (cont.) Fire and chemical reactivity data

Chemical Name	Flash Point, °F CC	Flash Point, °F OC	Flammable Limits in Air, % LEL	Flammable Limits in Air, % UEL	Extinguishing Agents	Extinguishing Agents NOT To Be Used	Behavior in Fire	Ignition Temp, °F	Water Reactive	Avoid Contact
Di-N-Butyl Ether	92		1.5	7.6	Dry chemical, "alcohol" foam, or carbon dioxide	Water may be ineffective	Vapor is heavier than air and may travel considerable distance to a source of ignition and flash back	382	No	
Di-N-Butyl Ketone	Data not available	Data not available	Data not available	Data not available	Foam, dry chemical, carbon dioxide	Water may be ineffective	Data not available	Data not available	No	
Dibutyl-phenol		>20	Not pertinent	Not pertinent	Dry chemical, carbon dioxide, foam	Water may be ineffective	Not pertinent	Data not available	No	
Dibutyl Phthalate	315	355	0.5	2.5	Dry, carbon dioxide, foam	Water or foam may cause frothing	Not pertinent	757	No	
O-Dichloro-benzene	155	165	2.2	9.2	Water, foam, dry chemical or carbon dioxide	Not pertinent	Not pertinent	1198	No	
P-Dichloro-benzene	150	165	Data not available	Data not available	Water, foam, carbon dioxide, or dry chemical	Not pertinent	Not pertinent	No data	No	

Fire Hazards and Fire Fighting | Chemical Reactivity

Fire and Chemical Reactivity Data

Di-(P-Chlorobenzoyl) Peroxide	Not pertinent	Not pertinent	Not pertinent	Not pertinent	Flood with water, or use dry chemical, foam, carbon dioxide	Not pertinent	Solid may explode	Data not available	No
Dichlorobutene	Data not available	Data not available	1.5	4	Water, foam, dry chemical, or carbon dioxide	Not pertinent	Not pertinent	Data not available	Reacts slowly to form hydrochloric acid
Dichlorodifluoromethane	Not flammable	Not flammable	Not flammable	Not flammable	Not pertinent	Not pertinent	Helps extinguish fire	Not flammable	No
1,2-Dichloroethylene	37	9.7	12.8	Dry chemical, foam, carbon dioxide	Water may be ineffective	Vapor is heavier than air and may travel considerable distance to a source of ignition and flash back	Data not available	No
Dichloroethyl Ether	131	180	Data not available	Data not available	Water, foam, dry chemical, carbon dioxide	Not pertinent	Not pertinent	696	No
Dichloromethane	Not flammable under conditions likely to be encountered	Not flammable under conditions likely to be encountered	12	19	Not pertinent	Not pertinent	Not pertinent	1184	No

Table 8.1 (cont.) Fire and chemical reactivity data

Chemical Name	Flash Point, °F CC	Flash Point, °F OC	Flammable Limits in Air, % LEL	Flammable Limits in Air, % UEL	Extinguishing Agents	Extinguishing Agents NOT To Be Used	Behavior in Fire	Ignition Temp, °F	Water Reactive	Avoid Contact
2,4-Dichlorophenol	237	200	Data not available	Data not available	Water, foam, carbon dioxide, dry chemical	Water or foam may cause frothing	Solid melts and burns	Data not available	No	
2,4-Dichlorophenoxyacetic Acid	Not pertinent (combustible solid)		Not pertinent	Not pertinent	Water, foam	Not pertinent	Not pertinent	Not pertinent	No	
Dichloropropane	60	70	3.4	14.5	Foam, carbon dioxide, dry chemical	Not pertinent	Not pertinent	1035	No	
Dichloropropene	95		Data not available	Data not available	Water, dry chemical, foam, carbon dioxide	Not pertinent	Not pertinent	Data not available	No	
Dicyclopentadiene		90	0.8	6.3	Foam, carbon dioxide, dry chemical, or water spray	Not pertinent	Not pertinent	941	No	
Dieldrin	Not flammable	Not flammable	Not flammable	Not flammable	Not pertinent	Not pertinent	Not pertinent	Data not available	No	

Fire and Chemical Reactivity Data

Diethanol-amine		305	1.6 (calc.)	9.8 (est.)	Water, alcohol foam, carbon dioxide, dry chemical	Addition of water may cause frothing	Not pertinent	1224	No
Diethylamine		5	1.8	9.1	Dry chemical, carbon dioxide, or alcohol foam	Data not available	Vapors are heavier than air and may travel a considerable distance to a source of ignition and flash back. Fires are difficult to control because of re-ignition	594	No
Diethyl-benzene	135		Data not available	Data not available	Foam, water, carbon dioxide, or dry chemical	Not pertinent	Not pertinent	743 (ortho)	No
Diethyl Carbonate	77	115	Data not available	Data not available	Foam, carbon dioxide, dry chemical	Water	Not pertinent	No data	Too slow to be hazardous
Diethylene Glycol	255		1.6	10.8	Alcohol foam, carbon dioxide, dry chemical	Water or foam may cause frothing	Not pertinent	444	No
Diethylene Glycol Dimethyl Ether		158	Data not available	Data not available	Dry chemical, foam, carbon dioxide	Not pertinent	Not pertinent	Data not available	No

Table 8.1 (cont.) Fire and chemical reactivity data

Chemical Name	Flash Point, °F CC	Flash Point, °F OC	Flammable Limits in Air, % LEL	Flammable Limits in Air, % UEL	Extinguishing Agents	Extinguishing Agents NOT To Be Used	Behavior in Fire	Ignition Temp, F	Water Reactive	Avoid Contact
Diethylene-glycol Monobutyl Ether	172	230	Not pertinent	Not pertinent	Water, "alcohol" foam, carbon dioxide, dry chemical	Not pertinent	Not pertinent	442	No	
Diethylene-glycol Monobutyl Ether Acetate		240	0.8	5	Water, alcohol foam, dry chemical, carbon dioxide	Not pertinent	Not pertinent	563	No	
Diethylene Glycol Monoethyl Ether	201	205	1.2	8.5 (est.)	Alcohol foam, dry liquid or carbon dioxide	Not pertinent	Not pertinent	400	No	
Diethylene-triamine		200	1	10	Water spray, alcohol foam, carbon dioxide or dry chemical	Water or foam may cause frothing	Not pertinent	676	No	
Di (2Ethyl-hexyl) Phosphoric Acid		385	Not pertinent	Not pertinent	Dry chemical, alcohol foam, carbon dioxide	Water or foam may cause frothing	Not pertinent	No data	No	

Fire Hazards and Fire Fighting | Chemical Reactivity

Fire and Chemical Reactivity Data 439

Diethyl Phthalate	305	0.75 (at 368°C)	Not pertinent	Dry chemical, foam, carbon dioxide	Water or foam may cause frothing	Data not available	855	No
Diethylzinc	Not pertinent (ignites spontaneously)			Dry chemical, sand, or powdered limestone	Water, foam, halogenated agents, carbon dioxide	Reacts spontaneously with air or oxygen, and violently with water, evolving flammable ethane gas. Contact with water applied to adjacent fires will intensify the fire	<0	Reacts violently to form flammable ethane gas
1,1 Difluoroethane	Not pertinent	3.7	18	Shut off gas source; use water to cool adjacent combustibles	Data not available	Containers may explode	Data not available	No
Difluorophosphoric Acid, Anhydrous	Not flammable	Not flammable	Not flammable	Not pertinent	Do not use water on adjacent fires	Not pertinent	Not pertinent	Reacts vigorously to form corrosive and toxic hydrofluoric acid
Diheptyl Phthalate	Data not available	Data not available	Data not available	Foam, dry chemical, carbon dioxide	Water may be ineffective	Data not available	Data not available	No

Table 8.1 (cont.) Fire and chemical reactivity data

Chemical Name	Flash Point, °F CC	Flash Point, °F OC	Flammable Limits in Air, % LEL	Flammable Limits in Air, % UEL	Extinguishing Agents	Extinguishing Agents NOT To Be Used	Behavior in Fire	Ignition Temp, °F	Water Reactive	Avoid Contact
Diisobutyl-carbinol	165	162	0.8	6.1	Carbon dioxide, dry chemical, alcohol foam	Not pertinent	Not pertinent	494 (calc)	No	
Diiso-butylene	35 (est.)		0.9 (est.)		Dry chemical, foam, or carbon dioxide	Water may be ineffective	Not pertinent	788	No	
Diisobutyl Ketone	120	131	0.81	7.1	Foam, dry chemical, carbon dioxide	Water may be ineffective	Data not available	745	No	
Diisodecyl Phthalate		450			Dry chemical, foam, carbon dioxide	Water may be ineffective	Data not available	755	No	
Diisoprpoa-nolamine		200	1.1 (calc.)	5.4 (est.)	Water, alcohol foam, dry chemical, or carbon dioxide	Water or foam may cause frothing	Not pertinent	580	No	
Diisopro-pylamine	35	20	0.8	7.1	"Alcohol" foam, dry chemical, carbon dioxide	Water may be ineffective	Vapor is heavier than air and may travel to a source of ignition and flash back	600	No	

Fire and Chemical Reactivity Data 441

Diisopropyl-benzene Hydro-peroxide	175		Data not available	Data not available	Foam, dry chemical, carbon dioxide	Water may be ineffective	Burns with a flare effect. Containers may explode	Data not available	No
Dimethyl-acetamide	158		1.5	11.5	Water, dry chemical, alcohol foam	Not pertinent	Not pertinent	914	No
Dimethyl-amine	20		2.8	14.4	Stop flow of gas. Use water spray, carbon dioxide, or dry chemical for fires in water solutions	Do not use foam	Not pertinent	756	No
Dimethyl Ether	Not pertinent (flammable gas)		2	50	Let fire burn; shut off gas flow; cool exposed surroundings with water	Not pertinent	Containers may explode. Vapors are heavier than air and may travel long distances to source of ignition and flash back	662	No
Dimethyl Sulfate	182	240	Data not available	Data not available	Water, foam, carbon dioxide, or dry chemical	Not pertinent	Not pertinent	370	Slow, non-hazardous reaction

Table 8.1 (cont.) Fire and chemical reactivity data

Chemical Name	Flash Point, °F CC	Flash Point, °F OC	Flammable Limits in Air, % LEL	Flammable Limits in Air, % UEL	Extinguishing Agents	Extinguishing Agents NOT To Be Used	Behavior in Fire	Ignition Temp, F	Water Reactive	Avoid Contact
Distillates: Flashed Feed Stocks	(a) <0; (b) 0–73; (c) 73–141		Data not available	Data not available	Foam, carbon dioxide, dry chemical	Water may be ineffective	Not pertinent	Data not available	No	
Dodecyltrichlorosilane	>150		Data not available		Dry chemical, carbon dioxide	Water, foam	Difficult to extinguish, reignition may occur. Contact with water applied to adjacent fires produces irritating hydrogen chloride fumes	Data not available	Generates hydrogen chloride (hydrochloric acid)	
Ethane	−211		2.9	13	Stop flow of gas	Data not available	Not pertinent	940	No	
Ethyl Acetate	24	55	2.2	9	Alcohol foam, carbon dioxide, or dry chemical	Not pertinent	Not pertinent	800	No	
Ethyl Alcohol	55	64	3.3	19	Carbon dioxide, dry chemical water spray, alcohol foam	None	Not pertinent	689	No	

FIRE AND CHEMICAL REACTIVITY DATA

Ethyl Butanol	61	128	1.9	8.8	Carbon dioxide or dry chemical for small fires; alcohol foam for large fires	Not pertinent	Not pertinent	580 (calc.)	No
Ethyl Chloroformate	82		Data not available	Data not available	Water, dry chemical, carbon dioxide	Not pertinent	Not pertinent	932	Slow reaction with water, evolving hydrogen chloride (hydrochloric acid)
Ethylene	−213		2.75	28.6	Stop flow of gas if possible. Use carbon dioxide, dry chemical, water fog	Not pertinent	Container may explode	842	No
Ethylene Dichloride	55	60	6.2	15.6	Foam, carbon dioxide, dry chemical	Water may be ineffective	Vapor is heavier than air and may travel considerable distance to a source of ignition and flash back	775	No
Ethyl Lactate	115	1.58	1.5	11.4	Water, dry chemical, alcohol foam, carbon dioxide	Not pertinent	Not pertinent	752	No

Table 8.1 (cont.) Fire and chemical reactivity data

Chemical Name	Flash Point, °F CC	Flash Point, °F OC	Flammable Limits in Air, % LEL	Flammable Limits in Air, % UEL	Extinguishing Agents	Extinguishing Agents NOT To Be Used	Behavior in Fire	Ignition Temp, F	Water Reactive	Avoid Contact
Ethyl Nitrate	−31		3	>50	Water dry chemical, carbon dioxide, water foam	Not pertinent	Vapors are heavier than air and may travel a considerable distance to a source of ignition and flash back, can decompose violently above 194°F; containers may explode in fire	194	No	
Ferric Ammonium Citrate	Not flammable	Not flammable	Not flammable	Not flammable	Not pertinent	Not pertinent	Not pertinent	Not pertinent	No	
Ferric Nitrate	Not flammable	Not flammable	Not flammable	Not flammable	Not pertinent	Not pertinent	In contact with combustible materials, will increase the intensity of a fire	Not pertinent	No	
Ferric Sulfate	Not flammable	Not flammable	Not flammable	Not flammable	Not pertinent	Not pertinent	Not pertinent	Not pertinent	No	

Fire Hazards and Fire Fighting / Chemical Reactivity

Fluorine	Not flammable	Not flammable	Not flammable	Not flammable	Not pertinent	Do not direct water onto fluorine leaks	Dangerously reactive gas. Ignites most combustibles	Not flammable	Reacts with water to form hydrogen fluoride, oxygen and oxygen difluoride
Formaldehyde Solution	182 (based on solution of 37% formaldehyde and Methanol free); 122 (based on a solution with 15% Methanol)		7	73	Water, dry chemical, carbon dioxide, alcohol foam	No data or recommendations not found	Not pertinent	806	No
Gallic Acid	Not pertinent. This is a combustible solid		Not pertinent	Not pertinent	Water, foam, dry chemical, or carbon dioxide	Not pertinent	No data	Not pertinent	No
Gasolines: Automotive	-36		1.4	7.4	Foam, carbon dioxide, dry chemical	Water may be ineffective	Vapor is heavier than air and may travel considerable distance to a source of ignition and flash back	853	No

Table 8.1 (cont.) Fire and chemical reactivity data

Chemical Name	Flash Point, °F CC	Flash Point, °F OC	Flammable Limits in Air, % LEL	Flammable Limits in Air, % UEL	Extinguishing Agents	Extinguishing Agents NOT To Be Used	Behavior in Fire	Ignition Temp, F	Water Reactive	Avoid Contact
Gasolines: Aviation	−50		1.2	7.1	Foam, carbon dioxide, or dry chemical	Water may be ineffective	Vapor is heavier than air and may travel considerable distance to a source of ignition and flash back	824	No	
Glycerine	320	350	Not pertinent	Not pertinent	Alcohol foam, dry chemical, carbon dioxide, water fog	Water or foam may cause frothing	Not pertinent	698	No	
Heptanol		170	No data	No data	Foam, carbon dioxide, or dry chemical	Not pertinent	Not pertinent	No data	No	
N-Hexaldehyde		90	No data	No data	Dry chemical, foam, or carbon dioxide	Water may be ineffective	Vapor is heavier than air and may travel considerable distance to a source of ignition and flash back	No data	No	
Hydrochloric Acid	Not flammable	Not flammable	Not flammable	Not flammable	Not pertinent	Not pertinent	Not pertinent	Not flammable	No	

Fire Hazards and Fire Fighting — Chemical Reactivity

Fire and Chemical Reactivity Data

Hydrogen Chloride	Not flammable	Not flammable	Not flammable	Not flammable	Not flammable	Not flammable	Not flammable	Not flammable	Moderate reaction with the evolution of heat
Hydrogen, Liquified	Not pertinent	Not pertinent	4		75	Let fire burn; shut off gas supply	Carbon dioxide	Burn with almost invisible flame	1065	Ambient temperature of water will cause vigorous vaporization of hydrogen
Isobutane	−117		1.8		8.4	Stop flow of gas	Not pertinent	Not pertinent	890	No
Isobutyl Alcohol	82	90	1.6		10.9	Alcohol foam, dry chemical, or carbon dioxide	Water may be ineffective	Not pertinent	800	No
Isodecaldehyde		185	No data		No data	Foam, dry chemical, or carbon dioxide	Not pertinent	Not pertinent	No	
Isohexane	−20		1.2		7.7	Foam, dry chemical, or carbon dioxide	Water may be ineffective	Not pertinent	585	No
Isopropyl Alcohol	53	65	2.3		12.7	Alcohol foam, dry chemical, or carbon dioxide	Water may be ineffective	Not pertinent	750	No

Table 8.1 (cont.) Fire and chemical reactivity data

Chemical Name	Flash Point, °F CC	Flash Point, °F OC	Flammable Limits in Air, % LEL	Flammable Limits in Air, % UEL	Extinguishing Agents	Extinguishing Agents NOT To Be Used	Behavior in Fire	Ignition Temp, F	Water Reactive	Avoid Contact
Isopropyl Mercaptan		−30	No data	No data	Dry chemical, alcohol foam, or carbon dioxide	Water may be ineffective	Vapor is heavier than air and may travel to a source of ignition and flash back	No data	No	
Kerosene	100		0.7	5	Foam, dry chemical, or carbon dioxide	Water may be ineffective	Not pertinent	444	No	
Lactic Acid	Not pertinent (Not flammable)	Not pertinent (Not flammable)	Not pertinent	Not pertinent	Water, foam, dry chemical, carbon dioxide	Not pertinent	Not pertinent	Not pertinent	No	
Lead arsenate	Not flammable	Not flammable	Not flammable	Not flammable	Not pertinent	Not pertinent	Not pertinent	Not pertinent	No	
Lean Iodide	Not flammable	Not flammable	Not flammable	Not flammable	Not pertinent	Not pertinent	Not pertinent	Not pertinent	No	
Linear Alcohols		180–285	Data not available	Data not available	Alcohol foam, dry chemical, or carbon dioxide	Water of foam may cause frothing	Not pertinent	Data not available	No	

Fire Hazards and Fire Fighting / Chemical Reactivity

Fire and Chemical Reactivity Data

Liquified Natural Gas	Flammable gas	Flammable gas	5.3	14	Do not extinguish large spill fires. Allow to burn while cooling adjacent equipment with water spray. Shut off leak if possible. Extinguish small fires with dry chemicals	Water	Not pertinent	999	No
Liquified Petroleum Gas	Propane: −156; Butane: −76		Propane: 2.2; Butane: 1.8	Propane: 9.5; Butane: 8.4	Allow to burn while cooling adjacent equipment with water spray. Extinguish small fires with dry chemicals. Shut off leak if possible	Water (let fire burn)	Containers may explode. Vapor is heavier than air and may travel a long distance to a source if ignition and flash back	No data	No

Table 8.1 (cont.) Fire and chemical reactivity data

| Chemical Name | Fire Hazards and Fire Fighting ||||||||| Chemical Reactivity ||
| | Flash Point, °F || Flammable Limits in Air, % || Extinguishing Agents | Extinguishing Agents NOT To Be Used | Behavior in Fire | Ignition Temp, F | Water Reactive | Avoid Contact |
	CC	OC	LEL	UEL						
Lithium, Metallic	Not pertinent (combustible solid)	Not pertinent (combustible solid)	Not pertinent	Not pertinent	Graphite, lithium chloride	Water, sand, halogenated hydrocarbons, carbon dioxide, soda acid, or dry chemical	Molten lithium is quite easily ignited and is then difficult to extinguish. Hot or burning lithium will react with all gases except those of the helium-argon group. It also reacts violently with concrete, wood, asphalt, sand, asbestos, and in fact nearly everything except metal. Do not apply water to adjacent fires. Hydrogen explosion may result	No data	Reacts violently to form flammable hydrogen gas and strong caustic solution. Ignition usually occurs	

Fire and Chemical Reactivity Data

Magnesium	Not pertinent (solid). Flammable when in the form of turnings or powder	Not pertinent (solid). Flammable when in the form of turnings or powder	Not pertinent	Not pertinent		Water, foam, halogenated agents, carbon dioxide	Forms dense white smoke. Flame is very bright	883	In finely divided form, reacts with water and acids to release flammable hydrogen gas
Mercuric Acetate	Not flammable	Not flammable	Not flammable	Not flammable	Not pertinent	Not pertinent	Not pertinent	Not pertinent	No
Mercuric Cyanide	Not flammable	Not flammable	Not flammable	Not flammable	Not pertinent	Not pertinent	Not pertinent	Not pertinent	No
Methane	Flammable gas	Flammable gas	5	15	Stop flow of gas	Water	Not pertinent	1004	No
Methyl Alcohol	54	61	6	36	Alcohol foam, dry chemical, carbon dioxide	Water may be ineffective	Containers may explode	867	No
Methyl Chloride	<32		8.1	17.2	Dry chemical or carbon dioxide. Stop flow of gas	Not pertinent	Containers may explode	1170	No
Methyl Isobutyl Carbinol	106	120–130	1	5.5	Alcohol foam, dry chemical, or carbon dioxide	Not pertinent	Not pertinent	Data not available	No
Alpha-Methyl-styrene	137		1.9	6.1	Dry chemical, foam, carbon dioxide	Water may be ineffective	Data not available	1066	No

Table 8.1 (cont.) Fire and chemical reactivity data

Chemical Name	Flash Point, °F CC	Flash Point, °F OC	Flammable Limits in Air, % LEL	Flammable Limits in Air, % UEL	Extinguishing Agents	Extinguishing Agents NOT To Be Used	Behavior in Fire	Ignition Temp, F	Water Reactive	Avoid Contact
Mineral Spirits	105–140, depending on grade		0.8	5	Foam, carbon dioxide, dry chemical	Do not use straight hose water stream	Not pertinent	540	No	
Nitrous Oxide	Not pertinent (non-flammable compressed gas)		Not pertinent	Not pertinent	Not pertinent	Not pertinent	Will support combustion but does not cause spontaneous combustion, and may increase intensity of fire. Containers may explode when heated	Not pertinent	No	
Nonanol	165	210	0.8	6.1	Alcohol foam, dry chemical, or carbon dioxide	Water may be ineffective	Not pertinent	Data not available	No	
Octane	56		1	6.5	Dry chemical, foam, carbon dioxide	Water may be ineffective	Vapor is heavier than air and may travel a considerable distance to source of ignition and flash back	428	No	

Fire Hazards and Fire Fighting / *Chemical Reactivity*

Fire and Chemical Reactivity Data

Oils: Clarified	Data not available	Data not available	Data not available	Data not available	Dry chemical, foam, or carbon dioxide	Water may be ineffective	Not pertinent	Data not available	No
Oils: Crude	Data not available	Data not available	Data not available	Data not available	Dry chemical, foam or carbon dioxide	Water may be ineffective	Not pertinent	Data not available	No
Oils: Diesel	(1-D) 100; (2-D) 125		1.3	6	Dry chemical, foam, or carbon dioxide	Water may be ineffective	Not pertinent	(1-D) 350–625, (2-D) 490–545	No
Oils, Edible: Castor	445		Data not available	Data not available	Dry chemical, foam, or carbon dioxide	Water or foam may cause frothing	Not pertinent	840	No
Oils, Edible: Coconut	420 (crude); 580 (refined)		Not pertinent	Not pertinent	Dry chemical, foam, or carbon dioxide	Water or foam may cause frothing; water may be ineffective	Not pertinent	Data not available	No
Oils, Edible: Cottonseed	486 (refined oil); 610 (cooking oil)		Data not available	Data not available	Dry chemical, foam, or carbon dioxide	Water or foam may cause frothing	Not pertinent	650 (refined oil)	No
Oils, Edible: Fish	420		Data not available	Data not available	Dry chemical, foam, or carbon dioxide	Water or foam may cause frothing	Not pertinent	Data not available	No
Oils, Edible: Lard	395		Not pertinent	Not pertinent	Dry chemical, foam, or carbon dioxide	Water or foam may cause frothing; water may be ineffective	Not pertinent	833	No

Table 8.1 (cont.) Fire and chemical reactivity data

Chemical Name	Flash Point, °F CC	Flash Point, °F OC	Flammable Limits in Air, % LEL	Flammable Limits in Air, % UEL	Extinguishing Agents	Extinguishing Agents NOT To Be Used	Behavior in Fire	Ignition Temp, F	Water Reactive	Avoid Contact
Oils, Fuel: 2	136		Data not available	Data not available	Dry chemical, foam, or carbon dioxide	Water may be ineffective	Not pertinent	494	No	
Oils, Fuel: 4	>130		1	5	Dry chemical, foam, or carbon dioxide	Water may be ineffective	Not pertinent	505	No	
Oils, Miscellaneous: Coal Tar	60–77		1.3	8	Dry chemical, foam, or carbon dioxide	Water may be ineffective	Not pertinent	Data not available	No	
Oils, Miscellaneous: Motor	275–600		Data not available	Data not available	Dry chemical, foam, or carbon dioxide	Water may be ineffective	Not pertinent	325–625	No	
Oils, Miscellaneous: Penetrating	295		Data not available	Data not available	Dry chemical, foam, or carbon dioxide	Water or foam may cause frothing	Not pertinent	Data not available	No	
Oils, Miscellaneous: Resin	255–290		Data not available	Data not available	Dry chemical, foam, or carbon dioxide	Water may be ineffective	Not pertinent	648	No	
Oils, Miscellaneous: Spray	140 (min)		0.6	4.6	Dry chemical, foam, or carbon dioxide	Water may be ineffective	Not pertinent	475	No	

FIRE AND CHEMICAL REACTIVITY DATA 455

Oils, Miscellaneous: Tanner's	Data not available	Data not available	Data not available	Data not available	Data not available	Dry chemical, foam, or carbon dioxide	Water may be ineffective	Not pertinent	Data not available	No
Oleic Acid		390–425	Data not available	Data not available	Data not available	Dry chemical, foam, or carbon dioxide	Water of foam may cause frothing	Data not available	685	No
Oleum	Not flammable	Not flammable	Not flammable	Not flammable	Not flammable	Not pertinent	Avoid water on adjacent fires	Not pertinent	Not flammable	Vigorous reaction with water. Splatters
Oxalic Acid	Not flammable	Not flammable	Not flammable	Not flammable	Not flammable	Not pertinent	Not pertinent	Not pertinent	Not pertinent	No
Oxygen, Liquified	Not flammable	Not flammable	Not flammable	Not flammable	Not flammable	Not pertinent	Not pertinent	Increases the intensity of the fire. Mixtures of liquid oxygen and any fuel are highly explosive	Not pertinent	Heat of water will vigorously vaporize liquid oxygen
Paraformaldeyde	160	199	7	73	Water, foam, dry chemical, carbon dioxide	Data not available	Changes to formaldehyde gas which is highly flammable	572	Forms water solution of formaldehyde
Pentaerythritol	Not pertinent	Not pertinent	Not pertinent	Not pertinent	Water, dry chemical, carbon dioxide	Not pertinent	Not pertinent	842	No

Table 8.1 (cont.) Fire and chemical reactivity data

Chemical Name	Flash Point, °F CC	Flash Point, °F OC	Flammable Limits in Air, % LEL	Flammable Limits in Air, % UEL	Extinguishing Agents	Extinguishing Agents NOT To Be Used	Behavior in Fire	Ignition Temp, °F	Water Reactive	Avoid Contact
Pentane	-57		1.4	8.3	Foam, dry chemical, carbon dioxide	Water may be ineffective	Containers may explode	544	No	
1-Pentene	-60	0	1.4	8.7	Foam, dry chemical, carbon dioxide. Stop the flow of vapor	Water may be ineffective	Containers may explode	527	No	
Peracetic Acid		104	Data not available	Data not available	Water	Not pertinent	Vapors are flammable and explosive. Liquid will detonate if concentration rises above 56% because of exaporation of acetic acid	392	No	
Petrolatum	360–430		Data not available	Data not available	Water, foam, dry chemicals, or carbon dioxide	Not pertinent	Not pertinent	Data not available	No	

456 A Guide to Safe Material and Chemical Handling

Fire and Chemical Reactivity Data

Phenol	175	185	1.7	8.6	Water fog, carbon dioxide, dry chemical, foam	Not pertinent	Yields flammable vapors when heated which form explosive mixtures in the air	1319	No
Phenyldichloroarsine, Liquid	Data not available	Data not available	Data not available	Data not available	Water	Not pertinent	Data not available	Data not available	Very slow reaction, considered non-hazardous. Hydrochloric acid is formed
Phosgene	Not flammable	Not flammable	Not flammable	Not flammable	Water to cool containers	Not pertinent	Not pertinent	Not pertinent	Decomposes but not vigorously
Phosphoric Acid	Not flammable	Not flammable	Not flammable	Not flammable	Not pertinent	Not pertinent	Not pertinent	Not flammable	Reacts with water to generate heat and form phosphoric acid. The reaction is not violent
Piperazine		225	Not pertinent	Not pertinent	Water, dry chemical, alcohol foam, carbon dioxide	Water may cause frothing	Data not available	851	No

Table 8.1 (cont.) Fire and chemical reactivity data

Chemical Name	Flash Point, °F CC	Flash Point, °F OC	Flammable Limits in Air, % LEL	Flammable Limits in Air, % UEL	Extinguishing Agents	Extinguishing Agents NOT To Be Used	Behavior in Fire	Ignition Temp, °F	Water Reactive	Avoid Contact
Polybutene	215–470		Data not available	Data not available	Carbon dioxide, dry chemical, or foam	Water may be ineffective	Not pertinent	Data not available	No	
Polychlorinated Biphenyl			Data not available	Data not available	Water, foam, dry chemical, or carbon dioxide	Not pertinent	Not pertinent	Data not available	No	
Poly-propylene	−162		2	1.1	Stop the flow of gas	Not pertinent	Containers may explode. Vapor is heavier than air and may travel a long distance to a source if ignition and flash back	927	No	

Fire and Chemical Reactivity Data 459

Potassium Cyanide	Not flammable	Not flammable	Not flammable	Not flammable	Not pertinent	Not pertinent	Not flammable	Not flammable	When potassium cyanide dissolves in water, a mild reaction occurs and poisonous hydrogen cyanide gas is released. The gas readily dissipates if it is collected in a confined space, then workers may be exposed to toxic levels. If the water is acidic, weak acids will result in the formation of deadly hydrogen cyanide gas
Potassium Iodide	Not flammable	Not flammable	Not flammable	Not flammable	Not pertinent	Not pertinent	Not flammable	Not flammable	No

Table 8.1 (cont.) Fire and chemical reactivity data

Chemical Name	Flash Point, °F CC	Flash Point, °F OC	Flammable Limits in Air, % LEL	Flammable Limits in Air, % UEL	Extinguishing Agents	Extinguishing Agents NOT To Be Used	Behavior in Fire	Ignition Temp, °F	Water Reactive	Avoid Contact
Propane	−156		2.1	9.5	Stop the flow of gas. For small fires use dry chemicals. Cool adjacent areas with water spray	Water	Containers may explode. Vapor is heavier than air and can travel considerable distances to a source of ignition and flash back	842	No	
Propionaldehyde	−22		2.6	16.1	On small fires use carbon dioxide or dry chemical. For large fires use alcohol type foam	Water may be ineffective	Vapor is heavier than air and can travel considerable distances to a source of ignition and flash back	545	Reacts slowly forming a weak propionic acid. The reaction is non-violent and non-hazardous	

Fire and Chemical Reactivity Data

Propylene Oxide	−35	−20	2.1	38.5	Carbon dioxide or dry chemical for small fires; alcohol or polymer foam for large fires	Water may be ineffective	Containers may explode. Vapor is heavier than air and can travel considerable distances to a source of ignition and flash back	869	No
Pyridine	68		1.8	12.4	Alcohol foam, dry chemical, or carbon dioxide	Water may be ineffective	Vapor is heavier than air and can travel considerable distances to a source of ignition and flash back	900	No
Pyrogallic Acid	Not pertinent	Not pertinent	Not pertinent	Not pertinent	Water, foam, dry chemical, or carbon dioxide	Not pertinent	Not pertinent	Data not available	No
Quinoline	225		Data not available	Data not available	Water, dry chemical, foam, or carbon dioxide	Not pertinent	Exposure to heat can result in pressure build-up in closed containers, resulting in bulging or even explosion	896	No

Table 8.1 (cont.) Fire and chemical reactivity data

Chemical Name	Flash Point, °F CC	Flash Point, °F OC	Flammable Limits in Air, % LEL	Flammable Limits in Air, % UEL	Extinguishing Agents	Extinguishing Agents NOT To Be Used	Behavior in Fire	Ignition Temp, F	Water Reactive	Avoid Contact
Salicylic Acid	Not pertinent	Not pertinent	Not pertinent	Not pertinent	Water, dry chemical, foam, or carbon dioxide	Application of water or foam may cause frothing	This product sublimes and forms vapor or dust that can explode	Data not available	No	
Selenium Dioxide	Not flammable	Not flammable	Not flammable	Not flammable	Not pertinent	Not pertinent	Not pertinent	Not pertinent	No	
Selenium Trioxide	Not flammable	Not flammable	Not flammable	Not flammable	Not pertinent	Not pertinent	Not pertinent	Not pertinent	Reacts vigorously with water forming selenic acid solution	
Silicon Tetrachloride	Not flammable	Not flammable	Not pertinent	Not pertinent	Not pertinent	Do not apply water or foam on adjacent fires	Contact with water or foam applied to adjacent fires results in the formation of toxic and irritating fumes of hydrogen chloride	Not pertinent	Reacts vigorously with water forming hydrogen chloride (hydrochloric acid)	
Silver Acetate	Data not available	Data not available	Data not available	Data not available	Data not available	Data not available	Data not available	Data not available	No	

Fire and Chemical Reactivity Data

Silver Carbonate	Data not available	Data not available	Data not available	Data not available	Data not available	Data not available	Decomposes to silver oxide, silver, and carbon dioxide	Not pertinent	No
Silver Fluoride	Data not available	Data not available	Data not available	Data not available	Data not available	Data not available	Not pertinent	Not pertinent	No
Silver Iodate	Data not available	Data not available	Data not available	Data not available	Data not available	Data not available	Not pertinent	Not pertinent	No
Silver Nitrate	Not flammable	Not flammable	Not flammable	Not flammable	Not pertinent	Not pertinent	Increases the flammability of combustible materials	Not pertinent	No
Silver Oxide	Not flammable	Not flammable	Not flammable	Not flammable	Not pertinent	Not pertinent	Decomposes into metallic silver and oxygen. If large amounts of the product are involved in a fire, the oxygen liverated may increase the intensity of the fire	Not pertinent	No
Silver Sulfate	Not flammable	Not flammable	Not flammable	Not flammable	Not pertinent	Not pertinent	Not pertinent	Not pertinent	No

Table 8.1 (cont.) Fire and chemical reactivity data

Chemical Name	Flash Point, °F CC	Flash Point, °F OC	Flammable Limits in Air, % LEL	Flammable Limits in Air, % UEL	Extinguishing Agents	Extinguishing Agents NOT To Be Used	Behavior in Fire	Ignition Temp, F	Water Reactive	Avoid Contact
Sodium	Not pertinent	Not pertinent	Not pertinent	Not pertinent	Dry soda ash, graphite, salt, or other approved dry powder such as dry limestone	Water, carbon dioxide, or halogenated extinguishing agents	Not pertinent	250	Reacts violently with water forming flammable hydrogen gas and caustic soda solution. Fire often accompanies the reaction	
Sodium Alkyl-benzene-sulfonates	Not flammable	Not flammable	Not flammable	Not flammable	Not pertinent	Not pertinent	Irritating vapors form in fires	Not pertinent	No	
Sodium Alkyl Sulfates	Not flammable	Not flammable	Not flammable	Not flammable	Not pertinent	Not pertinent	Not pertinent	Not pertinent	No	
Sodium Amide	Not pertinent	Not pertinent	Not pertinent	Not pertinent	Dry soda ash, graphite, salt, or other approved dry powder such as dry limestone	Water	Data not available	Not pertinent	Reacts violently and often bursts into flames	

Fire Hazards and Fire Fighting / Chemical Reactivity

Fire and Chemical Reactivity Data 465

Sodium Arsenate	Not flammable	Not flammable	Not flammable	Not flammable	Not pertinent	Not pertinent	Not pertinent	No
Sodium Arsenite	Not flammable	Not flammable	Not flammable	Not flammable	Not pertinent	Not pertinent	Not pertinent	No
Sodium Azide	Not flammable	Not flammable	Not flammable	Not flammable	Not pertinent	Not pertinent	Containers may explode	Dissolves to form an alkaline solution. The reaction is non-violent
Sodium Bisulfite	Not flammable	Not flammable	Not flammable	Not flammable	Not pertinent	Not pertinent	Not pertinent	No
Sodium Borate	Not flammable	Not flammable	Not flammable	Not flammable	Not pertinent	Not pertinent	Not pertinent	No
Sodium Borohydride	Not flammable	Not flammable	Not flammable	Not flammable	Graphite, soda ash, limestone, sodium chloride powders	Water, carbon dioxide, or halogenated extinguishing agents	Decomposes and produces highly flammable hydrogen gas	Reacts to form flammable hydrogen gas
Sodium Cacodylate	Not flammable	Not flammable	Not flammable	Not flammable	Not pertinent	Not pertinent	Not pertinent	No
Sodium Chlorate	Not flammable	Not flammable	Not pertinent	Not pertinent	Not pertinent	Fire blankets	The product melts and then decomposes solid or liquid with all organic matter and some metals	No

Table 8.1 (cont.) Fire and chemical reactivity data

| Chemical Name | Fire Hazards and Fire Fighting ||||||||| Chemical Reactivity ||
| | Flash Point, °F || Flammable Limits in Air, % || Extinguishing Agents | Extinguishing Agents NOT To Be Used | Behavior in Fire | Ignition Temp, F | Water Reactive | Avoid Contact |
	CC	OC	LEL	UEL						
Sodium Chromate	Not flammable	Not flammable	Not flammable	Not flammable	Not pertinent	Not pertinent	Can increase the intensity of fires when in contact with combustible materials	Not pertinent	No	
Sodium Cyanide	Not flammable	Not flammable	Not flammable	Not flammable	Not pertinent	Not pertinent	Not pertinent	Not pertinent	When sodium cyanide dissolves in water, a mild reaction occurs and some poisonous hydrogen cyanide gas is liberated. The gas is not generally a concern unless it is generated in an enclosed space. If the water is acidic, then large amounts of the toxic gas form rapidly	

Fire and Chemical Reactivity Data

Sodium Dichromate	Not flammable	Not flammable	Not flammable	Not flammable	Flood with large amounts of water	Not pertinent	Decomposes to produce oxygen upon heating. May ignite other combustibles upon contact	Not pertinent	No
Sodium Hydride	Not pertinent	Not pertinent	Not pertinent	Not pertinent	Powdered limestone and nitrogen-propelled dry powder	Water, soda ash, chemical foam, or carbon dioxide	Not pertinent	Data not available	Reacts vigorously with water with the release of flammable hydrogen gas
Sodium Hydrosulfide Solution	Not flammable	Not flammable	Not flammable	Not flammable	Not pertinent	Not pertinent	Not pertinent	Not pertinent	No
Sodium Hydroxide	Not flammable	Not flammable	Not flammable	Not flammable	Not pertinent	Not pertinent	Not pertinent	Not pertinent	Dissolves with the liveration of considerable heat. The reaction violently produces steam and agitation

Table 8.1 (cont.) Fire and chemical reactivity data

Chemical Name	Flash Point, °F CC	Flash Point, °F OC	Flammable Limits in Air, % LEL	Flammable Limits in Air, % UEL	Extinguishing Agents	Extinguishing Agents NOT To Be Used	Behavior in Fire	Ignition Temp, F	Water Reactive	Avoid Contact
Sodium Hypochlorite	Not flammable	Not flammable	Not flammable	Not flammable	Not pertinent	Not pertinent	May decompose, generating irritating chlorine gas	Not pertinent	No	
Sodium Methylate	Not pertinent	Not pertinent	Not pertinent	Not pertinent	Dry chemical, inert powders such as sand or limestone, or carbon dioxide	Water, foam	Contact with water or foam applied to adjacent fires will produce flammable methanol	Not pertinent	Produces a caustic soda solution and a solution of methyl alcohol. The reaction is not violent	
Sodium Oxalate	Not flammable	Not flammable	Not flammable	Not flammable	Not pertinent	Not pertinent	Not pertinent	Not pertinent	No	
Sodium Phosphate	Not flammable	Not flammable	Not flammable	Not flammable	Not pertinent	Not pertinent	May melt with the loss of steam	Not pertinent	All variations or grades of this chemical readily dissolve in water. ASPP and MSP form weakly acidic solutions. TSP forms a strong	

Fire Hazards and Fire Fighting | Chemical Reactivity

Fire and Chemical Reactivity Data

Sodium Silicate	Not flammable	Not flammable	Not flammable	Not pertinent	Not pertinent	Not pertinent	caustic solution, similar to soda lye. TSPP forms weakly alkali solution
Sodium Silicofluoride	Not flammable	Not flammable	Not flammable	Not pertinent	Not pertinent	Decomposes at red heat	Not pertinent	No
Sodium Sulfide	Not flammable	Not flammable	Not flammable	Water	Not pertinent	Not pertinent	Not pertinent	No
Sodium Sulfite	Not flammable	Not flammable	Not flammable	Not pertinent	Not pertinent	Not pertinent	Not pertinent	No
Sodium Thiocyanate	Not flammable	Not flammable	Not flammable	Not pertinent	Not pertinent	Not pertinent	Not pertinent	No
Sorbitol	Not flammable	Not flammable	Not flammable	Water	Data not available	Data not available	Data not available	No
Stearic Acid	Not pertinent	Not pertinent	Not pertinent	Foam, dry chemical, or carbon dioxide	Water or foam may cause frothing	Data not available	743	No
Styrene	88	93	1.1	6.1	Water fog, foam, carbon dioxide, or dry chemical	Water may be ineffective	Vapor is heavier than air and may travel considerable distance to a source of ignition and flash back	914	No

Table 8.1 (cont.) Fire and chemical reactivity data

Chemical Name	Flash Point, °F CC	Flash Point, °F OC	Flammable Limits in Air, % LEL	Flammable Limits in Air, % UEL	Extinguishing Agents	Extinguishing Agents NOT To Be Used	Behavior in Fire	Ignition Temp, °F	Water Reactive	Avoid Contact
Sucrose	Not pertinent	Not pertinent	Not pertinent	Not pertinent	Water	Not pertinent	The product melts and chars	Not pertinent	No	
Sulfolane	330	Data not available	Data not available	Data not available	Water, foam, dry chemicals, or carbon dioxide	Not pertinent	Data not available	Data not available	No	
Sulfer Dioxide	Not flammable	Not flammable	Not flammable	Not flammable	Not pertinent	Not pertinent	Containers may rupture releasing toxic and irritating sulfur dioxide	Not flammable	No	
Sulfuric Acid	Not flammable	Not flammable	Not flammable	Not flammable	Not pertinent	Water used on adjacent fires should be carefully handled	Not pertinent	Not flammable	Reacts violently with the exaolution of heat (exothermic reaction). Significant agitation and spattering occurs when water is added to the chemical	

Fire and Chemical Reactivity Data

Sulfuric Acid, Spent	Not flammable	Not flammable	Not flammable	Not pertinent	Not pertinent	Not pertinent	No reaction unless strength is above 80–90 %, in which case an exothermic reaction will occur
Titanium Tetrachloride	Not flammable	Not flammable	Not flammable	Not pertinent	Do not use water if it can directly contact this chemical	If containers leak a very dense white fume can form and obscure operations	Not flammable	Reacts with moisture in the air forming a dense white fume. Reaction with liquid water gives off heat and forms hydrochloric acid
Toluene	40	55	1.27	7	Carbon dioxide or dry chemical for small fire, or-dinary foam for large fires	Water may be ineffective	Vapor is heavier than air and may travel considerable distance to a source of ignition and flash back	997	No

Table 8.1 (cont.) Fire and chemical reactivity data

Chemical Name	Flash Point, °F CC	Flash Point, °F OC	Flammable Limits in Air, % LEL	Flammable Limits in Air, % UEL	Extinguishing Agents	Extinguishing Agents NOT To Be Used	Behavior in Fire	Ignition Temp, °F	Water Reactive	Avoid Contact
Toluene 2, 4-Diisocyante	270		0.9	9.5	Water, foam, dry chemicals, or carbon dioxide	Water or foam may cause frothing	Not pertinent	>300	A non-violent reaction occurs forming carbon dioxide gas and an organic base	
P-Toluenesulfonic Acid	Not pertinent	Not pertinent	Not pertinent	Not pertinent	Water	Not pertinent	Not pertinent	Not pertinent	No	
O-Toludine	85	167	Data not available	Data not available	Foam, dry chemical, or carbon dioxide	Water may be ineffective	Data not available	900	No	
Toxamphene	84		1.1	6.4	Foam, dry chemical, or carbon dioxide	Water may be ineffective	Solution in xylene may produce corrosive products when heated	986	No	
Trichloroethylene	90		8	10.5	Water fog	Not pertinent	Not pertinent	770	No	
Trichlorofluoromethane	Not flammable	Not flammable	Not flammable	Not flammable	Not flammable	Not pertinent	Not pertinent	Not pertinent	No	

Fire and Chemical Reactivity Data

Trichloro-silane	>−58	−18	1.2	90.5	Dry chemical, carbon dioxide	Water, foam	Difficult to extinguish, re-ignition may happen. Also vapor is heavier than air and can travel to a source of ignition and flash back	220	Reacts violently to form hydrogen chloride fumes (hydrochloric acid)
Tridecanol		250	Data not available	Data not available	Alcohol, dry chemical, water fog	Water or foam may cause frothing	Not pertinent	Data not available	No
1-Tridecene	Data not available	Data not available	Data not available	Data not available	Dry chemical, foam, or carbon dioxide	Water may be ineffective	Not pertinent	Data not available	No
Triethyl-aluminum	Spontaneously ignites in air at all temperatures		Not pertinent	Not pertinent	Inert powders such as limestone or sand, or dry chemical	Water, foam, halogenated extinguishing agents	Dense smoke of aluminum oxide is formed. Contact with water on adjacent fires causes violent reaction producing toxic and flammable gases	Not pertinent	Reacts violently to form flammable ethane gas

Table 8.1 (cont.) Fire and chemical reactivity data

Chemical Name	Flash Point, °F CC	Flash Point, °F OC	Flammable Limits in Air, % LEL	Flammable Limits in Air, % UEL	Extinguishing Agents	Extinguishing Agents NOT To Be Used	Behavior in Fire	Ignition Temp, °F	Water Reactive	Avoid Contact
Triethylamine		20	1.2	8	Carbon dioxide or dry chemical for small fire, alcohol foam for large fires	Water may be ineffective	Not pertinent	842	No	
Thiethylene Glycol	350		0.9	9.2	Alcohol foam, dry chemical, carbon dioxide	Water or foam may cause frothing	Not pertinent	700	No	
Tripropylene Glycol		285	0.8	5	Alcohol foam, dry chemical, carbon dioxide	Water may be ineffective	Data not available	Data not available	No	
Turpentine	95		0.8		Foam, dry chemical, or carbon dioxide	Water may be ineffective	Forms heavy black smoke and soot	488	No	
Undecanol		200	Data not available	Data not available	Foam, dry chemical, or carbon dioxide	Water or foam may cause frothing	Not pertinent	Data not available	No	
1-Undecene		160	Data not available	Data not available	Foam, dry chemical, or carbon dioxide	Water may be ineffective	Not pertinent	Data not available	No	

Fire and Chemical Reactivity Data

	N-Undecyl-benzene	Uranyl Acetate	Uranyl Nitrate	Uranyl Sulfate	Urea	Urea Peroxide	Vanadium Oxytri-chloride
	285	Not flammable	Not flammable	Not flammable	Not flammable	Not pertinent	Not flammable
	Not flammable	Not flammable	Not flammable	Not flammable	Not flammable	Not pertinent	Not flammable
	Data not available	Not flammable	Not flammable	Not flammable	Not flammable	Not pertinent	Not flammable
	Data not available	Not flammable	Not flammable	Not flammable	Not flammable	Not pertinent	Not flammable
	Foam, dry chemical, or carbon dioxide	Not pertinent	Apply flooding amounts of water	Not pertinent	Not pertinent	Inert powders such as limestone or sand, or water	Not pertinent
	Water may be ineffective	Not pertinent	Not pertinent	Not pertinent	Not pertinent	Not pertinent	Water, unless flooding amounts should not be used on adjacent fires
	Data not available	Not pertinent	Intensifies fires	Not pertinent	Melts and decomposes generating ammonia	Melts and decomposes giving off oxygen and ammonia	Data not available
	Data not available	Not pertinent	Not pertinent	Not pertinent	Not flammable	>680	Not pertinent
	No	No	Dissolves in water forming a weak solution of nitric acid. The reaction is non-hazardous	No	No	Forms solution of hydrogen peroxide. The reaction is non-hazardous	Reacts forming a solution of hydrochloric acid

Table 8.1 (cont.) Fire and chemical reactivity data

Chemical Name	Flash Point, °F CC	Flash Point, °F OC	Flammable Limits in Air, % LEL	Flammable Limits in Air, % UEL	Extinguishing Agents	Extinguishing Agents NOT To Be Used	Behavior in Fire	Ignition Temp, F	Chemical Reactivity Water Reactive	Chemical Reactivity Avoid Contact
Vanadium Pentoxide	Not flammable	Not flammable	Not flammable	Not flammable	Not pertinent	Not pertinent	May increase the intensity of fires	Not pertinent	No	
Vanadyl Sulfate	Not flammable	Not flammable	Not flammable	Not flammable	Not pertinent	Not pertinent	Data not available	Not pertinent	No	
Vinyl Acetate	18	23	2.6	13.4	Carbon dioxide or dry chemical for small fires, and ordinary foam for large fires	Water may be ineffective	Vapor is heavier than air and may travel considerable distance to a source of ignition and flash back causing product to polymerize and burst or explode containers	800	No	
Vinyl Chloride		−110	4	26	Carbon dioxide or dry chemical for small fires, and stop the flow of gas if feasible for large fires	Not pertinent	Containers may explode. Gas is heavier than air and may travel to a source of ignition and flash back	882	No	

Fire and Chemical Reactivity Data 477

Vinyl Fluoride, Inhibited	Not pertinent	Not pertinent	2.6	21.7	Allow fire to burn out, stop the flow of gas if feasible. Cool adjacent containers with water	Not pertinent	Vapor is heavier than air and can travel to a source of ignition and flash back. Containers may explode	725	No
Vinylidene Chloride, Inhibited		0	7.3	16	Foam, dry chemical, or carbon dioxide	Water may be ineffective	May explode due to polymerization. Vapor is heavier than air and can travel to a source of ignition and flash back	955–1031	No
Vinyl Methyl Ether, Inhibited		−69	2.6	39	Allow fire to burn out, stop the flow of gas if feasible. Extinguish small fires with dry chemical or carbon dioxide	Water may be ineffective	Containers may explode. Vapor is heavier than air and can travel to a source of ignition and flash back	Data not available	Reacts slowly to form acetaldehyde. The reaction is generally non-hazardous unless occurring in hot water or acids are present

Table 8.1 (cont.) Fire and chemical reactivity data

Chemical Name	Flash Point, °F CC	Flash Point, °F OC	Flammable Limits in Air, % LEL	Flammable Limits in Air, % UEL	Extinguishing Agents	Extinguishing Agents NOT To Be Used	Behavior in Fire	Ignition Temp, °F	Water Reactive	Avoid Contact
Vinyltoluene	125	137	0.8	11	Water fog, foam, carbon dioxide, or dry chemical	Not pertinent	Containers may explode or rupture due to polymerization	914	No	
Vinyltrichlorosilane	52	60	3		Dry chemical or carbon dioxide	Water, foam	Fire is difficult to extinguish because of ease in re-ignition. Contact with water applied to adjacent fires will result in the formation of irritating hydrogen chloride gas	505	Reacts vigorously, producing hydrogen chloride (hyrdochloric acid)	
Waxes: Carnauba	540		Not pertinent	Not pertinent	Water, foam, dry chemicals, or carbon dioxide	Water or foam may cause frothing	Not pertinent	Data not available	No	

Fire and Chemical Reactivity Data

Waxes: Paraffin	390	380–465	Not pertinent	Not pertinent	Not pertinent	Water, foam, dry chemicals, or carbon dioxide	Water or foam may cause frothing	Not pertinent	473	No
M-Xylene	84		1.1	6.4		Foam, dry chemical, or carbon dioxide	Water may be ineffective	Vapor is heavier than air and may travel a considerable distance to a source of ignition and flash back	986	No
Xylenol		1.4			Water, dry chemical, carbon dioxide, foam	Not pertinent	Not pertinent	Not pertinent	1110	No
Zinc Acetate	Not flammable	Not flammable	Not flammable	Not flammable		Not pertinent	Not pertinent	Not pertinent	Not pertinent	No
Zinc Ammonium Chloride	Not flammable	Not flammable	Not flammable	Not flammable		Not pertinent	Not pertinent	Not pertinent	Not pertinent	No
Zinc Bromide	Not flammable	Not flammable	Not flammable	Not flammable		Not pertinent	Not pertinent	Not pertinent	Not pertinent	No
Zinc Chloride	Not flammable	Not flammable	Not flammable	Not flammable		Not pertinent	Not pertinent	Not pertinent	Not pertinent	No
Zinc Chromate	Not flammable	Not flammable	Not flammable	Not flammable		Not pertinent	Not pertinent	Not pertinent	Not pertinent	No

Table 8.1 Fire and chemical reactivity data

Chemical Name	Fire Hazards and Fire Fighting									Chemical Reactivity	
	Flash Point, °F		Flammable Limits in Air, %		Extinguishing Agents	Extinguishing Agents NOT To Be Used	Behavior in Fire	Ignition Temp, F	Water Reactive	Avoid Contact
	CC	OC	LEL	UEL						
Zinc Fluoroborate	Not flammable	Not flammable	Not flammable	Not flammable	Not pertinent	Not pertinent	Not pertinent	Not pertinent	No	
Zinc Nitrate	Not flammable	Not flammable	Not flammable	Not flammable	Not pertinent	Not pertinent	Not pertinent	Not pertinent	No	
Zinc Sulfate	Not flammable	Not flammable	Not flammable	Not flammable	Not pertinent	Not pertinent	Not pertinent	Not pertinent	No	
Zirconium Acetate	Not flammable	Not flammable	Not flammable	Not flammable	Not pertinent	Not pertinent	Not pertinent	Not pertinent	No	
Zirconium Nitrate	Not flammable	Not flammable	Not flammable	Not flammable	Not pertinent	Not pertinent	May increase the intensity of fires	Not pertinent	Dissolves to give an acid solution	